Oxidative Stress in Applied Basic Research and Clinical Practice

Editor-in-Chief
Donald Armstrong

For other titles published in this series, go to
http://www.springer.com/series/8145

Note from the Editor-in-Chief

All books in this series illustrate point-of-care testing and critically evaluate the potential of antioxidant supplementation in various medical disorders associated with oxidative stress. Future volumes will be updated as warranted by emerging new technology, or from studies reporting clinical trials.

Donald Armstrong
Editor-in-Chief

Natan Gadoth • Hans Hilmar Göbel
Editors

Oxidative Stress and Free Radical Damage in Neurology

Humana Press

Editors
Natan Gadoth
Sackler Faculty of Medicine
Tel Aviv University
Tel Aviv 69978, Israel
and
Department of Neurology
Mayanei Hayeshua Medical Center
Bnei-Brak 51544
gadoth@post.tau.ac.il

Hans Hilmar Göbel
Johannes Gutenberg University
Medical Center
Mainz, Germany
goebel@neuropatho.klinik.uni-mainz.de

ISBN 978-1-60327-513-2 e-ISBN 978-1-60327-514-9
DOI 10.1007/978-1-60327-514-9
Springer New York Dordrecht Heidelberg London

Library of Congress Control Number: 2010937792

Printed on acid-free paper

Humana Press is part of Springer Science+Business Media (www.springer.com)

Preface

The possible role of free radicals and oxidative stress in neurological disorders was not recognized until recently and thus is relatively unknown to practicing as well as academically minded clinical neurologists. This fact can be appreciated as one goes through the major textbooks recommended and used by residents in neurology looking, in vain, for information on this topic. One may find only short reminders of the possible association of superoxidase dismutase with familial amyotrophic lateral sclerosis and nitrous oxide with migraine. We may also find bits of information regarding the possible role of free radicals in the pathogenesis of traumatic brain injury.

This limitation is not at all surprising, because the biological role of free radicals was discovered and scientifically accepted only recently. Denham Harman proposed the free radical theory of aging more than 50 years ago, but acceptance of this theory was slow until only a few years ago. During the late 1940s to early 1950s, retrolental fibroplasia in premature newborns was recognized as being caused by oxygen toxicity. The presence of free radicals in biological systems was not generally considered likely until the discovery of superoxide dismutase in 1969 by Irwin Fridovich, although in the 1950s, the basis of oxygen toxicity and X-irradiation was proposed to be a common free radical mechanism.

Thus, the goal of introducing this topic to neurology for teaching and practice led us to compile this volume with the help of a number of colleagues, who agreed to cover several aspects of clinical and research issues regarding free radicals and the nervous system. Familiarity with the various antioxidants and their possible therapeutic role in some neurological disorders, which is described in the relevant chapters throughout this volume, is an additional important and unique contribution of this volume.

We hope that clinicians as well as basic researchers in neuroscience will find this volume of help in expanding their understanding, not only of the widespread involvement of free radicals in the central nervous system, but also of some ambiguous issues related to answering the question of the primary or secondary role of free radical damage in neurology.

Tel-Aviv, Israel — Natan Gadoth
Mainz, Germany — Hans H. Göbel

Contents

Chapter 1
The Role of Free Radicals in the Nervous System

Joseph Friedman

Abstract This chapter is an introduction to the biology of reactive oxygen species (ROS) in the brain. In healthy aerobes, there is a balance between the production of various ROS and antioxidant defenses. Living organisms have not only adapted to coexistence with free radicals but have developed various mechanisms for the advantageous use of free radicals in various physiological functions. Infectious diseases were a powerful driver of natural selection in early human civilizations. Indeed, ROS participate directly in defense against infection. In a normal situation, microglia, which are resident macrophages of the brain, fight against infection by ROS. ROS are well recognized for playing a dual role, having both deleterious and beneficial effects, which in most cases depend on concentration. At high ROS concentrations there are harmful effects, and in a low–moderate concentration ROS are involved in physiological roles in cellular response to noxious stimuli. It was suggested that the main effects of ROS on cells are through their actions on signaling pathways rather than causing nonspecific damage. With aging, when these pathways deteriorate, accumulation of higher concentrations of ROS occurs in amounts beyond the capacity of antioxidants to cope. This deterioration results in the age-associated neurodegenerative disorders such as stroke and central nervous system (CNS) trauma as well as Parkinson's and Alzheimer's disease. Some of the CNS-evolved specific signaling pathways are described.

Keywords Neuron · Microglia · Mitochondria · Complex I · ROS · OS · NADPH oxidase · Excitoxicity · Antioxidants · BBB · Signaling pathways · Brain stroke · Brain trauma · Parkinson's disease · Alzheimer's disease

---

J. Friedman (✉)
Life Science Laboratory, Weizmann Science Park, Ness-Ziona, Israel
e-mail: dr.joseph.friedman@gmail.com

N. Gadoth and H.H. Göbel (eds.), *Oxidative Stress and Free Radical Damage in Neurology*, Oxidative Stress in Applied Basic Research and Clinical Practice,
DOI 10.1007/978-1-60327-514-9_1,

1 Introduction to ROS

In the structure of atoms and molecules, electrons usually associate in pairs, each moving within a defined region of space around the nucleus. This space is referred to as the atomic or molecular orbital. One electron in each pair has a spin quantum number of +½, and the other, –½. The process of removing electrons is called oxidation, and the substance receiving electrons becomes reduced. The reactions involved in electron transfer are called redox (reduction–oxidation) reactions. A free radical is any species capable of independent existence (hence the term free) that contains one or more unpaired electrons [1, 2]. Free radicals are sometimes reactive, although the chemical reactivity of radicals varies over a wide spectrum. Consideration of our broad definition shows that there are many free radicals in chemistry and biology. The simplest is the atomic hydrogen with one proton and a single electron, which must therefore be unpaired. Hence, removal of a hydrogen atom from a biological molecule leaves behind an unpaired electron on the atom to which hydrogen was originally attached. The diatomic oxygen molecule O_2 qualifies as a radical inasmuch as it possesses two unpaired electrons, each located in a different orbital but both having the same spin quantum number: that is the reason why O_2 itself has a relatively low reactivity in contrast to other radicals, which can be highly reactive. Radicals can be formed by the loss of a single electron from a nonradical: $x \rightarrow \bar{e} + x^{+}$, or by the gain of a single electron by a nonradical: $y + \bar{e} \rightarrow y^{-}$. Radicals can react with other molecules in a number of ways. Thus, if two radicals meet, they can combine their unpaired electrons and join to form a covalent bond: $x\bullet + y\bullet \rightarrow x - y$.

A radical might donate its unpaired electron to another molecule: $x\bullet + y \rightarrow x^{+} + y^{-}$. Or, it might catch an electron from another molecule to pair: $x\bullet + y \rightarrow x^{-} + y^{-}$.

There are reactive oxygen molecules such as H_2O_2 that do not fit the definition of free radicals, so all the reactive species, radicals or not, are called reactive oxygen species (ROS).

The first organic free radical, identified by Moses Gomberg in 1900, was the triphenyl methyl radical. Until 40–50 years ago, ROS were not thought to occur in biological systems because of their high reactivity and lack of selectivity. This point of view was changed in 1956 by Denham Harman, who proposed the concept of free radicals playing a role in aging. A second epoch of research regarding free radicals in biological systems was reached in 1969 when McCord and Fridovich discovered the enzyme superoxide dismutase (SOD) and thus provided convincing evidence about the importance of free radicals in living systems.

More than 90% of the oxygen that enters human cells is used for the production of energy. Mitochondria produce more than 80% of the adenosine triphosphate (ATP) needed by the mammalian cells. During this process, four electrons are added to each O_2 molecule, resulting in the formation of two molecules of water. An estimated 1–5% of the O_2 taken into cells, however, forms partially reduced O_2 species, the ROS. Some of them contain an unpaired electron and are therefore referred to as free radicals:

$$O_2 \xrightarrow{+\bar{e}} \dot{\bar{O}}_2 \xrightarrow[+2H^+]{+\bar{e}} H_2O_2 \xrightarrow{+\bar{e}} OH^\bullet + O\bar{H} \xrightarrow[+2H^+]{+\bar{e}} 2H_2O$$

The intermediates have various degrees of reactivity with nonradical species.

1.1 The Superoxide Anion Radical

$\dot{\bar{O}}_2$ is produced in vivo in a variety of ways, the major source being the electron chain in mitochondria [3, 4]. $\dot{\bar{O}}_2$ is not highly reactive with biological substrates in an aqueous environment. Once formed, $\dot{\bar{O}}_2$ quickly undergoes dismutation to generate H_2O_2. This reaction is markedly accelerated by a family of enzymes, the SOD [5]. Thus, SOD is generally considered an important antioxidative enzyme [6]. In addition to SOD, several other enzymes that generate H_2O_2 also exist in human tissues.

1.2 Hydrogen Peroxide

Hydrogen peroxide (H_2O_2) itself is not especially toxic unless it is present in high concentrations within cells. H_2O_2 readily diffuses through cellular membranes and can thereby reach sites distant from where it was generated. Also, in the presence of transition metals, mainly Fe^{2+} but also Cu^{1+}, H_2O_2 is reduced to the hydroxyl radical ($OH^\bullet$) via either the Haber–Weiss or Fenton reactions [7–9]. In most cells, H_2O_2 is converted to innocuous products by the actions of two important antioxidant enzymes, that is, catalase and selenium-dependent glutathione peroxidase (GPx). GPx utilizes H_2O_2 hydroperoxides as substrates during the conversion of reduced glutathione (GSH) to its sulfide (GSSG) [10].

1.3 Hydroxyl Radical

There is universal agreement that, once formed, the $OH^\bullet$ radical reacts rapidly with any molecule within a few angstroms (Å) from the site where it is produced. Because of its high reactivity, its estimated half-life at 37°C is of the order of 10^{-9} s. The $OH^\bullet$ readily damages nuclear and mitochondrial DNA, membrane lipids, and carbohydrates [11]. There are at least two ways in which DNA damage is achieved. In many cases the mutilated DNA is created because H_2O_2 reacts with either Fe^{2+} or Cu^{1+} that is bound to molecules in the immediate vicinity of DNA. So, when toxic $OH^\bullet$ is formed, its first target is the adjacent nucleic acids [12]. Alternatively, during excitatory neurotransmitter stimulation of neurons, the large increase in intracellular free Ca^{2+} activates nuclear enzymes, which results in formation of $OH^\bullet$, which subsequently leads to DNA damage [13]. $OH^\bullet$ also interacts with membrane lipids to initiate lipid peroxidation. This reaction is accomplished when $OH^{\bullet}$ removes the allelic H^+ from a polyunsaturated fatty acid (PUFA), which results in a radical chain reaction wherein lipid peroxidation is self-propagated [8].

1.4 Singlet Oxygen

Singlet oxygen ($^1O_2$) is formed under photoxidative conditions when energy is transferred to O_2 from photoexcited sensitizers. $^1O_2$ is not a diradical as is molecular oxygen, and it is highly reactive toward most olefins; thus, it can subtract H^+ from a PUFA to initiate lipid peroxidation [1].

1.5 Peroxyl Radical (LOO•)

The breakdown of lipids in cellular membranes during which peroxyl radical ($LOO^{\bullet}$) is formed is perhaps the most thoroughly studied of all oxidative processes [14]. The process referred to as lipid peroxidation is extremely complex and can be self-propagating, which means that once initiated it would theoretically lead to the oxidation of all the lipids in a cell; thus, it can be highly destructive. Besides $OH^{\bullet}$, other radicals such as $^1O_2$ and $ONOO^-$ can also initiate the process of lipid destruction [15]. A number of toxic products are generated during the decomposition of fatty acids, including lipid hydroperoxides and peroxyl radicals. The $LOO^{\bullet}$ can then attack a nearby PUFA and reinitiate (propagate) the process. Vitamin E (α-tocopherol) is the premier $LOO^{\bullet}$ scavenger and chain-breaking antioxidant.

1.6 Nitric Oxide and Peroxynitrite Anion

$NO^{\bullet}$ is often characterized as a double-edged sword. $NO^{\bullet}$ rapidly combines with $\dot{\bar{O}}_2$ to generate the peroxynitrite anion ($ONOO^-$) [16]. It is this latter molecule that accounts for much of the toxicity of $NO^{\bullet}$. The reactivity of ONOO− is roughly the same as that of $OH^{\bullet}$. The toxicity of ONOO− derives from its ability to directly nitrate and hydroxylate the aromatic rings of amino acid residues [17], to react with sulfhydryls [16] and with zinc-thiolate moieties [18] as well as with lipids [19] and proteins [20].

2 Physiological Roles of ROS

In healthy aerobes, there is a balance between the production of ROS and antioxidant defenses. In health, the cell has become well equipped to cope with the normal production of ROS. Indeed, continuous low concentrations of ROS induce expression of antioxidant enzymes and related defense mechanisms. A large body of evidence has been accumulated that living organisms have not only adapted to a coexistence with free radicals but have developed various mechanisms for the advantageous use of free radicals in various physiological functions [1, 21]. Infectious diseases were a powerful driver of natural selection in early human civilizations. Indeed, ROS participate directly in defense against infection and also are important coordinators of the inflammation. , resident macrophages of the brain in the normal situation, fight against infection by means of ROS. ROS are well recognized for playing both deleterious and beneficial roles, which in most cases depend on their concentration. At high ROS concentrations there are harmful effects, and in a low–moderate concentration ROS are involved in physiological roles in cellular response to noxious stimuli. It was suggested that the main effects of ROS on cells are through their actions on signaling pathways rather than causing nonspecific damage to macromolecules [22]. I suggest that normally ROS participate in many signal transduction pathways that are essential for the many functions of the brain such as memory and learning. With aging, when these pathways deteriorate, there is an accumulation of high concentrations of ROS, which cause age-associated neurodegenerative disorders. The CNS thus evolved specific signaling pathways [23], a number of which are described herein.

2.1 ROS and Protein Kinase C Activation

Protein kinase C (PKC) activation protects nerve cells from oxidative stress (OS)-induced cell death [24]. PKC expression has been previously coupled with the preservation of cell survival and the formation and consolidation of different types of memory. The induction of PKC activity in neurons is thought to be a prerequisite for neuroprotection against several exogenous insults.

2.2 ROS and the PI3K/AKT Pathway

The PI3K/AKT pathway has been shown to play control functions in neuronal protection against a variety of extracellular insults and thus to be essential for neuronal differentiation and survival [25].

2.3 ROS and NF-κB

Activation of NF-κB by H_2O_2 or hydroperoxides in neurons has been shown to have an antiapoptotic effect and is protective against glutamate exposure, glucose deprivation, hypoxia, and low K^+ [26, 27].

2.4 *ROS and Matrix Metalloproteinase Activation*

Cerebral ischemia activates matrix metalloproteinases (MMPs), which attack the integrity of the blood–brain barrier (BBB) by breakdown of the extracellular matrix around cerebral blood vessels and neurons [28].

2.5 *ROS Effects on Apoptotic Pathways*

Damage induced by ROS can trigger cell death by influencing pathways that reduce the survival potential of cells. It has been suggested [29] that $NO^{\bullet}$ and $\bar{O}_2$ may contribute to damage to nuclear genetic material through the formation of peroxinitrite. Low levels of these species trigger apoptosis, but higher levels cause the cell to undergo necrosis. Possible mechanisms by which the apoptotic effects are mediated could involve mitochondria, DNA-repair enzymes, and death membrane receptors [30].

ROS can thus affect a variety of pathways in the cell, and these pathways determine the survival status of the cell. Pro-apoptotic pathways involve p38, JNK, JAK, ATM, and p53, and anti-apoptotic pathways include ERK1/2, PI3K/AKT, and heat shock factor-1 (HSF-1) [23, 31]. The hormetic properties of ROS may increase the expression and activity of vitagenes, indicating that ROS may be "good" for the cells and thus have a clear physiological benefit for the maintenance of cellular homeostasis, particularly when the cells are challenged by the normal insults that occur in cellular physiology.

Heat shock proteins (HSP) serve as chaperones that bind to other proteins and regulate their conformation. Under conditions of cell stress, a major risk to cell survival is represented by misfolded and aggregated proteins. HSPs prevent protein misfolding and oligomerization. Therefore, not surprisingly, HSPs are induced under conditions of cellular stress [32]. Molecules that activate this defense mechanism are therefore possible candidates for cytoprotection [33]. After a variety of CNS insults, HSP70 is synthesized at high levels [34, 35]. Denatured proteins are thought to serve as stimulators for induction of HSPs [36]. Cells have evolved mechanisms for rescuing and recycling misfolded proteins, but these systems are not perfect. Chaperones can rescue misfolded proteins by breaking up aggregates and assisting the refolding process. Proteins that cannot be rescued by refolding can be delivered by chaperones to the proteosome to be recycled. Furthermore, the compelling evidence of the vitagenes network [37] as a defense system operating in the brain during times of ROS stress offers new perspectives in the treatment of Alzheimer's disease (AD).

3 Neurogenerative Disorders: General

The various neurodegenerative diseases (diseases in which neurons degenerate and die) have a variety of different symptoms, affect different parts of the brain, and have different causes. They have in common impaired mitochondrial function, increased oxidative damage, defects in the ubiquitin-proteosome system, presence of abnormal aggregated proteins, changes in iron metabolism, and some involvement of excitoxicity and of inflammation. It seems likely that all these events are involved in a vicious cycle and that any of them could initiate neuronal cell death, rapidly recruiting the others to its destructive purpose. Oxidized proteins are usually removed by the proteosome. Inhibition of the proteosome allows abnormal proteins to accumulate and produces OS, but how this is done is still unclear. Finally, ROS-producing agents could initiate neurodegeneration, because ROS damage mitochondria, cause rise in Ca^{2+}, and may inhibit proteosome function. The iron content of most brain areas increases with age, and iron and other metals promote the aggregation of several proteins. How do neurons die in these various diseases? Sometimes they die by necrosis, as in excitoxicity, and sometimes, probably, by apoptosis. However, as more studies are done, the role of intermediate types of cell death, with features of both necrosis and apoptosis, is becoming more prominent [1].

4 Brain Ischemia: Stroke

Ischemic injury to neurons is mainly caused by interruption of blood flow, hypoxia, ATP depletion, and subsequent reoxygenation of the ischemic brain by reperfusion. It has been observed that ROS are predominantly involved in the pathogenesis of stroke. They also play an important role in exacerbation of the phase following stroke by triggering pro-apoptotic pathways that reduce the survival chances of the neurons [28, 38]. Multiple agents have been shown to decrease neuronal loss after ischemia in animal models, including Ca^{2+} blockers, inhibitors of neuronal nitric oxide synthase (nNOS), xanthene oxidase (XO), or phospholipase A2, agents that prevent binding of glutamate to its receptors, and antioxidants such as as PEG-SOD (CuZn-SOD conjugated to the polymer polyethylene glycol, PEG), some plant phenols, *N*-acetyl-cysteine (NAC), and related compounds such as NXY-059 and proteosomal inhibitors. It is possible that the latter can dampen inflammation by decreasing activation of NF-κB [39]. For example, α-tocopherol-deficient animals suffer more damage after ischemia-reperfusion [40]. The fact that these varieties of interventions can work in animals suggests that cell death is not caused by a single event, but by multiple events acting in parallel or synergistically. Suggestions as to why some neurons are more vulnerable than others include the levels of various glutamate receptors, basal levels of OS, propensity to suffer proteosomal dysfunction, and the level of intracellular free Ca^{2+} and mitochondrial damage [41]. The ability to activate signal pathways that promote neuronal survival such as the PI3K/AKT pathway may also be important. Sadly, none of the agents mentioned here has yet given convincing clinical benefit in human stroke patients, although NXY-059 seems promising. In contrast to the highly controlled animal models, human patients vary in their degree of hypoxia and in other factors. There may also be a gender effect, a trend to better outcome in females compared to males, which has been speculated to be the result of better antioxidant protection in females [42].

NO• is not necessarily bad: it can decrease levels of adhesion molecules, improve blood flow (by vasoconstriction), and inhibit lipid peroxidation, for example. Ischemia-reperfusion also affects the BBB, causing increased permeability and raising levels of adhesion molecules, and attracting lymphocytes and neutrophils [43]. ROS can damage the barrier directly (e.g., by causing swelling of astrocytes and damaging the endothelial tight junction) and by activating matrix metalloproteinases (MMPs) [1].

5 Brain Trauma

Traumatic injury to the brain or spinal cord involves both direct tissue damage (crushing, tearing, bruising) and secondary damage involving many of the events relevant to stroke such as release of transition metal ions, Zn^{2+} and hemoglobin (Hb), increased NO• production, phospholipase and calpain activation, release of fatty acids, and increase of glutamate [44]. For example, damage to the rat spinal cord caused extracellular iron release, accompanied by OH• formation [45]. Blood vessels ruptured by the trauma leak blood, and the parts of the brain they normally supply will become hypoxic. The vasospasm that can occur several days after intracranial bleeding seem to involve oxidative damage, for example, by OH• production [46] and by end products of lipid peroxidation. During the first 5 days after an intracranial bleed, the erythrocytes in the CSF slowly hemolyze, and glial cells gather at the site to take up the iron and eventually store it as ferritin or hemosiderin. Much of Hb is converted to bilirubin by heme oxygenase (HO), but part undergoes nonenzymatic oxidation and degradation [47]. Injury or ischemia increases HO-1 levels, especially in glia. Degradation of heme removes one potential pro-oxidant but causes release of another iron ion. Transgenic mice overexpressing HO-1 were less sensitive to ischemic brain injury, apparently as a consequence of increased ferritin synthesis secondary to the increased HO [48]. Traumatic injury can directly damage the BBB, allowing entry of agents that do not normally cross the BBB into the brain. The inflammatory immune response after injury may be good, bad, or, more likely, both (as in the case of stroke). Many agents tested for the treatment of stroke have also been examined in models of traumatic injury. For example, PEG-CuZnSOD appears to be beneficial when administered to rats after percussive brain injury; it probably enters the brain only because the BBB is damaged. The damage by ROS can become more widespread because of the weakened cellular antioxidant defense system. Moreover, acute brain injury increases the levels of excitotoxic amino acids (such as glutamate) that also produce ROS, thereby promoting parenchymal destruction. The principal pathophysiological processes in acute CNS injury are extremely complex and involve pathological permeability of BBB, energy failure, loss of cell ion homeostasis, acidosis, increased intracellular Ca^{2+}, excitotoxicity, and ROS-mediated toxicity. These events can lead to ischemic necrosis, which occurs in the severely ischemic regions and is associated with loss of calcium and glutamate homeostasis. It can also lead to apoptosis, which is more likely to occur in the moderately ischemic regions, evolves more slowly, and depends on the activation of a sequence of genes [38].

6 Parkinson's Disease

The initiation and progression of the neurodegeneration process in Parkinson's disease (PD) are still obscure. The underlying pathophysiology is complex and in most cases probably multifactorial, differing among the individuals affected. A large number of studies have provided evidence that redox imbalance contributes to all forms of PD, although it is not yet established whether OS is a primary event or a consequence of other pathogenetic factors. Nevertheless, overproduction of ROS is unquestionably an important mediator of cell death in PD [49]. It has been suggested that the pathogenesis of PD may involve two processes: a disease-specific insult combined with damage associated with normal aging. PD is the result of neurodegeneration occurring in specific brain areas (substantia nigra pars compacta and striatum), resulting in dopamine depletion [50]. Factors including dopamine, neuromelanin, increased iron deposition in substantia nigra, a decrease in ferritin and glutathione (GSH), a defect in the mitochondrial respiratory chain function involving complex I, mitochondrial dysfunction, and excitoxicity may be a cause or result of OS. Environmental toxins including MPTP, paraquat, and rotenone have been shown to increase the risk of PD in humans. Studies with cell and animal models reveal oxidative and inflammatory properties of these toxins and their ability to activate glial cells that subsequently destroy neighboring dopaminergic neurons. Mitochondrial complex I activity is deficient in the substantia nigra in PD and may be associated with a genetic abnormality of complex I. However, the role of a postulated mitochondrial DNA defect remains uncertain. It has been demonstrated that glutathione depletion in dopaminergic cells in culture results in a selective decrease in complex I activity (a major hallmark of PD) and marked reduction in mitochondrial function. Current evidence suggests that complex I inhibition may be the central cause of sporadic PD and that derangements in complex I cause α-synuclein aggregation, which contributes to the demise of dopamine neurons. Complex I is inhibited by production of nitric oxide, which can interact with the proteins within complex I and thereby inhibit its activity. Treatment of glutathione-depleted cultured dopaminergic cells with inhibitors of NOS, the enzyme that makes NO•, prevents mitochondrial complex I inhibition [1, 51].

6.1 Neurotoxins Producing PD-Like Symptoms

6-Hydroxy dopamine (6-OHDA) was the first dopaminergic neurotoxin discovered and has been used experimentally in models of PD for 30 years [52, 53]. This compound is not only toxic to neurons but also can induce activation of glial cells [53]. Interestingly, accumulation of endogenous 6-OHDA has been shown in PD patients [54]. Upon transport to the neurons, 6-OHDA is oxidized similarly to dopamine to generate free radicals and quinones [53, 54]. 6-OHDA inhibits mitochondrial complex I and produces superoxide and hydroxyl radicals through participation in the Fenton reaction [54]. Although administration of 6-OHDA produces many symptoms resembling those of PD, it is worth noting that treatment with this compound does not cause the formation of Lewy bodies, the characteristic pathological hallmark of PD [52, 53].

6.2 1-Methyl-1-4-Phenyl-1,2,3,6-Tetrahydropyridine

1-Methyl-1-4-Phenyl-1,2,3,6-Tetrahydropyridine (MPTP) was the first exogenous toxin that may be involved in the pathophysiology of PD [55]. As in PD, autopsy of patients exposed to MPTP showed the presence of damage to the dopaminergic system in the substantia nigra, but in contrast to PD, Lewy bodies were not found [53, 54, 56]. MPTP is not harmful to dopaminergic neurons. In astrocytes, MAO (monoamine oxidase) B first converts MPTP into 1-methyl-4-phenyl-2,3-dihydropyridinium (MPDP) and then oxidizes it to a toxic pyridinium ion (MPP^+) [54, 55, 57]. MPP^+ is taken up by dopaminergic neurons [54]. Normal dopaminergic neurons accumulate MPP^+ to high levels in mitochondria, where it inhibits the electron transport chain of complex I and in turn produces ROS. In dopaminergic cell lines, MPP^+ produces ROS in mitochondria, which then trigger the classic apoptosis cascade including caspase activation and DNA fragmentation [58]. Although MPTP toxicity shares many similarities with PD, it is not associated with the formation of Lewy bodies, as already mentioned [53, 54, 56].

6.3 Paraquat

Paraquat is a widely used herbicide. There is significant overlap between geographic areas where paraquat is used and areas where PD is prevalent [59]. Paraquat is structurally similar to MPP^+, the active metabolite of MPTP. Similarly to MPTP, paraquat also depletes dopaminergic neurons in substantia nigra and causes the symptoms of parkinsonism. Within neurons, paraquat can produce intracellular ROS, leading to production of malonaldehyde, protein carbonyls, and DNA fragmentation [60–62]. Paraquat inhibits mitochondrial complex I and perturbs the mitochondrial respiration chain, causing impaired energy metabolism, proteosomal dysfunction, and intracellular ROS production [61, 63–64].

6.4 Rotenone

One of the most recent approaches to develop cell models to study PD results from the use of the pesticide rotenone. Since the 1950s it has been believed that rotenone acts as an inhibitor of the mitochondrial respiratory chain [58]. It can enter organelles, including double membrane-encased mitochondria, where it inhibits with high affinity complex I of the mitochondrial electron transport chain. Fortunately, most cells are able to detoxify the oxidative damage caused by rotenone through their antioxidant protective enzyme pathways. Unfortunately, chronic inhibition of this mitochondrial complex can cause selective degeneration of dopaminergic neurons despite being uniformly distributed throughout the brain [52]. This point raises the questions why dopaminergic cells are targeted for destruction and whether rotenone in the environment contributes to the etiology of PD in certain individuals. For reasons of its inhibition of complex I, rotenone is known to cause ROS generation, ATP depletion, and cell death in neurons [52]. In agreement with the involvement of ROS in rotenone toxicity, antioxidants are capable of attenuating rotenone-mediated cell death [52]. Similar to PD, treatment with rotenone leads to the aggregation of α-synuclein and formation of Lewy bodies. Another similarity to PD is that rotenone depletes glutathione, a primary cell antioxidant [52].

6.5 Involvement of Microglia in Oxidative Stress

All known dopaminergic neurotoxins exhibit a common characteristic: they inhibit either mitochondrial complex I or complex III in neurons [54]. Additionally, OS in the CNS comes not only from mitochondria-generated ROS in neurons but also from activated microglia [65]. Reactive microglia are known to play a role in several neurodegenerative disorders, including PD [66–68]. Indeed, PD patients can have more than six times the number of reactive microglia as compared to controls. However, it is unknown whether these microglia initiate or aggravate neurodegeneration [68]. Activated microglia also produce superoxide and nitric oxide [69]. To survive in the environment of these oxidative generators, it is not surprising that microglia also contain high levels of endogenous antioxidants. Activated microglia in the substantia nigra are found in several models of PD, including exposure to MPTP, rotenone, substance P, and methamphetamine. In many instances, activation of microglia and generation of ROS coincide with neurochemical changes such as the decrease in dopamine synthesis [70]. Microglia may play a role in the initiation and progression of PD and enhance neurotoxicity elicited by neurotoxins [71, 72]. Inhibition of microglial activation by the antibiotic drug minocycline could attenuate the neurotoxicity of 6-OHDA, MPTP, and rotenone [72, 73]. A number of models are available for PD, but each has its strengths and weakness in mimicking the pathophysiological features of the disease.

There are models that mimic mitochondrial dysfunction and raise OS; both features have been shown in postmortem studies of PD patients. Microglia play an active role in generation of ROS in PD. and NOS are known to play a role in inflammation and generation of ROS, and inhibition of these enzymes offers protection to dopaminergic neurons. These models suggest that, in addition to the genetic component, environmental factors also play a role in progression of PD. Furthermore, exposure of neurons in the substantia nigra to neurotoxins can cause dopaminergic cell loss similar to that seen in PD. Further investigations to develop therapeutics for PD and its prevention should focus on antioxidant and antiinflammatory compounds that may inhibit ROS formation and suppress microglial activation [74].

7 Alzheimer's Disease

Alzheimer's disease (AD) is a complex, progressive degeneration of the neurons of the neocortex, characterized by memory loss and deterioration of higher cognitive functions. It is the most common form of adult-onset dementia [75]. Histopathologically, AD is characterized by synaptic and nerve cell loss, extracellular deposition of β-amyloid protein forming senile plaques, and intracellular precipitation of *tau* protein [75, 76]. At present, available treatments are unable to stop the progression of AD, which makes the identification of novel treatment for prevention and neuroprotection a pressing scientific concern [77]. The exact biochemical mechanism responsible for the pathogenesis of AD is still unknown, but much attention has been given to the role of the massive loss of the neurotransmitter acetylcholine (necessary for cognition and memory) and to the possible implication of OS in AD by postmortem findings of enhanced lipid peroxidation in specific areas of the brain. AD is essentially a disorder in which there is an acceleration of the aging process in affected brain regions that become progressively more damaged by ROS. In the initial phase of AD, β-amyloid deposition and *tau* protein precipitation may function as compensatory response and downstream adaptations to ensure that neuronal cells do not succumb to oxidative injury. However, during the progression of the disease, the antioxidant activity of both β-amyloid protein and *tau* evolves pro-oxidant activity representing a typical gain-of-function transformation, which can result from increase in ROS and a decrease in clearance mechanisms [76]. The complex nature and genesis of oxidative damage in AD can be partly answered by mitochondrial and redox-active metal abnormalities. Mitochondrial involvement in the pathogenesis of AD and its relationship to OS are repeatedly hypothesized in research and clinical literature. Direct evidence supporting increased OS in AD brain has been described [75]. Although circumstantial evidence for an involvement of ROS in the formation of β-amyloidosis is abundant [78], there is also direct evidence for a link between amyloid aggregation, OS generation, and neurotoxicity [79]. It was reported [80] that the aggregation of β-amyloid peptide ($A_{\beta}P$) and amyloid precursor protein (APP) fragments could be induced by metal-catalyzed ROS-generating systems and that ROS scavengers prevented the aggregation process. $A_\beta P$ can generate free radical peptides and produce ROS [81]. Therefore, $A_\beta P$ neurotoxicity appears to involve complex free radical chemistry. Furthermore, it was also reported that a number of antioxidants can protect neuronal cells from $A_\beta P$ cytotoxicity and that $A_\beta P$ increases ROS in exposed cells [82]. All these data provide strong support to the hypothesis that the cytotoxic action of $A_\beta P$ results, at least in part, from ROS-mediated oxidative damage to susceptible cells. How can $A_\beta P$ lead to ROS production? First, it can form ROS itself [83]. Mixture of $A_\beta P$ and Cu^{1+} can cause oxidation of ascorbate, L-DOPA, dopamine, or cholesterol accompanied by H_2O_2 production [83]. Second, $A_\beta P$ may stimulate ROS production by brain cells [84]. Microglia activation has been implicated in the progressive nature of AD. Microglia can become deleteriously activated in response to disease-specific stimuli (amyloid-β (Aβ) oligomers, Aβ fibrils, and senile plaques) to produce neuronal toxins such as ROS and cytokines. In addition to disease-specific proinflammatory stimuli, neuronal damage/death can also activate microglia to produce these toxic factors. This ongoing and self-perpetuating cycle of neuronal damage/death followed by microglial activation is commonly referred to as microgliosis and may be an underlying mechanism of the progressive nature of diverse neurodegenerative diseases, including AD. NADPH oxidase has been implicated as a key mechanism through which microglia damage neurons in response to A_β and neuron damage/death [77]. As for OS, there is strong evidence to support its role in AD. This evidence is stronger than for any other neurodegenerative disease. It was speculated [85] that melatonin, because of its free radical scavenging activity, neutralized the radicals that were generated extracellularly by $A_\beta P$ as well as those produced intracellularly by the markedly increased calcium, thereby reducing the neurotoxicity of the $A_\beta P$ and leading to increased cellular survival. These observations, coupled with the known reduction in melatonin [86] in the aged, raise the possibility that the loss of this antioxidant in the later stages of life may predispose individuals to the neurotoxic effects of $A_\beta P$ and to AD. Antioxidants in general are receiving increased interest as agents that may defer the onset of this devastating disease.

References

1. Halliwell B, Gutteridge JMC. Free radicals in biology and medicine. 4th edn. Oxford University Press, Oxford, 2007.
2. Halliwell B. Oxidative stress and neurodegeneration: where are we now? J Neurochem. 2006;97:1634–58.
3. Nohl H, Hegner D. Do mitochondria produce oxygen free radicals in vivo? Eur J Biochem. 1978;82:563–7.
4. Sohal RS. Mitochondria generate superoxide anion radicals and hydrogen peroxide. FASEB J. 1997;11:1269–70.
5. Fridovich I. Superoxide dismutases. An adaptation to a paramagnetic gas. J Biol Chem. 1989;264:7761–4.

6. Touati D. The molecular genetics of superoxide dismutase in *E. coli*. An approach to understanding the biological role and regulation of SOD in relation to other elements of the defense system against oxygen toxicity. Free Radic Biol Med. 1989;8:1–9.
7. Imlay JA, Chin SM, Linn S. Toxic DNA damage by hydrogen peroxide through the Fenton reaction in vivo and in vitro. Science. 1988;240:1302–9.
8. Halliwell B, Gutteridge JM. Role of free radicals and catalytic metal ions in human disease: an overview. Methods Enzymol. 1990;186:1–85.
9. Yamazaki I, Piette LH. EPR spin trapping study on the oxidizing species formed in the reaction of the ferrous ion with hydrogen peroxide. J Am Chem Soc. 1991;113:7588–93.
10. Jain A, Martensson J, Stole E, et al. Glutathione deficiency leads to mitochondrial damage in brain. Proc Natl Acad Sci USA. 1991;88:1913–7.
11. Cochrane CG. Mechanisms of oxidant injury of cells. Mol Aspects Med. 1991;12:137–47.
12. Halliwell B, Aruoma OI. DNA damage by oxygen-derived species. Its mechanism and measurement in mammalian systems. FEBS Lett. 1991;28:9–19.
13. Orrenius S, McConkey DJ, Bellomo G, et al. Role of Ca^{2+} in toxic cell killing. Trends Pharmacol Sci. 1989;10:281–5.
14. Asano T, Matsui T, Takuwa Y. Lipid peroxidation, protein kinase C and cerebral ischemia. Crit Rev Neurosurg. 1991;1:361–79.
15. Sevanian A, McLeod LL. Formation and biological reactivity of lipid peroxidation. In: Wallace KB, editors. Free radical toxicology. Washington, DC, Taylor & Francis, 1997:47–70.
16. Radi R, Beckman JS, Bush KM, et al. Peroxynitrite oxidation of sulfhydryls. The cytotoxic potential of superoxide and nitric oxide. J Biol Chem. 1991;266:4244–50.
17. Beckman JS, Ischiropoulos H, Zhu L, et al. Kinetics of superoxide dismutase- and iron-catalyzed nitration of phenolics by peroxynitrite. Arch Biochem Biophys. 1992;298:438–55.
18. Crow JP, Beckman NJS, McCord JM. Sensitivity of the zinc-thiolate moiety of yeast alcohol dehydrogenase to hypochlorite and perxoinitrite. Biochemistry. 1995;34:3544–52.
19. Radi R, Beckman JS, Bush KM, et al. Peroxynitrite induced membrane lipid peroxidation. The cytotoxic potential of superoxide and nitric oxide. Arch Biochem Biophys. 1991;288:481–7.
20. Moreno JJ, Pryor WA. Inactivation of alpha 1-proteinase inhibitor by peroxynitrite. Chem Res Toxic. 1992;5:425–31.
21. Kirkwood TB. Understanding the odd science of aging. Cell. 2005;120:437–47.
22. Maher P, Schubert D. Signaling by reactive oxygen species in the nervous system. Cell Mol Life Sci. 2000;57:1287–305.
23. Sugawara T, Fujimura M, Noshita N, et al. Neuronal death/survival signaling pathways in cerebral ischemia. NeuroRx. 2004;1:17–25.
24. Maher P. How protein kinase C activation protects nerve cells from oxidative stress-induced cell death. J Neurosci. 2001;21:2929–38.
25. Kamada H, Nito C, Endo H, et al. Bad as a converging signaling molecule between survival PI3-K/Akt and death JNK in neurons after transient focal cerebral ischemia in rats. J Cereb Blood Flow Metab. 2006;27:521–33.
26. Adibhatla RM, Hatcher JF, Dempsey RJ. Citicoline: neuroprotective mechanisms in cerebral ischemia. J Neurochem. 2002;80:12–23.
27. Saeed SA, Shad KF, Saleem T, et al. Some new prospects in the understanding of the molecular basis of the pathogenesis of stroke. Exp Brain Res. 2007;182:1–10.
28. Taylor JM, Crack PJ. Impact of oxidative stress on neuronal survival. Clin Exp Pharmacol Physiol. 2004;31:397–406.
29. Hewett SJ, Uliasz TF, Vidwans AS, et al. Cyclooxygenase-2 contributes to *N*-methyl-D-aspartate-mediated neuronal cell death in primary cortical cell culture. J Pharmacol Exp Ther. 2000;293:417–25.
30. Toescu EC. Hypoxia sensing and pathways of cytosolic Ca^{2+} increases. Cell Calcium. 2004;36:187–99.

31. Chen RM, Chen TL, Chiu WT, et al. Molecular mechanism of nitric oxide-induced osteoblast apoptosis. J Orthop Res. 2005;23:462–8.
32. Maines MD. The heme oxygenase system: update. Antioxid Redox Signal. 2005;7:1761–6.
33. Poon HF, Calabrese V, Scapagnini G, et al. Free radicals: key to brain aging and heme oxygenase as a cellular response to oxidative stress. J Gerontol A Biol Sci Med Sci. 2004;59:478–93.
34. Alam J, Cook JL. Transcriptional regulation of the heme oxygenase-1 gene via the stress response element pathway. Curr Pharm Des. 2003;9:2499–511.
35. Franklin TB, Krueger-Naug AM, Clarke DB, et al. The role of heat shock proteins Hsp70 and Hsp27 in cellular protection of the central nervous system. Int J Hyperthermia. 2005;21:379–92.
36. Igarashi K, Sun J. The heme-Bach1 pathway in the regulation of oxidative stress response and erythroid differentiation. Antioxid Redox Signal. 2006;8:107–18.
37. Calabrese V, Guagliano E, Sapienza M, et al. Redox regulation of cellular stress response in aging and neurodegenerative disorders: role of vitagenes. Neurochem Res. 2007;32:757–73.
38. Gilgun-Sherki Y, Rosenbaum Z, Melamed E et al. Antioxidant therapy in acute central nervous system injury: current state. Pharmacol Rev. 2002;54:271–84.
39. Williams AJ, Berti R, Dave JR, et al. Delayed treatment of ischemia/reperfusion brain injury: Stroke. 2004;35:1186–91.
40. Van der Worp HB, Bar PR, Kappelle LJ, et al. Dietary vitamin E levels affect outcome of permanent focal cerebral ischemia in rats. Stroke. 1998;29:1002–5.
41. Asai A, Tanahashi N, Qiu JH, et al. Selective proteasomal dysfunction in the hippocampal CA1 region after transient forebrain ischemia. J Cereb Blood Flow Metab. 2002;22:705–10.
42. Bayir H, Marion DW, Puccio AM, et al. Marked gender effect on lipid peroxidation after severe traumatic brain injury in adult patients. J Neurotrauma. 2004;21:1–8.
43. Ballabh P, Braun A, Nedergaard M. The blood–brain barrier: an overview: structure, regulation, and clinical implications. Neurobiol Dis. 2004;16:1–13.
44. Phillis JW, O'Regan MH. A potentially critical role of phospholipases in central nervous system ischemic, traumatic, and neurodegenerative disorders. Brain Res Rev. 2004;44:13–47.
45. Liu D, Liu J, Sun D, et al. Spinal cord injury increases iron levels: catalytic production of hydroxyl radicals. Free Radic Biol Med. 2003;34:64–71.
46. Macdonald RL, Marton LS, Andrus PK, et al. Time course of production of hydroxyl free radical after subarachnoid hemorrhage in dogs. Life Sci. 2004;75:979–89.
47. Wagner KR, Sharp FR, Ardizzone TD, et al. Heme and iron metabolism: role in cerebral hemorrhage. J Cereb Blood Flow Metab. 2003;23:629–59.
48. Chang EF, Wong RJ, Vreman HJ, et al. Heme oxygenase-2 protects against lipid peroxidation-mediated cell loss and impaired motor recovery after traumatic brain injury. J Neurosci. 2003;23:3689–96.
49. Berg D, Youdim MB, Riederer P. Redox imbalance. Cell Tissue Res. 2004;318:201–13.
50. Cui K, Luo X, Xu K, et al. Role of oxidative stress in neurodegeneration: recent developments in assay methods for oxidative stress and nutraceutical antioxidants. Prog Neuro-psychopharmacol Biol Psychiatry. 2004;28:771–99.
51. Fato R, Bergamini C, Leoni S, et al. Generation of reactive oxygen species by mitochondrial complex I: implications in neurodegeneration. Neurochem Res. 2008;33:2487–501.
52. Uversky VN. Neurotoxicant-induced animal models of Parkinson's disease: understanding the role of rotenone, maneb and paraquat in neurodegeneration. Cell Tissue Res. 2004;318:225–41.
53. Bove J, Prou D, Perier C, et al. Toxin-induced models of Parkinson's disease. NeuroRx. 2005;2:484–94.
54. Schober A. Classic toxin-induced animal models of Parkinson's disease: 6-OHDA and MPTP. Cell Tissue Res. 2004;318:215–24.
55. Wersinger C, Sidhu A. An inflammatory pathomechanism for Parkinson's disease? Curr Med Chem. 2006;13:591–602.

56. Landrigan PJ, Sonawane B, Butler RN, et al. Early environmental origins of neurodegenerative disease in later life. Environ Health Perspect. 2005;113:1230–3.
57. Maguire-Zeiss KA, Short DW, Federoff HJ. Synuclein, dopamine and oxidative stress: co-conspirators in Parkinson's disease? Brain Res Mol Brain Res. 2005;134:18–23.
58. Kotake Y, Ohta S. MPP^+ analogs acting on mitochondria and inducing neuro-degeneration. Curr Med Chem. 2003;10:2507–16.
59. Shimizu K, Matsubara K, Ohtaki K, et al. Paraquat leads to dopaminergic neural vulnerability in organotypic midbrain culture. Neurosci Res. 2003;46:523–32.
60. Peng J, Stevenson FF, Doctrow SR, et al. Superoxide dismutase/catalase mimetics are neuroprotective against selective paraquat-mediated dopaminergic neuron death in the substantial nigra: implications for Parkinson disease. J Biol Chem. 2005;280:2194–8.
61. Yang W, Tiffany-Castiglioni E. The bipyridyl herbicide paraquat produces oxidative stress-mediated toxicity in human neuroblastoma SH-SY5Y cells: relevance to the dopaminergic pathogenesis. J Toxicol Environ Health Part A. 2005;68:1939–61.
62. Thiruchelvam M, Prokopenko O, Cory-Slechta DA, et al. Overexpression of superoxide dismutase or glutathione peroxidase protects against the paraquat + maneb-induced Parkinson disease phenotype. J Biol Chem. 2005;280:22530–9.
63. Shimizu K, Matsubara K, Ohtaki K, et al. Paraquat induces long-lasting dopamine overflow through the excitotoxic pathway in the striatum of freely moving rats. Brain Res. 2003;976:243–52.
64. Bretaud S, Lee S, Guo S. Sensitivity of zebrafish to environmental toxins implicated in Parkinson's disease. Neurotoxicol Teratol. 2004;26:857–64.
65. Chun HS, Gibson GE, DeGiorgio LA, et al. Dopaminergic cell death induced by MPP (+), oxidant and specific neurotoxicants shares the common molecular mechanism. J Neurochem. 2001;76:1010–21.
66. Block ML, Li G, Qin L, et al. Potent regulation of microglia-derived oxidative stress and dopaminergic neuron survival: substance P vs. dynorphin. FASEB J. 2006;20:251–8.
67. Li J, Baud O, Vartanian T, et al. Peroxynitrite generated by inducible nitric oxide synthase and NADPH oxidase mediates microglial toxicity to oligodendrocytes. Proc Natl Acad Sci USA. 2005;102:9936–41.
68. Zhang W, Wang T, Pei Z, et al. Aggregated alpha-synuclein activates microglia: a process leading to disease progression in Parkinson's disease. FASEB J. 2005;19:533–42.
69. Dringen R. Oxidative and antioxidative potential of brain microglial cells. Antioxid Redox Signal. 2005;7:1223–33.
70. Scheller C, Sopper S, Jenuwein M, et al. Early impairment in dopaminergic neurotransmission in brains of SIV-infected rhesus monkeys due to microglia activation. J Neurochem. 2005;95:377–87.
71. Gao HM, Hong JS, Zhang W, et al. Distinct role for microglia in rotenone-induced degeneration of dopaminergic neurons. J Neurosci. 2002;22:782–90.
72. Casarejos MJ, Menendez J, Solano RM, et al. Susceptibility to rotenone is increased in neurons from parkin null mice and is reduced by minocycline. J Neurochem. 2006;97:934–46.
73. Croisier E, Moran LB, Dexter DT, et al. Microglial inflammation in the parkinsonian substantia nigra: relationship to alpha-synuclein deposition. J Neuroinflammation. 2005; 2:14–22.
74. Miller RL, James-Kracke M, Sun GY, et al. Oxidative and inflammatory pathways in Parkinson's disease. Neurochem Res. 2009;34:55–65.
75. Markesbery WR. Oxidative stress hypothesis in Alzheimer's disease. Free Radic Biol Med. 1997;23:134–47.
76. Zhu X, Smith MA, Perry G, et al. Mitochondrial failures in Alzheimer's disease. Am J Alzheimer's Dis Other Demen. 2004;19:345–52.
77. Block ML. NADPH oxidase as a therapeutic target in Alzheimer's disease. BMC Neurosci. 2008;9:S8.

78. Friedlich AL, Butcher LL. Involvement of free oxygen radicals in beta-amyloidosis: an hypothesis. Neurobiol Aging. 1994;15:443–55.
79. Hashimoto M, Rockenstein E, Crews L, et al. Role of protein aggregation in mitochondrial dysfunction and neurodegeneration in Alzheimer's and Parkinson's diseases. Neuromolecular Med. 2003;4:21–36.
80. Dyrks T, Dyrks E, Hartmann T, et al. Amyloidogenicity of beta A4 and beta A4-bearing amyloid protein precursor fragments by metal-catalyzed oxidation. J Biol Chem. 1992;267:18210–7.
81. Hensley K, Carney J, Hall N, et al. Electron paramagnetic resonance investigations of free radical-induced alterations in neocortical synaptosomal membrane protein infrastructure. Free Radic Biol Med. 1994;17:321–31.
82. Behl C, Davis JB, Lesley R, et al. Hydrogen peroxide mediates amyloid beta protein toxicity. Cell. 1994;77:817–27.
83. Barnham KJ, Masters CL, Bush AI. Neurodegenerative diseases and oxidative stress. Nat Rev Drug Disc. 2004;3:205–14.
84. Abramov AY, Canevari L, Duchen MR. Beta-amyloid peptides induce mitochondrial dysfunction and oxidative stress in astrocytes and death of neurons through activation of NADPH oxidase. J Neurosci. 2004;24:565–75.
85. Pappolla MA, Sos M, Omar RA, et al. Melatonin prevents death of neuroblastoma cells exposed to the Alzheimer amyloid peptide. J Neurosci. 1997;17:1683–90.
86. Reiter RJ. The ageing pineal gland and its physiological consequences. Bio Essays. 1992;14:169–75.

Chapter 2
Why Is the Nervous System Vulnerable to Oxidative Stress?

Joseph Friedman

Abstract The nervous system is especially vulnerable to reactive oxygen species (ROS)-mediated injury for the following reasons. (1) High oxygen consumption of the brain for high energy needs, that is, high O_2 consumption, results in excessive ROS produced. (2) Neuronal membranes are rich in polyunsaturated fatty acids (PUFA), which are particularly vulnerable to free radical attack. (3) The ratio of membrane surface area to cytoplasmic volume is high. (4) Specialized neuronal conduction and synaptic transmission activity depend on efficient membrane function. (5) Extended axonal morphology is prone to peripheral injury. (6) Neuronal anatomic network is vulnerable to disruptions. (7) The excitotoxic glutamate is the major effector that causes oxidative stress (OS). (8) The high Ca^{2+} traffic across neuronal membranes and interference of ion transport increase intracellular Ca^{2+}, often leading to OS. (9) Auto-oxidation of neurotransmitters can generate O_2 and quinones that reduce glutathione. (10) Iron is formed throughout the brain, and brain damage readily releases iron ions capable of catalyzing free radical reactions. (11) Antioxidant defense mechanisms are modest, in particular, low levels of catalase, glutathione peroxidase, and vitamin E. (12) ROS directly downregulate proteins of tight junctions and indirectly activate matrix metalloproteinases (MMP) that contribute to open the blood–brain barrier (BBB). (13) Activated microglia produce ROS and cytokines in a perpetual process. (14) Cytochrome P450 produces ROS. (15) Loss of trophic support can activate NADPH oxidase, which increases ROS. (16) The presence of hemoglobin within the neural tissues secondary to spontaneous, iatrogenic, or traumatic causes is neurotoxic. Heme and iron are released and promote ROS. (17) Neuronal mitochondria generate O_2. (18) The interaction of NO with superoxide can be implicated also in neuronal degeneration. (19) Neuronal cells are nonreplicating and thus are sensitive to ROS. In comparison with other organs, the neuronal network may be especially

J. Friedman (✉)
Weizmann Science Park, Ness-Ziona, Israel
e-mail: dr.josephfriedman@gmail.com

N. Gadoth and H.H. Göbel (eds.), *Oxidative Stress and Free Radical Damage in Neurology*, Oxidative Stress in Applied Basic Research and Clinical Practice,
DOI 10.1007/978-1-60327-514-9_2,

vulnerable to ROS-mediated injury because of the following anatomic, physiological, and biochemical properties of the brain.

Keywords Neuron · Microglia · Mitochondria · ROS · OS · NO · O_2 consumption · PUFA · Excitoxicity · Ca^{2+} · Iron · NADPH oxidase · Auto-oxidation · Neurotransmitters · Antioxidants · BBB

1 High O_2 Consumption [1, 2]

The brain, which accounts for only 2% of body weight, consumes 20% of the total oxygen inspired. The brain processes much O_2 per unit tissue mass. The discrepancy is even more striking in young children who have much smaller bodies but without proportionally smaller brains. A major reason for the high O_2 uptake by the brain is the vast amounts of adenosine triphosphate (ATP) needed for its normal activity (4×10^{12} ATP molecules every minute). The large amount of ATP is necessary to maintain neuronal intracellular ion homeostasis in face of all the openings and closings of ion channels associated with propagation of action potentials and neurosecretion. The heavy reliance of the brain on oxidative phosphorylation makes the brain vulnerable to interruptions of O_2 supply. Similarly, inhibitors of mitochondrial ATP synthesis readily cause neuronal death. Examples include cyanide 3-nitropropionic acid (an inhibitor of mitochondrial succinate dehydrogenase) and rotenone (an inhibitor of complex I). Because approximately 5% of oxygen consumed by cells is estimated to be reduced to reactive oxygen species (ROS), a relatively higher amount of ROS may be generated in the brain as compared to other tissues that use less oxygen.

2 Neuronal Membranes Rich in PUFA [3, 4]

Neuronal membrane lipids are rich in polyunsaturated fatty acids (PUFA) side chains, especially those of eicosapentanoic ($C_{20:5}$) and decosahexanoic ($C_{22:6}$) acids. PUFAs are particularly vulnerable to free radical attack because of the double bonds, within the membrane allowing easy removal of hydrogen ions by ROS such as $OH^{\bullet}$. Homogenization of brain tissue causes rapid lipid peroxidation, which can be largely inhibited by ion-chelating agents such as desferrioxamine. In addition, products of lipid peroxidation can injure the brain. 4-Hydroxy-nonenal, which is one of those products, is especially cytotoxic to neurons, increasing Ca^{2+} levels, inactivating glutamate transporters, and damaging neurofilament proteins. It can also inactivate α-ketoglutarate dehydrogenase (α-KGDH), a key enzyme of the tricarboxylic acid cycle. 4-Oxo-2-nonenal may be equally or more toxic, and isoprostanes (IPs) may act as vasoconstrictive agents in brain and can damage developing oligodendrocytes in premature babies. Other products of the IP pathway may also be neurotoxic by damaging the proteosome, for example.

3 High Ratio of Membrane Surface Area to Cytoplasmic Volume

4 Specialized Neuronal Conduction and Synaptic Transmission Activity Depend on Efficient Membrane Function

5 Extended Axonal Morphology Is Prone to Peripheral Injury

6 Neuronal Anatomic Network Is Vulnerable to Disruption

7 Oxidative Stress and Excitotoxic Amino Acids [5, 6]

Although multiple factors can precipitate intracellular oxidative stress (OS), the neurotransmitter glutamate is the major effector of this process in the brain, primarily through activation of its ionotropic receptors. Brain extracellular concentrations of glutamate are normally low (<μM). Neuronal death or collapse of normal ion gradients (e.g., owing to severe energy depletion) in neurons cause massive glutamate release. Glutamate and related excitatory amino acids account for most of the excitatory synaptic activity in the mammalian central nervous system (CNS) and are released by as many as 40% of all synapses.

The *N*-methyl-D-aspartate (NMDA), α-amino-3-hydroxy-5-methyl-4-isoxasol-propionic acid (AMPA), and kaineic acid (KA) ionotropic receptors can be distinguished by their pharmacologic and electrophysiological properties. The binding to neuronal receptors leads to excessive and prolonged increases in $\bar{O}_2$ and NO formation and intracellular Ca^{2+}. Neurons treated with excess glutamate or other excitatory toxins swell rapidly and die, usually by necrosis. OS can damage neurons and promote the release of excitatory amino acids, generating a vicious cycle of events. Several ROS are able to decrease glutamate uptake by glial cells and to inactivate glutamine synthetase, preventing conversion of glutamate to glutamine. This enzyme is inactivated in Alzheimer's disease. The early drop in cellular glutathione (GSH) levels observed together with oxidative glutamate toxicity is very similar to the changes seen in vivo within neurons responding to both acute and chronic injury.

8 High Ca^{2+} Traffic Across Neuronal Membranes [7, 8]

The high Ca^{2+} traffic across neuronal membranes interferes with ion transport (e.g., by disruption of energy metabolism) and can produce rapid rises in intracellular free Ca^{2+}, often leading to OS. Increase in intracellular free Ca^{2+} activates neuronal nitric oxide synthase (nNOS), phospholipase A_2, and calpain.

9 Autooxidation of Neurotransmitters [9–11]

Several neurotransmitters can be autooxidized. Dopamine, its precursor L-DOPA, serotonin, and norepinephrine can react with O_2 to generate not only O_2 but also quinones/semiquinones that can deplete reduced glutathione (GSH) and bind to protein sulfhydryl (SH) groups. Oxidation can be catalyzed by transition metal ions, but when $\bar{O}_2$ is in excess, it can react with norepinephrine, dopamine, and serotonin to initiate their oxidation, which then continues with production of more ROS, quinones, etc. Monoamine oxidase (MAO) exists in two forms, MAO-A and -B. MAO-A preferentially oxidizes hydroxylated amines such as serotonin and noradrenalin and is found primarily in catecholaminergic neurons. MAO-B preferentially oxidizes nonhydroxylated amines and is located in serotoninergic neurons. Both oxidases are present in glial cells and are able to oxidize dopamine.

10 Iron [12, 13]

Iron is found throughout the brain. Important iron-containing proteins include cytochromes, ferritin, aconitases, mitochondrial non-heme iron proteins, cytochrome P450, and iron is found in both tyrosine and txyptophan hydroxylase enzymes, which catalyze the first steps in the synthesis of dopamine and serotonin, respectively. Several brain areas (e.g., substantia nigra, caudate nucleus, putamen, and globus pallidus) have high iron content. Ferritin contains a major portion of the healthy brain iron, and smaller amounts are found in hemosiderine. The iron content of the brain is low at birth and rapidly increases during early life to reach the adult level at the age of 30 years. If insufficient dietary iron is present in infants and children, it may cause impairment of brain function. Transferrin delivers most of the required iron across the blood–brain barrier (BBB), utilizing receptors located at the brain microvasculature. However, when the brain is damaged, it readily releases iron (and copper) ions in forms capable of catalyzing such free radical reactions as $OH^{\bullet}$ formation from H_2O_2, lipid peroxidation, and autooxidation of neurotransmitters. Injection of iron salts into the brains of animals can produce epileptic seizures, dopamine depletion, and death of neurons accompanied by lipid peroxidation, oxidative damage, and $OH^{\bullet}$ generation. Catalytic iron released by brain damage can persist because the cerebrospinal fluid (CSF), which surrounds the brain and spinal cord, has little or no iron-binding capacity. Total iron values in CSF from normal humans range, according to various reports, from 0.29 to 1.11 μM. The transferrin content is 0.24 μM. Because 1 mol of transferrin binds 2 mol of iron ions, these data suggest that CSF transferrin is often at, or close to, its iron saturation.

11 Antioxidant Defenses are Modest [14, 15]

Brain antioxidant defenses are modest. In particular, catalase levels are low in most brain regions; levels are higher in hypothalamus and substantia nigra than in the cortex or cerebellum. Brain catalase is located in small peroxisomes (microperoxisomes), and its activity in rat or mouse brain is rapidly inhibited if aminotriazole is administered to the animals. This agent inhibits only catalase of complex I, confirming that the brain generates H_2O_2 in vivo and that at least some of it reaches catalase. The catalase probably cannot handle H_2O_2 generated in other subcellular compartments. The brain contains less catalase, glutathione peroxidase (GPx), and vitamin E as compared to liver. Ascorbic acid, which can act as an antioxidant as well as a pro-oxidant, is present at elevated levels in both white and gray matter. Transport systems exist in the choroid plexus and neurons that serve to concentrate ascorbic acid into brain cells and the CSF. Ascorbic acid acts as a pro-oxidant when the free iron of the brain increases as a result of intracerebral hemorrhage. The developing brain may be particularly susceptible to free radical injury during ischemia because of a relative developmental deficiency in its antioxidant enzymes. The activity of those enzymes has not yet been determined in the cerebral cortex of developing humans, although expression levels and cellular localization have been reported in the cortex, basal ganglia, and brainstem nuclei by immunohistochemical methods. The premature infant has low circulating levels of glutathione (GSH) and relative inability to sequester iron because of low transferrin levels. It is well known that glial cells are more resistant to OS than neurons, probably the result of transcriptional upregulation of glutathione synthesis.

12 The Blood–Brain Barrier [16, 17]

The brain needs a barrier that can selectively and efficiently prevent harmful molecules from entering the brain from the blood, thus enabling the rigorous control of the brain microenvironment that is necessary for complex neuronal signaling. ROS, both directly (e.g., by downregulating the synthesis of proteins involved in tight junctions between cells) and/or by activation of matrix metalloproteinases (MMP), can contribute to "open up" the BBB, allowing the entry of neurotoxins and inflammatory cells. The adult brain neurons are protected by the BBB, and the blood–axon barrier acts similarly in the peripheral nervous system (PNS). In the immature nervous system, these barriers are not fully formed, and thus it is more vulnerable. In elderly people, the combination of accumulating vascular damage with associated breakdown of those barriers may render them sensitive to exposure to ROS and toxins. Taken together, oxidative stress emerges as a common underlying cause of BBB dysfunction.

13 Microglia and ROS [18, 19]

The microglia are resident macrophage-type cells that arise from monocytes entering the brain during embryonic development. Normally, they help clear cellular debris (including apoptotic cells) and are "alert" to threats to neurons. However, microglia can become activated to produce $\bar{O}_2$, H_2O_2, and cytokines such as interleukin (IL)-1, IL-6, and tumor necrosis factor-alpha (TNF-α). Subsequently such cytokines can cause microglia to generate more ROS and to produce iNOS (induced NO synthase) and hence excess of $NO^{\bullet}$ (nitric oxide). Cytokines can additionally be produced by activated astrocytes, and the latter may again respond to cytokines by iNOS induction. Thus, microglia and astrocytes are major mediators in brain inflammation.

14 Cytochrome P450 [20, 21]

Cytochrome P450 (CYPs), such as CYP46 that metabolizes cholesterol, CYP2D6, and CYP2E1 are present in several brain regions. Because CYP2E1 leaks electrons readily during its catalytic cycle, it produces more ROS than most other CYPs. Thus, it is another possible source of oxidative stress, although its magnitude may be small because brain CYPs levels are low compared to their levels in the liver. However, CYP2E1 metabolizes ethanol, acetone, halothane, and organic solvents such as CCl_4 and $CHCl_3$ and may be inducible in human brain by ethanol and smoking. Thus, it could contribute to solvent neurotoxicity.

15 Loss of Trophic Support [22, 23]

Loss of trophic support can lead to oxidative stress and apoptosis in neurons, in part by excess activation of NADPH (nicotinamide adenine dinucleotide phosphate) oxidase (NOX). The latter were first detected in phagocytes, but are now known to be widely distributed in animal tissues, seemingly producing $\bar{O}_2$ for defense and for signaling purposes. Neuronal NOX enzymes may promote necessary apoptosis during development of the nervous system, but if trophic support is lost in the developing brain, they are overactivated, leading to neuronal death.

16 Hemoglobin Is Neurotoxic [24–27]

Hemoglobin (Hb) is neurotoxic. This protein is normally safely transported in erythrocytes, which are rich in antioxidant enzymes. However, isolated Hb is degraded when exposed to excess H_2O_2, with release of pro-oxidant iron ions from the heme ring. Heme can also be released and is a powerful promoter of lipid peroxidation. In addition, Hb reacts with H_2O_2 and other peroxides to form oxidizing species capable of stimulating lipid peroxidation. Hb also binds NO avidly, producing vasoconstriction.

17 Mitochondria [28]

Neuronal mitochondria generate $\dot{\bar{O}}_2$, mostly from complex I. The levels of 8-hydroxydeoxyguanosine (8-OHDG) mutations and deletions increase in brain mitochondrial DNA with aging.

18 Nitric Oxide

NO, also referred to as endothelium-derived relaxing factor because of its production in the endothelial cells, is also formed in neurons by the enzyme nitric oxide synthase (NOS), which is widely distributed in the brain and is activated by calmodulin. The interaction of NO with superoxide radicals is believed to be implicated not only in the normal metabolism of the neuron but also in its degradation.

19 Neurons are Nonreplicating [29]

Neurons are generally postmitotic cells, and thus the brain is sensitive to loss of function if neurons die. Neurons are very active and are highly susceptible to a reduction in the supply of oxygen and glucose. The insidious reduction in the number of neurons and their synaptic connections eventually compromises virtually all CNS functions. Fortunately, many parts of the brain have considerable redundancy and plasticity. Neurons in culture are prone to injury, often leading to necrosis or apoptosis if treated with toxins that interfere with energy metabolism, or exposed to ROS, or if neurotrophic factors are withdrawn from the culture medium.

In this chapter, 19 reasons were mentioned why the nervous system is especially vulnerable to ROS-mediated injury. It was suggested that the main effects of ROS on cells are through their actions on signaling pathways rather than by causing nonspecific damage. I suggested that with aging these pathways deteriorate, causing accumulation of higher concentrations of ROS in amounts beyond the capacity of the brain's antioxidant system to handle. This deterioration results in age-associated neurodegenerative disorders such as stroke and CNS trauma, as well as Parkinson's disease, Alzheimer's disease, and other neurodegenerative disorders of aging.

References

1. Carey J, editor. Brain facts: a primer on the brain and nervous system. Washington, DC: The Society for Neuroscience; 2002.
2. Halliwell B. Oxidative stress and neurodegeneration: where are we now? J Neurochem. 2006;97:1634–58.
3. Chen CT, Green JT, Orr SK, Bazinet RP. Regulation of brain polyunsaturated fatty acid uptake and turnover. Prostaglandins Leukot Essent Fatty Acids. 2008;79:85–91.
4. Dyall SC, Michael-Titus AT. Neurological benefits of omega-3 fatty acids. Neuromolecular Med. 2008;10:219–35.
5. Mailly F, Marin P, Israël M, Glowinski J, Prémont J. Increases in external glutamate and NMDA receptor activation contribute to H_2O_2-induced neuronal apoptosis. J Neurochem. 1999;73:1181–88.
6. Phillis JW, Oregan MH. Characterization of modes of release of amino acids in the ischemic/reperfused rat cerebral cortex. Neurochem Int. 2003;43:461–7.
7. Vannucci RC, Brucklaher RM, Vannucci SJ. Intracellular calcium accumulation during the evolution of hypoxic-ischemic brain damage in the immature rat. Brain Res Dev Brain Res. 2001;126:117–20.
8. González A, Schmid A, Salido GM, Camello PJ, Pariente JA. XOD-catalyzed ROS generation mobilizes calcium from intracellular stores in mouse pancreatic acinar cells. Cell Signal. 2002;14:153–9.
9. Ameijeiras AH, Alvarez EM, Sanchez S. Autoxidation of MAO-mediated metabolism of dopamine as a potential cause of oxidative stress: role of ferrous and ferric ions. Neurochem Int. 2004;45:103–16.
10. Troadec JD, Marien M, Darios F, Hartmann A, Ruberg M, Colpaert F, et al. Noradrenaline provides long-term protection to dopaminergic neurons by reducing oxidative stress. J Neurochem. 2001;79:200–10.
11. Spencer JP, Jenner P, Daniel SE, Lees AJ, Marsden DC, Halliwell B. Conjugates of catecholamines with cysteine and GSH in Parkinson's disease: possible mechanisms of formation involving reactive oxygen species. J Neurochem. 1998;71:2112–22.
12. Stankiewicz JM, Brass SD. Role of iron in neurotoxicity: a cause for concern in the elderly? Curr Opin Clin Nutr Metab Care. 2009;12:22–9.
13. Hider RC, Ma Y, Molina-Holgado F, Gaeta A, Roy S. Iron chelation as a potential therapy for neurodegenerative disease. Biochem Soc Trans. 2008;36:1304–8.
14. Asha Devi S. Aging brain: prevention of oxidative stress by vitamin E and exercise. ScientificWorldJournal. 2009;9:366–72.

15. Valko M, Leibfritz D, Moncol J, Cronin MT, Mazur M, Telser J. Free radicals and antioxidants in normal physiological functions and human disease. Int J Biochem Cell Biol. 2007;39:44–84.
16. Kim GW, Gasche Y, Grzeschik S, Copin JC, Maier CM, Chan PH. Neurodegeneration in striatum induced by the mitochondrial toxin 3-nitropropionic acid: role of matrix metalloproteinase-9 in early blood–brain barrier disruption. J Neurosci. 2003;23:8733–42.
17. Pun PB, Lu J, Moochhala S. Involvement of ROS in BBB dysfunction. Free Radic Res. 2009;43:348–64.
18. Nimmerjahn A, Kirchhoff F, Helmchen F. Resting microglial cells are highly dynamic surveillants of brain parenchyma in vivo. Science. 2005;308:1314–8.
19. Duncan AJ, Heales SJ. Nitric oxide and neurological disorders. Mol Aspects Med. 2005;26:67–96.
20. Miksys S, Tyndale RF. The unique regulation of brain cytochrome P450 2 (CYP2) family enzymes by drugs and genetics. Drug Metab Rev. 2004;36:313–33.
21. Howard LA, Miksys S, Hoffmann E, Mash D, Tyndale RF. Brain CYP2E1 is induced by nicotine and ethanol in rat and is higher in smokers and alcoholics. Br J Pharmacol. 2003;138:1376–86.
22. Puntambekar P, Mukherjea D, Jajoo S, Ramkumar V. Essential role of Rac1/NADPH oxidase in nerve growth factor induction of TRPV1 expression. J Neurochem. 2005;95:1689–703.
23. Sánchez-Carbente MR, Castro-Obregón S, Covarrubias L, Narváez V. Motoneuronal death during spinal cord development is mediated by oxidative stress. Cell Death Differ. 2005;12:279–91.
24. Phumala N, Porasuphatana S, Unchern S, Pootrakul P, Fucharoen S, Chantharaksri U. Hemin: a possible cause of oxidative stress in blood circulation of beta-thalassemia/hemoglobin E disease. Free Radic Res. 2003;37:129–35.
25. Alash AI. Oxygen therapeutics: can we tame hemoglobin? Nat Rev Drug Discov. 2004;3:152–9.
26. Asaeda M, Sakamoto M, Kurosaki M, Tabuchi S, Kamitani H, Yokota M, et al. A nonenzymatic derived arachidonyl peroxide, 8-iso-prostaglandin F2-alpha, in cerebrospinal fluid of patients with aneurysmal subarachnoid hemorrhage participates in the pathogenesis of delayed cerebral vasospasm. Neurosci Lett. 2005;373:222–5.
27. Chen-Roetling J, Li Z, Chen M, Awe OO, Regan RF. Heme oxygenase activity and hemoglobin neurotoxicity are attenuated by inhibitors of the MEK/ERK pathway. Neuropharmacology. 2009;56:922–8.
28. Kudin AP, Debska-Vielhaber G, Kunz WS. Characterization of superoxide production sites in isolated rat brain and skeletal muscle mitochondria. Biomed Pharmacother. 2005;59:163–8.
29. Fishel ML, Vasko MR, Kelley MR. DNA repair in neurons: so if they don't divide what's to repair? Mutat Res. 2007;614:24–36.

Chapter 3
Perinatal Hypoxic-Ischemic Neural Injury and Free Radical Injury: A Fine Balance

M. Perlman and Po-Yin Cheung

Abstract During the transition from fetal to neonatal life, neonates are exposed to risks of hypoxic and ischemic insults. Since the 1980s, evidence has accumulated of reperfusion or reoxygenation injury in the human fetus and neonate, and its relationship to oxygen-derived free radical generation has been described. The first of these types of oxygen-related cerebral injury was described more than half a century ago when retinopathy of prematurity was found to be caused by uncontrolled administration of oxygen. Oxygen-derived free radicals are generated during and following neonatal resuscitation and lead to cerebral damage with cellular dysfunction and/or death. In this chapter, birth asphyxia and hypoxic-ischemic brain injury in term and preterm neonates are defined and described, as well as the clinical aspects of neurological oxygen-derived free radical disease of the newborn. The management of cerebral hypoxic-ischemic insult was limited to supportive measures until therapeutic hypothermia was recently proven to be beneficial. Based on the results of controlled studies of room air vs. 100% oxygen in neonatal resuscitation and on the relationship between the blood PO_2 and neurodevelopmental outcome in infants with hypoxic-ischemic encephalopathy, hyperoxia-hypocapnia should be rigorously avoided when resuscitating the asphyxiated neonate. The therapeutic potential of antioxidant interventions warrants further investigation.

Keywords Neonate · Hypoxia · Ischemia · Asphyxia · Encephalopathy · Free radicals · Hyperoxia · Antioxidants · Oxygen · Oxidative stress · Brain injury · Retinopathy of prematurity · Hypothermia

---

M. Perlman (✉)
Department of Pediatrics, University of Toronto, Toronto, ON, Canada
and
Division of Neonatology, Hospital for Sick Children, Toronto, ON, Canada
e-mail: max.perlman@sympatico.ca

N. Gadoth and H.H. Göbel (eds.), *Oxidative Stress and Free Radical Damage in Neurology*, Oxidative Stress in Applied Basic Research and Clinical Practice,
DOI 10.1007/978-1-60327-514-9_3,

1 Introduction

Oxygen-derived free radicals (OFR) are known to also play an important pathogenetic postinsult role in fetal and neonatal hypoxic ischemic brain injuries. Enormous developmental changes occur in this “age group”; a potentially viable 23-week gestational age fetus (or newly born infant of the age) may be further removed developmentally from a full-term neonate than the full-term neonate is from the fully mature adult. Moreover, unique demands are made on the newly born infant in making the transition from the airless intrauterine environment to ambient air. The transition carries risks of hypoxic and ischemic insults. Birth asphyxia of term and premature neonates, and the more prolonged postnatal respiratory insufficiency of premature neonates, keep neonatal intensive care units (NICUs) busy.

In adult medicine, evidence has accumulated that, in the postinsult period, “reperfusion injury” or OFR occurs and this may lead to multiorgan injury and dysfunction [1]. The risk of brain injury in asphyxiated neonates is reduced by antioxidant therapy or increased by inappropriate oxygen therapy [2–4]. The term “oxygen free radical disease of the newborn” was popularized by Dr. Saugstad to cover this pathogenesis and the various types of injury observed [5].

The first recognized association between oxygen therapy and neonatal disease was the epidemic of blindness in premature infants in the 1940s as a consequence of “retinopathy of prematurity” (ROP). Overgenerous oxygen therapy was proven to be the cause in 1954. Blindness caused by ROP was then virtually eliminated by more prudent oxygen therapy. A small resurgence occurred in the 1970s and 1980s when intensive care was extended to include extremely premature neonates, resulting in their increasing survival. Again, with more vigilant oxygen therapy the condition became less common. The premature neonate was recognized to be especially vulnerable to oxygen toxicity. The stage was set to discover other neonatal diseases with similar pathogenesis and to provide potential therapeutic strategies to prevent oxidative injury associated with OFR.

This chapter on the clinical aspects of neurological oxygen free radical disease of the newborn complements Chap. 4 in this volume on the basic science. In this chapter we also define and describe birth asphyxia and hypoxic ischemic brain injury in some detail for the reader unfamiliar with the fetus and newborn infant, as confusion exists about these subjects. In addition to the retina and central nervous system, OFR has been implicated in the pathogenesis of secondary lung injury of neonates treated for respiratory failure [6] and in infant carcinogenesis [7, 8].

Knowledge of the role of free radical formation in the pathogenesis of human fetal and neonatal disease has been based on cohort studies, randomized controlled trials, and the measurement of OFR, metabolites, oxidative states, antioxidant levels, and factors that increase or diminish antioxidant activity. Research subjects have included normal neonates, neonates during resuscitation and recovering from intrapartum hypoxic-ischemic insults of varying severity, and premature neonates with respiratory insufficiency of varying duration. Organs studied include the brain, eye, lung, kidney, and intestine. The outcomes studied have included mortality, and isolated short-term and long-term morbidities. A variety of animals with a broad range of levels of maturation at birth have also been studied. Before summarizing the literature on the role of OFR in the pathogenesis of fetal and neonatal neurological disorders, we define the conditions in which oxidative stress is thought to play a role.

2 Asphyxia and Hypoxic-Ischemic Encephalopathy

2.1 Pathophysiology During the Asphyxial Insult

Asphyxia starts with reduced oxygen and carbon dioxide exchange across the pulmonary circulation or, in the fetus, the placental circulation. *Hypoxia* (lack of oxygen at the tissue level) and *hypercapnia* (and a "respiratory acidosis") result. The third essential component of "asphyxia" is *metabolic acidosis*, reflecting the accrual of organic acids produced by anaerobic metabolism in conditions of oxygen lack [9]. According to this pathophysiological definition (hypoxia, hypercapnia, and metabolic acidosis), mild "asphyxia" occurs in most normal vaginal deliveries and causes no clinical dysfunction. Neonates born with moderate asphyxia may have *dysfunction* of one or more systems but rarely sustain *irreversible injury*. Severe asphyxia generally causes multisystem dysfunction and not infrequently causes permanent injury. The brain is the most common site of injury; other organs are rarely injured irreversibly in survivors. Recently the term asphyxia has been supplanted by hypoxic-ischemic encephalopathy (HIE), which is defined by a clinical pattern that includes encephalopathy and other organ dysfunction as well as the typical features of severe asphyxia (see below under Sect. 2.3).

In early asphyxia the heart rate and blood pressure rise, accompanied by apnea ("primary apnea"). As asphyxia persists, gasping appears and the heart rate and blood pressure fall gradually. After bradycardia has been present for some time, "secondary apnea" supervenes. From this stage, recovery of breathing and the circulation require resuscitation. Early in asphyxia, blood is shunted from the so-called nonessential to the essential organs, that is, heart, adrenal glands, and central nervous system (the so-called diving reflex is a compensatory mechanism). With progression, the essential organs become ischemic. The term hypoxia ischemia describes this process. As the hypoxia continues, cerebral, myocardial and adrenal ischemia supervene, which plays a critical pathogenetic role in brain injury. In addition to the foregoing changes, body stores of glycogen are utilized and even depleted to generate energy under anaerobic conditions.

Asphyxia can occur in the human fetus during pregnancy, labor, or after birth. A great variety of pathological conditions cause difficulty in fetal gas exchange. The sites of pathology vary from the inspired air (e.g., carbon monoxide poisoning) to the fetal cerebral circulation (e.g., hemorrhagic or septic shock), and, in order of oxygen transfer, include maternal respiratory and cardiovascular failure, severe dysfunction or disease of the maternal uterine vasculature, uterine tetany, premature placental separation, and umbilical vascular compression or obstruction. Cerebral ischemia without "asphyxia" may be caused by shock, thromboembolism ("neonatal stroke"), and severe fetal head compression that reduces and even arrests cerebral blood flow and oxygenation [10].

Postnatal asphyxia may be caused by maladaptation at birth or by neonatal respiratory, cardiovascular, or nervous system failure resulting from congenital anomalies, fetal disease, or birth injury. Shock is most commonly caused by acute hemorrhage before, during, or after birth, or by septic shock from peripartum infections. The pathophysiological and clinical effects of shock and its sequelae are similar to those of asphyxia.

In the in vivo models of asphyxia and asphyxial brain injury in a wide range of experimental fetal and neonatal animals, the "last gasp" (the onset of secondary apnea) was used as an easily identified endpoint in the natural history of asphyxia. In the rhesus monkey, brain injury starts at about the time of this last gasp. Boyle and Le Gallois (cited by Dawes) showed that the more mature the victim of asphyxia, the earlier the last gasp occurred in asphyxiated neonatal rabbits and guinea pig [9]. Le Gallois also discovered that interruption of the circulation shortened the time to the last gasp. The compensatory role of redistribution of the circulation is mentioned above; in addition, the circulating blood acts as a waste basin for the organic acids produced by anaerobic glycolysis within cells.

2.2 Brain Pathophysiology During Resuscitation and Recovery Post Hypoxia-Ischemia

Oxygen free radical (OFR) generation and related brain injury occur during both the asphyxiating insult and the reoxygenation-reperfusion process that follows [1, 2]. In the latter phase, increased oxidative stress and OFR generation may destroy various cellular components, leading to cell damage and death with apoptosis [11, 12], which may be a potential target of therapy. Regarding OFR generation, cerebral concentrations of hydrogen peroxide and nitric oxide were found to increase following hypoxia-reoxygenation and ischemia-reperfusion [13, 14]. Nitric oxide is also involved in the pathogenesis of cerebral hypoxia-reoxygenation injury [15], which may be secondary to its inhibitory effect on mitochondrial respiration, or the reaction with superoxide anion to generate peroxynitrite, a potent oxidant. Significant correlations between OFR and markers of oxidative stress and lipid peroxidation have been reported.

The relationship between OFR generation and HIE in the clinical context was supported by the clinical report of Kumar et al. [16]. They studied 43 newborn infants with asphyxia who subsequently developed HIE, and related the increased malonyldialdehyde and nitric oxide levels in plasma to the extent of the OFR injury. In addition, an elevated ratio of cerebrospinal fluid to plasma albumin was attributed to increased blood–brain barrier permeability and considered to contribute to the progression of HIE.

2.3 Clinical Definition and Manifestations of Severe Asphyxia and HIE

The clinical definitions of asphyxia and HIE are controversial, in flux and confusing. Not all cases of severe asphyxia develop HIE, and not all HIE is caused by asphyxia (e.g., shock). The American College of Obstetricians and Gynecologists (ACOG) and others defined asphyxia and, more recently, HIE by a list of phenomena that are part of a clinical syndrome, some of which are deemed essential [17]. The consensus-based definitions of HIE are restricted to intrapartum causes, although HIE can be caused by asphyxia during the antepartum and postnatal periods. Moreover, the diagnosis of HIE often cannot be made immediately after birth, as some neonates born after severe asphyxia are free of encephalopathic signs for some hours post-insult; hypoxia-ischemia describes this period better than HIE. The "ischemia" of HIE is also ambiguous, as ischemia may be caused by other causes, as already mentioned, or is part of the natural history of asphyxia severe enough to cause HIE. Common to the definitions are the clinical and laboratory phenomena described below, as adapted from ACOG.

1. *Obstetrical features*: Obstetrical complications associated with sentinel events cause acute near-total asphyxia, whereas less acute placental insufficiency causes prolonged partial asphyxia. In general, these two insult types are associated with aberrant but different fetal heart rate patterns.
2. *Depression of vital signs (cardiorespiratory and neurological) at birth*: The magnitude of such a depression is expressed by Apgar scoring based on five vital signs. Scoring is routinely done at 1 and 5 min after birth and then at 5-min intervals until normal. Scoring is partly subjective and has poor interobserver reproducibility. Depressed scores are also caused by conditions other than asphyxia. The time of onset of spontaneous breathing (the first gasp and the establishment of regular breathing) were closely related to the severity of the asphyxia in the classical rhesus monkey model. Patients with severe brain-damaging late intrapartum asphyxia require resuscitation at birth. The Apgar scores, unless extreme, are only approximately correlated with the likelihood of permanent brain injury.
3. *Blood gas abnormalities (fetal scalp blood or umbilical cord arterial blood sampled at delivery)*: These include low pH, elevated PCO_2, and reduced base excess (or elevated base deficit) (the PO_2 is idiosyncratic). The base excess or deficit is considered the best measure of asphyxia and represents the "oxygen debt" [18]. Measurements may be spurious as a result of venous sampling, especially in cases with umbilical cord compression. In subjects with potentially brain-damaging asphyxia, base excess values usually return to normal by 2–4 h post insult [19].
4. *Encephalopathy*: The brain illness of intrapartum asphyxia has a fairly typical pattern and time-course, starting within a few hours of birth. Abnormalities of muscle tone, responsiveness, irritability, and a variety of involuntary movements

including subtle and clonic-tonic seizures are observed [20]. The Sarnat staging system (I–III) reflects severity. Characteristic EEG and brain imaging changes, and evolution of changes over the first several days, are present in cases likely to have permanent injury.

5. *Multisystem dysfunction*: As defined as central nervous system and at least one other system, is usually present in HIE and requires complete testing for detection [21].
6. *Long-term outcome*: Outcome may be normal. Adverse outcomes vary from minor cognitive deficits to death in the neonatal period or later. Acute total or near-total asphyxial insults typically result in central gray matter injury causing dystonic quadriplegia, athetosis, and expressive speech defects with relatively preserved intellectual function. Prolonged partial insults typically result in cortical and subcortical white matter injury causing spastic quadriplegia with cognitive deficit, often with microcephaly and central visual deficits. "Mixed" insults cause a combination of the above (i.e., mixed cerebral palsy, etc.). Auditory deficits occasionally occur. Cerebral palsy is a mandatory part of the definition of HIE in consensus statements; however, a plethora of recent evidence indicates that the spectrum of outcomes of HIE includes isolated learning impairments [22].

2.4 The Timing of Asphyxial Insults

Timing of insult refers to two different phenomena: the time during gestation or delivery of the causative asphyxial insult, or the duration characteristics of the insult (in cases with brain injury, the acute near-total insults last from about 10 to 45 min, whereas prolonged partial insults last from about 1 h to a few hours) [23]. Intrapartum refers to the period between the onset of labor and birth, when asphyxiating obstetrical accidents are probably most common, the fetus is most vulnerable owing to the presence of labor contractions, and obstetrical and postnatal interventions to minimize brain injury are most pertinent. Asphyxia at near-term or full-term pregnancy (37 completed gestational weeks) is associated with the types of brain injury already mentioned, whereas preterm HI insults in the fetus or prematurely born neonate tend to cause periventricular leukomalacia and cerebral diplegia.

Antepartum asphyxia occurs any time before labor onset, characteristically but not exclusively in the fetus with growth restriction and oligohydramnios. The insult is associated with aberrant fetal heart rate patterns (not usually detected at the time) and altered fetal movements (including breathing movements detectable by fetal ultrasound examination), during and post insult [24]. The insult may be lethal ("stillbirth") or, if less severe and limited (either by spontaneous resolution, or by prompt delivery and resuscitation) the fetus may survive with or without brain injury. The type of brain injury depends on fetal maturation at the time of the asphyxial insult; in the preterm fetus periventricular leukomalacia usually results.

If the fetus remains in utero for long enough postinsult (e.g., more than 1–2 weeks), the fetus may show muscle contractures (from reduced spontaneous movements), microcephaly (because of brain atrophy), and the appearance of "old" established brain lesions on imaging before or after birth.

Intrapartum asphyxia, often erroneously called "perinatal" asphyxia, is detected by fetal heart rate changes on monitoring, electronically or by auscultation, more often during the late intrapartum period. If detection is timely, obstetrical intervention is prompted to expedite delivery and enable resuscitation. If undetected, intrapartum stillbirth can occur but is rare; with "in utero resuscitation", the fetus may survive with or without brain injury.

Postnatal asphyxia may be caused by cardiorespiratory or central nervous system maladaptation secondary to intrapartum asphyxia (e.g., myocardial ischemia, persistent pulmonary hypertension, and meconium aspiration syndrome), flawed resuscitation and iatrogenic causes (e.g., tension pneumothorax). Postnatal asphyxia is expressed, among other ways, by delayed resolution of the metabolic acidosis present at birth. As already mentioned, respiratory, cardiovascular or central nervous system failure from a variety of causes can result in postnatal asphyxia unrelated to intrapartum events.

2.5 Asphyxia in Preterm Neonates

Determining the presence or absence of asphyxia in the preterm neonate is difficult for a number of reasons. First, vital signs are often depressed in preterm neonates without asphyxia. Second, acute brain injuries that are thought to be hypoxic-ischemic in origin (especially periventricular leukomalacia detected by brain imaging) often appear to be unassociated with recognized obstetrical or neonatal hypoxic ischemic events. Third, the clinical expression of brain dysfunction (encephalopathy) in the preterm neonate is limited.

3 Oxygen in Neonatal Resuscitation

The first step in treating apnea at birth is to initiate gas exchange by artificial ventilation. Most neonates resuscitated at birth are in primary apnea and start to breathe independently within a very few minutes [9]. They are not threatened by significant hypoxia. Given the doubts that have existed for decades about the safety of breathing high fractions of oxygen, it is hard to justify the use of 100% oxygen for resuscitation of mild asphyxia, a condition that becomes evident within a minute or two of birth. On the other hand, many clinicians hesitate to provide less than 100% oxygen in resuscitating severely asphyxiated neonates. Research on oxygen therapy in resuscitation is summarized briefly here, emphasizing the effects on the brain.

3.1 Animal Studies

As detailed in Chap. 4, numerous studies in the past 15 years have compared the effects of 21 and 100% oxygen in the resuscitation for asphyxia of newborn piglets, rat pups, and fetal lambs. In general, most studies reported significant increases in tissue oxidative stress markers in asphyxiated or hypoxic animals resuscitated with 100% oxygen compared to 21% oxygen. Aberrant glutathione content, and increased malonyldialdehyde levels were shown in the brain of animals resuscitated with 100% oxygen. This increased oxidative state, associated with increased apoptotic markers [25], may cause white matter injury and activation of matrix metalloproteinases, a family of endoproteases that have been implicated in cerebral inflammatory conditions including stroke, cerebral infarction, and hemorrhage [3, 26]. Further, in fetal lambs subjected to asphyxia by cord occlusion, Markus et al. [27] demonstrated that hyperoxic resuscitation at birth induced a cerebral pro-inflammatory response (increased tissue expressions of IL-1beta and IL-12p40 in the cortex and subcortex), which in turn may lead to increased tissue damage after exposure to hyperoxemia at birth. However, there is little definitive information available regarding the clinical significance and consequences of the increased oxidative stress and the induced injury in the brain when newborn subjects were resuscitated with 100% oxygen. Presti et al. [28] reported better neurobehavioral outcome in newborn rat pups with HIE that were resuscitated with 100% oxygen. However, the mortality rate in the 100% oxygen resuscitation group was modestly increased, as observed in human studies.

The adverse effects of hyperoxemia on the brain may be particularly hazardous to the brain during the period of reperfusion following severe asphyxia and are mediated primarily by increased OFR [29]. In animal models of asphyxia, oxygen supplementation increases OFR formation [30, 31], decreases cerebral perfusion [26, 32], and adversely affects short-term neurological outcome [33]. As already indicated, similar short- and long-term pathophysiological effects were observed in human neonates exposed to 100% oxygen during resuscitation [34]. That oxidative stress may continue for 28 days suggests the potency of the effects of hyperoxia [29].

3.2 Human Studies

3.2.1 Healthy Neonates

In the healthy neonate and adult, hyperoxia and consequent hyperoxemia cause initial suppression followed by subsequent augmentation of ventilation [35]. A proposed mechanism based on in vitro studies is brainstem excitation by OFR [36]. Augmented ventilation in turn causes hypocapnia and reduced cerebral blood flow. Additional adverse effects of hyperoxemia include reduced CO_2^- hemoglobin binding (the Haldane effect) and increased affinity of hemoglobin for O_2, causing reduced unloading of O_2 to tissues. Macey et al. [37] studied physiological responses to breathing room air, 100% oxygen, and a 5% CO_2–95% O_2 mixture in 14 children aged 8–15 years (study of neonates was impracticable). Cardiovascular and breathing functions were observed by functional magnetic resonance imaging. Hyperoxic breathing had adverse effects on the control of autonomic and hormonal function, whereas the admixture of 5% CO_2 with O_2 greatly mitigated these adverse effects.

3.2.2 Asphyxiated Term Neonates

Neonates born with asphyxia in centers with fewer resources for monitoring and controlling these parameters may be exposed to hyperoxemia and/or hypocapnia even for hours after birth, as we observed in neonates referred to our care [4]. In neonates with postasphyxial HIE, we found a significant association between episodes of severe hyperoxemia ($PaO_2 > 26.6$ kPa) and/or severe hypocapnia ($PaCO_2 < 2.6$ kPa), and adverse outcome (defined as death, severe cerebral palsy, or any cerebral palsy with blindness, deafness, or developmental delay at age 24 months). The risk of severe adverse outcome, adjusted by multivariate analysis, was increased approximately two- to fourfold if hyperoxemia [odds ratio, 3.85; 95% confidence interval (CI), 1.67–8.88; $P = 0.002$], hypocapnia (odds ratio, 2.34; 95% CI, 1.02–5.37; $P = 0.044$), or both (odds ratio, 4.56; 95% CI, 1.4–14.9; $P = 0.012$) were present [4].

3.3 Therapeutic Trials

In resuscitating severely asphyxiated neonates after birth, it is difficult immediately to achieve a fine balance between hyperoxemia and hypoxemia and between hypocapnia and hypercapnia. Oxygen supplementation and/or mechanical ventilation may inevitably be inadequate or excessive for a number of reasons. Postnatal cardiorespiratory adaptation varies considerably depending on the variable response to therapy and whether complications of asphyxia are present. It is often not possible to monitor arterial blood levels of oxygen and carbon dioxide partial pressures accurately and to implement corrective action rapidly. Aggressive treatment in relationship to the severity of the respiratory illness results in hyperoxemia or hypocapnia, possibly increasing the risk of brain insult and injury [4]. The clinical results of randomized controlled trials of 100% vs. 21% oxygen in neonatal resuscitation for asphyxia are summarized below (Table 3.1).

1. *The onset of respiration and the need for resuscitation immediately after birth.* Neonates resuscitated with room air had higher 5-min Apgar scores and earlier first breath [38].
2. *Early organ dysfunction/disease.* A larger proportion of subjects receiving 100% oxygen had (or tended to have) evidence of heart or kidney dysfunction than those receiving room air [39].

Table 1 Clinical studies (published as of November 2008 in English) to examine the effect of oxygen in the neonatal resuscitation

References[a]	Sample size	Major findings
Ramji et al. [74]	84	Room air is as effective as 100% oxygen regarding mortality and hypoxic-ischemic encephalopathy (HIE)
Saugstad et al. [75]	591[b]	Room air is as effective as 100% oxygen regarding mortality and HIE
Vento et al. [76]	537	1. Room air is as effective as 100% oxygen regarding mortality and HIE, but more quickly as assessed by Apgar scores, time to the first cry, and the sustained pattern of respiration 2. 100% resuscitated group has increased blood oxidative stress markers 28 days after birth
Vento et al. [77]	106	Asphyxia causes oxidative stress in the perinatal period, and resuscitation with 100% oxygen causes hyperoxemia and increased oxidative stress
Ramji et al. [78]	418[b]	Room air is as effective as 100% oxygen regarding mortality and HIE, but has significantly shorter onset for first cry and duration of resuscitation
Vento et al. [79]	39	The use of room air on resuscitation causes less oxidative stress and damage to heart and kidney than pure oxygen
Bajaj et al. [80]	204	Room air is as effective as 100% oxygen regarding mortality and HIE
Total	1,979	1. Significant reduction in mortality ($P < 0.01$) 2. No significant effect on stage 2 and 3 HIE

[a]Clinical studies
[b]Denotes multicenter clinical trials

3. *Mortality and HIE rates.* A recent meta-analysis revealed reduced mortality and a trend for a lower rate of HIE stages 2 and 3 [40].
4. *Long-term outcome.* No intergroup differences were observed in somatic growth rates, or the rate of developmental delay at 18–24 months of age, but more than 30% of subjects were lost to follow-up [41].

Clearly, oxygen supplementation and ventilation should be rigorously controlled during the first minutes and hours of life. Despite the evidence from controlled trials of the superiority of room air over 100% oxygen in the initial resuscitation, this issue remains controversial. The latest International Consensus Statement, for example, appears indecisive [42]. The Statement relates to oxygen *saturation* (SO_2) in its conclusion, a pragmatic but in our opinion potentially misleading choice, as this standard of measurement of oxygen status reflects possible bias (although SO_2 is the better measure of the risk of hypoxia or oxygen deficiency, PO_2 is the better measure of risk of hyperoxia or oxygen toxicity). Readers do not need to be reminded of the extremely wide range of hyperoxemia (from about 100 to more than 500 mmHg) that is associated with high but "acceptable" levels (in term neonates) of SO_2 of about 95–100%.

3.4 Oxygen Therapy in Preterm Neonates with Prolonged Respiratory Insufficiency

Oxygen has been administered with caution to premature infants with respiratory insufficiency since the discovery in the early 1950s that hyperoxia was the primary cause of the epidemic of blindness. How fine is the balance between oxygen toxicity and oxygen deficiency and their respective potential risks? The available human data suggest that above the low limit of SO_2 or PO_2 used in NICUs this threat was overestimated. Answers to these questions, beyond the scope of this chapter, will be answered in the next few years as the results of ongoing clinical trials become available.

What about the other extreme: the potential effects of hyperoxemia and hypocapnia? This question was investigated retrospectively in preterm and low birth weight neonates weighing less than 2,000 g at birth. The 1,105 subjects were derived from a single population base and born in 1984–1987 with a mean gestational age of 31 weeks [43]. The study outcome was disabling cerebral palsy diagnosed by age 2 years. Analyses were conducted of associations between the outcome and, among other ventilation-related risk factors, cumulative exposures to hypocapnia (defined as a $PCO_2 < 35$ mmHg) and hyperoxemia (defined as a $PO_2 > 60$ mmHg) during the first 8 days after birth. The risk of disabling cerebral palsy was increased two- to threefold if either of these blood gas aberrations was present for a longer period (highest quintile of cumulative exposure). When adjusted by multivariate analysis, these results were unchanged. In another observational study of the cumulative exposure of low birth weight infants to hypocapnia and hyperoxia, a somewhat different result was obtained. As in the study just described, more neonates in the highest quartile of exposure to hypocapnia had periventricular leukomalacia than neonates without hypocapnia, also after statistical adjustment. On the other hand, no association was found between hyperoxia and this brain injury [44].

4 Therapeutic Interventions to Alleviate Oxygen-Derived Free Radical Injury of the Neonatal Brain

4.1 Enzymatic and Nonenzymatic Antioxidants and Metal Ions in the Newborn Infant

Enzymatic antioxidants including superoxide dismutase, catalase, and glutathione peroxidase reduce the formation of hydroxyl radicals and remove hydrogen and lipid peroxides, preventing the Haber–Weiss reaction. The nonenzymatic antioxidants can neutralize very reactive radical couples to less reactive ones. α-Tocopherol (vitamin E) breaks the chain reaction of lipid peroxidation with the formation of the less reactive α-tocopherol radical, which is then neutralized by ascorbic acid (vitamin C), a plasma and intracellular antioxidant, and the resultant formation of the less reactive and water-soluble ascorbate radical, thus removing radicals from the vulnerable lipid membrane [45, 46]. Therefore, the vitamin E:vitamin C ratio is important in maintaining this process and thus may explain the controversial effects of vitamin E and C supplementation on oxidative stress-related conditions in the neonate [47]. Further, ascorbate at low concentrations can oxidize metals such as copper and iron due to its position in the redox potential hierarchy [47]. Interestingly, the neonatal brain has significant iron content and thus is predisposed to oxidative stress-induced injury. Neuroprotective effects of iron chelation and erythropoietin have been suggested in neonatal rodent models of HIE [48, 49]. Erythropoietin trials have been carried out in human preterm neonates for reducing blood transfusions; meta-analysis of mortality and morbidity have not shown any advantage or disadvantage for neonates receiving the treatment [50]. However, the interaction between these nonenzymatic antioxidants and metal ions in the system is complex and thus affects the pathogenesis of oxidative stress-induced injury and the therapeutic effect of antioxidant and other interventions in ways that are difficult to predict. Nonetheless, there is increasing evidence supporting the relationship between perinatal asphyxia, oxidative stress, and inflammatory reaction in the developing brain that leads to white matter injury [51].

4.2 Antioxidants and Conditions that Influence Antioxidant Defense Mechanisms

Antioxidants were defined as "substances that can delay or prevent oxidation of a particular substrate" [52]. Compounds that prevent OFR cellular damage may prevent their formation and/or scavenge OFR when formed. From the physiological perspective, reactive oxygen species or a pro-oxidant state may play specific roles in normal physiological conditions ranging from cell differentiation and proliferation in potential target organs such as lung, brain, intestine, and retina to bacterial killing by neutrophils [53]. The cellular redox state as governed by the balance between oxidants and antioxidants will modulate the activities of transcription factors that can affect gene expression and signaling pathways.

4.2.1 Endogenous Antioxidants: Bilirubin and Neonatal Jaundice

Most newborn infants have elevated serum bilirubin concentrations. A fine balance may exist between the neurotoxicity of unconjugated bilirubin and its powerful antioxidant role. Associations between low serum bilirubin levels and blindness caused by ROP, the classical neonatal free radical disease, have been reported and refuted [54–56]. It has also been proposed that phototherapy for jaundice not only reduces bilirubin by converting bilirubin to bilirubin photoisomer but also increases the production of free radicals from other substrates (photooxidation of the lipid and vitamin contents of parenteral solutions in translucent intravenous administration sets have been demonstrated to be major contributors to in vitro generation of peroxides) [57]. In managing jaundiced preterm neonates we again are confronted by a fine balance between undertreatment and overtreatment: the risk of the former being bilirubin encephalopathy and that of the latter endogenous bilirubin antioxidant with consequent free radical injury of the retina and nervous system.

4.2.2 Therapy to Arrest Postinsult Pathophysiological Processes of Asphyxia and Hypoxic-Ischemic Encephalopathy

Hypothermia Therapy

The potential benefit of low environmental temperature for animals subjected to experimental asphyxia was recognized by Le Gallois in 1812 (cited by Dawes) [9]. Bench-to-bedside translation was realized in 2005 when the results of two large clinical trials were published. Our meta-analysis that included eight trials showed a reduced incidence of the comprehensive outcome, mortality or major neurodevelopmental disability, in patients receiving hypothermia (relative risk, 0.76; 95% CI, 0.65–0.88) [58].

There are many possible explanations for the cerebral protection afforded by hypothermia. In addition to its effect on free radicals and nitric oxide generation [59], the neuroprotective actions of hypothermia may involve excitatory amino acids [60], adenosine [61], cerebral hemodynamics [62], energetics and metabolism [63], and apoptosis [64]. Although Kil et al. [65] demonstrated that temperature-related changes occur in hydroxyl radical production during cerebral ischemia and reperfusion in rats, mechanisms of the effect of hypothermia on free radical generation remain to be elucidated.

Pharmaceutical Antioxidant Therapies: Potential Benefits and Risks

A number of antioxidants and antiapoptosis agents have been examined regarding the possibility of clinical use in newborn subjects [66, 67]. As mentioned earlier, the administration of superoxide dismutase (an enzymatic antioxidant with large molecular size and lipophilic properties) was effective in alleviating OFR-induced injury in the brain of neonatal lambs [2, 32]. However, poor bioavailability limits the effectiveness of some therapeutic agents. According to the most recent reappraisal of published trials, vitamin E may reduce the risk of severe ROP, but more trials are needed [68]. On the other hand, vitamin E administration has been associated with an increased incidence of sepsis, necrotizing enterocolitis, and mortality [69, 70]. Another antioxidant that was found to be effective in experimental animals is the xanthine-oxidase inhibitor allopurinol. A meta-analysis of three human studies was inconclusive, owing to insufficient statistical power [71]. *N*-Acetylcysteine is a water-soluble, thiol-containing compound that has antioxidative properties (by restoring cellular redox-status) as a free radical scavenger and as a precursor to the main intracellular antioxidant, glutathione. The preliminary findings suggest neuroprotective effects of *N*-acetylcysteine in experimental asphyxia [66]. The mechanism of action and the neuroprotective effects of *N*-acetylcysteine are discussed in Chap 4.

5 Conclusions Regarding the Prevention of Oxygen Free Radical Diseases of the Neonate

The competing risks of diseases resulting from OFR generation and oxygen deficiency are best prevented by avoiding the circumstances that primarily give rise to oxygen lack. This task represents major challenges for obstetricians and neonatologists. When concerned with asphyxia and respiratory insufficiency, one can only attempt to maintain a fine balance between hypoxemia and hyperoxemia. From the standpoint of oxygen toxicity, continuous monitoring of preductal arterial blood PaO_2 is the ideal method (preductal blood supplies the brain and eye). However, this tactic is limited by our ignorance of the normal ranges of PO_2 and PCO_2, even for term neonates at different postnatal ages, not to mention at different gestational ages and postconceptional ages of preterm neonates. We also lack knowledge regarding the PO_2 and PCO_2 ranges that are associated with minimal short-term and long-term mortality and morbidity in these critically ill neonates. Thus, using indwelling arterial catheters to monitor blood gases could be associated with daunting risks. On the other hand, pulse SO_2 is frequently used as a "surrogate indicator of O_2 tension" for pragmatic reasons. This technique, however, is less likely to prevent hyperoxia when saturations exceed 90–95%, for logistic reasons similar to those that exist for PO_2, and in addition to the characteristics of the oxygen dissociation curve of fetal hemoglobin. The comment by Sola et al. [72] that "the saturation limits of 85 and 94% may be more than adequate in the majority of cases" seems a sensible approach to us. Apart from meticulous control of the PO_2 or SO_2, hyperthermia should be strictly avoided [73]. Pharmacologic approaches to prevent OFR have, similar to oxygen itself and, at the other end of the spectrum, antioxidants, have so far proved to be two-edged swords. Further research is warranted to provide answers to some of the foregoing questions in OFR-induced injury to the neonatal brain.

References

1. McCord JM. Oxygen-derived free radicals in postischemic tissue injury. N Engl J Med. 1985;312:159–63.
2. Thiringer K, Hrbek A, Karlsson K, Rosén KG, Kjellmer I. Postasphyxial cerebral survival in newborn sheep after treatment with oxygen free radical scavengers and a calcium antagonist. Pediatr Res. 1987;22:62–6.

3. Munkeby BH, Børke WB, Bjørnland K, Sikkeland LI, Borge GI, Halvorsen B, et al. Resuscitation with 100% O_2 increases cerebral injury in hypoxemic piglets. Pediatr Res. 2004;56:783–90.
4. Klinger G, Beyene J, Shah P, Perlman M. Do hyperoxaemia or hypocapnia add to the risk of brain injury after intrapartum asphyxia? Arch Dis Child Fetal Neonatal Ed. 2005;90:F49–52.
5. Saugstad OD. Hypoxanthine as an indicator of hypoxia: its role in health and disease through free radical production. Pediatr Res. 1988;23:143–50.
6. Bancalari E, Gerhardt T. Bronchopulmonary dysplasia. Pediatr Clin North Am. 1986;33:1–23.
7. Naumurg E, Bellocco R, Cnattinguis S, Jonzon A, Ekbom A. Supplementary oxygen and risk of childhood lymphatic leukaemia. Acta Paediatr. 2002;91:1328–33.
8. Spector LG, Klebanoff MA, Feusner JH, Georgieff MK, Ross JA. Childhood cancer following neonatal oxygen supplementation. J Pediatr. 2005;147:27–31.
9. Dawes GS. Birth asphyxia, resuscitation, and brain damage. In: Foetal and neonatal physiology: a comparative study of the changes at birth. Chicago: Year Book; 1967. p. 141–59[AU5].
10. Misrahy GA, Beran AV, Spradley JF, Garwood VP. Fetal brain oxygen. Am J Physiol. 1960;199:959–64.
11. Li C, Jackson RM. Reactive species mechanisms of cellular hypoxia-reoxygenation injury. Am J Physiol Cell Physiol. 2002;282:C227–41.
12. Rodrigo J, Fernandez AP, Serrano J, Peinado MA, Martinez A. The role of free radicals in cerebral hypoxia and ischemia. Free Radical Biol Med. 2005;39:26–50.
13. Hyslop PA, Zhang Z, Pearson DV, Phebus LA. Measurement of striatal H_2O_2 by microdialysis following global forebrain ischemia and reperfusion in the rat: correlation with cytotoxic potential of H_2O_2 in vitro. Brain Res. 1995;671:181–6.
14. Kutzsche S, Kirkeby OJ, Rise IR, Saugstad OD. Effects of hypoxia and reoxygenation with 21% and 100% oxygen on cerebral nitric oxide concentration and microcirculation in newborn piglets. Biol Neonate. 1999;76:153–67.
15. Matsui T, Nagafuji T, Kumanishi T, Asano T. Role of nitric oxide in pathogenesis underlying ischemic cerebral damage. Cell Mol Neurobiol. 1999;19:177–89.
16. Kumar A, Mittal R, Khanna HD, Basu S. Free radical injury and blood–brain barrier permeability in hypoxic-ischemic encephalopathy. Pediatrics. 2008;122:e722–7. doi:10.1542/peds.2008-0269.
17. The American College of Obstetricians and Gynecologists and the American Academy of Pediatrics. Neonatal encephalopathy and cerebral palsy. Defining the pathogenesis and pathophysiology. Washington, DC: American College of Obstetricians and Gynecologists and the American Academy of Pediatrics; 2003
18. ACOG Committee Opinion Number 348. Umbilical cord blood gas and acid–base analysis. Obstet Gynecol. 2006;108:1319–22.
19. Shah PR, Raju NV, Beyene J, Perlman M. Recovery of metabolic acidosis in term infants with postasphyxial hypoxic-ischemic encephalopathy. Acta Paediatr. 2003;92:941–7.
20. Volpe JJ. Neurology of the newborn. 4th ed. Philadelphia: Saunders; 2000.
21. Shah P, Perlman M. Birth asphyxia and multi-organ dysfunction. Recent Adv Paediatr. 2006;23:113–26.
22. Gonzalez FF, Miller SP. Does perinatal asphyxia impair cognitive function without cerebral palsy? Arch Dis Child Fetal Neonatal Ed. 2006;91:F454–9.
23. Brann AW Jr, Myers RE. Central nervous system findings in the newborn monkey following severe in utero partial asphyxia. Neurology. 1975;25:327–38.
24. Phelan JP, Ahn MO. Perinatal observations in forty-eight neurologically impaired term infants. Am J Obstet Gynecol. 1994;171:424–31.
25. Gerstner B, DeSilva TM, Genz K, Armstrong A, Brehmer F, Neve RL, et al. Hyperoxia causes maturation-dependent cell death in the developing white matter. J Neurosci. 2008;28: 1236–45.
26. Richards JG, Todd KG, Emara M, Haase E, Cooper SL, Bigam DL, et al. A dose–response study of graded reoxygenation on the carotid haemodynamics, matrix metalloproteinase-2

activities and amino acid concentrations in the brain of asphyxiated newborn piglets. Resuscitation. 2006;69:319–27.
27. Markus T, Hansson S, Amer-Wåhlin I, Hellström-Westas L, Saugstad OD, Ley D. Cerebral inflammatory response after fetal asphyxia and hyperoxic resuscitation in newborn sheep. Pediatr Res. 2007;62:71–7.
28. Presti AL, Kishkurno SV, Slinko SK, Randis TM, Ratner VI, Polin RA, et al. Reoxygenation with 100% oxygen versus room air: late neuroanatomical and neurofunctional outcome in neonatal mice with hypoxic-ischemic brain injury. Pediatr Res. 2006;60:55–9.
29. Vento M, Asensi M, Sastre J, Garcia-Gala F, Pallardo FV, Vina J. Resuscitation with room air instead of 100% oxygen prevents oxidative stress in moderately asphyxiated term neonates. Pediatrics. 2001;107:642–7.
30. Capani F, Loidl CF, Aguirre F, Piehl L, Facorro G, Hager A, et al. Changes in reactive oxygen species (ROS) production in rat brain during global perinatal asphyxia: an ESR study. Brain Res. 2001;914:204–7.
31. Lièvre V, Becuwe P, Bianchi A, Bossenmeyer-Pourié C, Koziel V, Franck P, et al. Intracellular generation of free radicals and modifications of detoxifying enzymes in cultured neurons from the developing rat forebrain in response to transient hypoxia. Neuroscience. 2001;105: 287–97.
32. Rosenberg AA, Murdaugh E, White CW. The role of oxygen free radicals in postasphyxia cerebral hypoperfusion in newborn lambs. Pediatr Res. 1989;26:215–9.
33. Temesvári P, Karg E, Bódi I, Németh I, Pintér S, Lazics K, et al. Impaired early neurologic outcome in newborn piglets reoxygenated with 100 percent oxygen compared with room air after pneumothorax-induced asphyxia. Pediatr Res. 2001;49:812–9.
34. Lundstrom KE, Pryds O, Greisen G. Oxygen at birth and prolonged cerebral vasoconstriction in preterm infants. Arch Dis Child Fetal Neonatal Ed. 1995;73:F81–6.
35. Cross KW, Warner P. The effect of inhalation of high and low oxygen concentrations on the respiration of the newborn infant. J Physiol. 1951;114:283–95.
36. Mulkey DK, Henderson RAIII , Putnam RW, Dean JB. Hyperbaric oxygen and chemical oxidants stimulateCO_2/H^+-senstive neurons in rat brain slices. J Appl Physiol. 2003;95: 910–21.
37. Macey PM, Woo MA, Harper RM. Hyperoxic brain effects are normalized by addition of CO_2. PLoS Med. 2007;4:e173.
38. Saugstad OD, Ramji S, Vento M. Resuscitation of depressed newborn infants with ambient air or pure oxygen: a meta-analysis. Biol Neonate. 2005;87:22–34.
39. Vento M, Sastre J, Asensi MA, Viña J. Room-air resuscitation causes less damage to heart and kidney than 100% oxygen. Am J Respir Crit Care Med. 2005;172:1393–98.
40. Saugstad OD, Ramji S, Soll RF, Vento M. Resuscitation of newborn infants with 21% or 100% oxygen: an updated systematic review and meta-analysis. Neonatology. 2008;94:176–82.
41. Saugstad OD, Ramji S, Irani SF, El-Meneza S, Hernandez EA, Vento M, et al. Resuscitation of newborn infants with 21% or 100% oxygen: follow-up at 18 to 24 months. Pediatrics. 2003;112:296–300.
42. International Liaison Committee on Resuscitation. The International Liaison Committee on Resuscitation (ILCOR) consensus on science with treatment recommendations for pediatric and neonatal patients: neonatal resuscitation. Pediatrics. 2006;117: e978–88.
43. Collins MP, Lorenz JM, Jetton JR, Paneth N. Hypocapnia and other ventilation-related risk factors for cerebral palsy in low birth weight infants. Pediatr Res. 2001;50:712–9.
44. Shankaran S, Langer JC, Kazzi SN, Laptook AR, Walsh M, National Institute of Child Health and Human Development Neonatal Research Network. Cumulative index of exposure to hypocarbia and hyperoxia as risk factors for periventricular leukomalacia in low birth weight infants. Pediatrics. 2006;118:1654–9.
45. Caraceni P, Yao T, Degli Esposti S, Gasbarrini A, Bowie BT, Zern M, et al. Effect of vitamin E on reoxygenation injury experienced by isolated rat hepatocytes. Life Sci. 1994;55:1427–32.

46. Vardareli E, Saricam T, Koken T, Degirmenci I, Aral E, Erenoglu E. The effect of alpha-tocopherol and pentoxyfilline on ischemia-reperfusion induced liver injury in rats. Hepatogastroenterology. 1998;45:1505–8.
47. Buettner GR, Jurkiewicz BA. Catalytic metals, ascorbate and free radicals: combinations to avoid. Radiat Res. 1996;145:532–41.
48. Sarco DP, Becker J, Palmer C, Sheldon RA, Ferriero DM. The neuroprotective effect of deferoxamine in the hypoxic-ischemic immature mouse brain. Neurosci Lett. 2000;282:113–6.
49. Aydin A, Genç K, Akhisaroglu M, Yorukoglu K, Gokmen N, Gonullu E. Erythropoietin exerts neuroprotective effect in neonatal rat model of hypoxic-ischemic brain injury. Brain Dev. 2003;25:494–8.
50. Aher S, Ohlsson A. Late erythropoietin for preventing red blood cell transfusion in preterm and/or low birth weight infants. Cochrane Database Syst Rev. 2006;3:CD004868.
51. Rezaie P, Dean A. Periventricular leukomalacia, inflammation and white matter lesions within the developing nervous system. Neuropathology. 2002;22:106–32.
52. Halliwell B. The antioxidant paradox. Lancet. 2000;355:1179–80.
53. Jankov RP, Negus A, Tanswell AK. Antioxidants as therapy in the newborn: some words of caution. Pediatr Res. 2001;50:681–7.
54. Yeo KL, Perlman M, Hao Y. Outcomes of extremely premature infants related to their peak serum bilirubin concentrations and exposure to phototherapy. Pediatrics. 1998;102:1426–31.
55. DeJong MH, Khuntia A, Maisels J, Bandagi A. Bilirubin levels and severe retinopathy of prematurity in infants with estimated gestational ages of 23 to 26 weeks. J Pediatr. 1999;135:102–4.
56. Oh W, Tyson JE, Fanaroff AA, Vohr BR, Perritt R, Stoll BJ, et al. Association between peak serum bilirubin and neurodevelopmental outcomes in extremely low birth weight infants. Pediatrics. 2003;112:773–9.
57. Lavoie JC, Bélanger S, Spalinger M, Chessex P. Admixture of a multivitamin preparation to parenteral nutrition: the major contributor to in vitro generation of peroxides. Pediatrics. 99:1997;e6.
58. Perlman M, Shah P. Time to adopt cooling for neonatal hypoxic-ischemic encephalopathy. Response to a previous commentary. Pediatrics. 2008;121:616–8.
59. Kader A, Frazzini VI, Baker CJ, Solomon RA, Trifiletti RR. Effect of mild hypothermia on nitric oxide synthesis during focal cerebral ischemia. Neurosurgery. 1994;35:272–7.
60. Thoresen M, Satas S, Puka-Sundvall M, Whitelaw A, Hallström A, Løberg EM, et al. Post-hypoxic hypothermia reduces cerebrocortical release of NO and excitotoxins. Neuroreport. 1997;8:3359–62.
61. Barros RC, Branco LG, Cárnio EC. Respiratory and body temperature modulation by adenosine A1 receptors in the anteroventral preoptic region during normoxia and hypoxia. Respir Physiol Neurobiol. 2006;153:115–25.
62. Baldwin WA, Kirsch JR, Hurn PD, Toung WS, Traystman RJ. Hypothermic cerebral reperfusion and recovery from ischemia. Am J Physiol. 1991;261(3 pt 2):H774–81.
63. Thoresen M, Penrice J, Lorek A, Cady EB, Wylezinska M, Kirkbride V, et al. Mild hypothermia after severe transient hypoxia-ischemia ameliorates delayed cerebral energy failure in the newborn piglet. Pediatr Res. 1995;37:667–70.
64. Edwards AD, Yue X, Squier MV, Thoresen M, Cady EB, Penrice J, et al. Specific inhibition of apoptosis after cerebral hypoxia-ischaemia by moderate post-insult hypothermia. Biochem Biophys Res Commun. 1995;217:1193–9.
65. Kil HY, Zhang J, Piantadosi CA. Brain temperature alters hydroxyl radical production during cerebral ischemia/reperfusion in rats. J Cereb Blood Flow Metab. 1996;16:100–6.
66. Jatana M, Singh I, Singh AK, Jenkins D. Combination of systemic hypothermia and *N*-acetylcysteine attenuates hypoxic-ischemic brain injury in neonatal rats. Pediatr Res. 2006;59:684–9.

67. Yesilirmak DC, Kumral A, Tugyan K, Cilaker S, Baskin H, Yilmaz O, et al. Effects of activated protein C on neonatal hypoxic ischemic brain injury. Brain Res. 2008;19:56–62.
68. Raju TN, Langenberg P, Bhutani V, Quinn GE. Vitamin E prophylaxis to reduce retinopathy of prematurity: a reappraisal of published trials. J Pediatr. 1997;131:844–50.
69. Johnson L, Bowen FW Jr, Abbasi S, Herrmann N, Weston M, Sacks L, et al. Relationship of prolonged pharmacologic serum levels of vitamin E to incidence of sepsis and necrotizing enterocolitis in infants with birth weight 1,500 grams or less. Pediatrics. 1985;75:619–38.
70. Lorch V, Murphy D, Hoersten LR, Harris E, Fitzgerald J, Sinha SN. Unusual syndrome among premature infants: association with a new intravenous vitamin E product. Pediatrics. 1985;75:598–602.
71. Chaudhari T, McGuire W. Allopurinol for the prevention of mortality and morbidity in newborn infants with suspected hypoxic-ischaemic encephalopathy. Cochrane Database Syst Rev. 2008;3:CD006817.
72. Sola A, Rogido MR, Deulofeut R. Oxygen as a neonatal health hazard: call for détente in clinical practice. Acta Paediatr. 2007;96:801–12.
73. Laptook A, Tyson J, Shankaran S, McDonald S, Ehrenkranz R, Fanaroff A, et al. Elevated temperature after hypoxic-ischemic encephalopathy: risk factor for adverse outcomes. Pediatrics. 2008;122:491–9.
74. Ramji S, Ahuja S, Thirupuram S, Rootwelt T, Rooth G, Saugstad OD. Resuscitation of asphyxic newborn infants with room air or 100% oxygen. Pediatr Res. 1993;34:809–12.
75. Saugstad OD, Rootwelt T, Aalen O. Resuscitation of asphyxiated newborn infants with room air or oxygen: an international controlled trial. The Resair 2 study. Pediatrics. 1998;102:e1.
76. Vento M, Asensi M, Sastre J, García-Sala F, Viña J. Six years of experience with the use of room air for the resuscitation of asphyxiated newly born term infants. Biol Neonate. 2001;79:261–7.
77. Vento M, Asensi M, Sastre J, Lloret A, García-Sala F, Viña J. Oxidative stress in asphyxiated term infants resuscitated with 100% oxygen. J Pediatr. 2003;142:240–6.
78. Ramji S, Rasaily R, Mishra PK, Narang A, Jayam S, Kapoor AN, et al. Resuscitation of asphyxiated newborns with room air or 100% oxygen at birth: a multicentric clinical trial. Indian Pediatr. 2003;40:510–7.
79. Vento M, Sastre J, Asensi MA, Viña J. Roomair resuscitation causes less damage to heart and kidney than 100% oxygen. Am J Respir Crit Care Med. 2005;172:1393–8.
80. Bajaj N, Udani RH, Nanavati RN. Room air vs. 100 per cent oxygen for neonatal resuscitation: a controlled clinical trial. J Trop Pediatr. 2005;51:206–11.

Chapter 4
Oxidative Stress in Neonatal Hypoxic-Ischemic Encephalopathy[1]

K.G. Todd, L.L. Jantzie, and Po-Yin Cheung

Abstract Despite improvements in the care of asphyxiated neonates, neonatal or perinatal hypoxia-ischemia remains a challenge to clinical practitioners. In this multisystem dysfunction, hypoxic-ischemic encephalopathy contributes to short- and long-term morbidity of these critically ill neonates. In the developing brain of premature and term neonates, there are immature responses to hypoxia-ischemia in the context of selective vulnerability of different brain structures and neural cells at different stages of development. This difference explains at least in part the diversity of clinical presentations and sequelae of neonatal hypoxic-ischemic brain injury. In addition to hypoxic-ischemic damage, cerebral reoxygenation or reperfusion injury plays an important role in the pathophysiology of hypoxic-ischemic brain injury. Mechanisms of cell death and apoptosis operate at multiple levels including oxygen-derived free radical damage and excitotoxicity. Cerebral oxidative stress and neurochemical changes are related in hypoxia-ischemia of the neonatal brain. Controlled reoxygenation to avoid hyperoxia and its related cerebral damage, and novel antioxidative agents such as *N*-acetylcysteine, are potential therapeutic interventions that may show promise in the improvement of clinical outcome of these asphyxiated neonates with cerebral hypoxic-ischemic injury. The effects of controlled reoxygenation and *N*-acetylcysteine on the neurochemistry of asphyxiated neonatal brain are discussed.

Keywords Neonate · Hypoxia · Ischemia · Asphyxia · Encephalopathy · Free radicals · Hyperoxia · Antioxidants · Oxidative stress · Brain injury · Oxygen · Periventricular leukomalacia · Oligodendrocyte · Prematurity · Excitotoxicity · Apoptosis · *N*-Acetylcysteine

---

[1]P.-Y. Cheung is a Clinical Investigator of the Canadian Institutes of Health Research and the Alberta Heritage Foundation for Medical Research (AHFMR).

P.-Y. Cheung (✉)
Department of Pediatrics and Pharmacology, University of Alberta, Edmonton, AB, Canada
e-mail: poyin@ualberta.ca

N. Gadoth and H.H. Göbel (eds.), *Oxidative Stress and Free Radical Damage in Neurology*, Oxidative Stress in Applied Basic Research and Clinical Practice,
DOI 10.1007/978-1-60327-514-9_4, © Springer Science+Business Media, LLC 2011

1 Introduction

In the past decade there have been significant improvements in obstetrical and neonatal care, including regionalization, that help to increase the survival rate and reduce the morbidity of asphyxiated neonates. Neonatal or perinatal hypoxia-ischemia (HI) is a common clinical occurrence, accounting for at least 16% of deaths in Canadian neonatal intensive care units (NICUs) [1]. Fetal (before birth), perinatal (20 weeks gestation up to postnatal day 7), and neonatal (birth up to 28 days post delivery) HI has a wide range of presentations and manifestations and affects the heart, lungs, gastrointestinal tract, kidneys, and brain most severely. Neonates that survive HI often exhibit a complex multisystem disorder with multi-organ dysfunction and failure. In addition to significant changes in hemodynamics, cardiovascular shock, papillary muscle necrosis, respiratory failure, pulmonary hypertension, pulmonary and gastrointestinal hemorrhage, necrotizing enterocolitis, acute tubular necrosis, and hypoxic-ischemic encephalopathy are commonly diagnosed post HI [2, 3].

1.1 Defining Neonatal Hypoxic-Ischemic Brain Injury

Brain injury that results from perinatal hypoxia remains a common clinical occurrence. Hypoxic-ischemic brain injury is specifically defined as neuropathology attributable to hypoxia and/or ischemia as evidenced by biochemical (serum creatine kinase-brain bound (CK-BB)), electrophysiological data (EEG), neuroimaging (magnetic resonance imaging or computed tomography), or postmortem abnormalities [4]. Although there has been a marked reduction in morbidity and mortality rates over the past four decades, HI in the perinatal period leading to major motor and cognitive disabilities continues to be a significant health problem worldwide [3]. Despite advances in fetal monitoring including biophysical profiling, two- and three-dimensional ultrasonography, and in utero magnetic resonance imaging, a large number of neonates with HI-caused neurological abnormalities are born each year, many caused by problems encountered during pregnancy and the birthing process [3].

2 Neonatal Cerebral Hypoxia-Ischemia

2.1 Neurological Sequelae

Regardless of the type or nature, it is well established that cerebral HI is the single most common cause of neurological and intellectual handicaps in children [3]. Many neonates that survive HI injury have significant neurological complications that are diverse in their progression, presentation, and contribution to long-term disability. Many psychiatric, neuropsychiatric, and neurological disorders such as cerebral palsy, mental retardation, epilepsy, attention deficit hyperactivity disorder, hearing and vision loss, and learning disabilities (significant psycho-educational handicaps at school age such as learning difficulties in mathematics and spelling, disturbances of language and speech, and soft neurological signs) are commonly linked to HI events before, during, or after birth [5]. Schizophrenia, conduct disorders, and developmental diseases such as neuronal migration disorders, lissencephaly, and heterotopias also likely have some degree of cerebral HI linked to their etiology [6–8].

2.2 Etiology

In general, HI is most easily understood to result from acute or chronic impaired gas exchange caused by disruption of the maternal–placental–fetal system, leading to inadequate provision of oxygen and removal of carbon dioxide and H^+ from the fetus [4]. Sentinel events in pregnancy and parturition that acutely cause fetal brain injury include placental and umbilical cord complications, acute maternal and feto-neonatal hemorrhage, and/or any condition causing abrupt decreases in maternal cardiac output or blood flow to the fetus [3, 4]. Specific conditions of pregnancy can also render a fetus more vulnerable to HI. For example, there is a greater risk for HI in multiple gestation pregnancies than singleton pregnancies, and twin–twin transfusion syndrome is paramount in these settings. Chronic HI insults, often associated with intrauterine growth retardation, or chronic HI, are commonly the major consequence of placental insufficiency. Critically ill neonates with cardiopulmonary diseases are commonly observed in NICUs and may develop secondary postnatal HI. These neonates often have multiple medical problems, and medications and instrumentation are required for life support. Critically ill neonates are more susceptible to fluctuations in systemic blood pressure, have increased propensity to form clots, and have impaired cerebral autoregulation, thus rendering them more vulnerable to HI [9]. Details of various antenatal, perinatal, and postnatal etiologies for HI are described in Chap. 3.

2.3 Developmental Vulnerability to Cerebral Hypoxia-Ischemia

Hypoxia-ischemia injury alters the trajectory of normal developmental events and initiates a cascade of altered maturation. Part of the reason why HI is so detrimental is that it is an injury which occurs in developing organs, on a background of natural and necessary programmed cell death. Further, molecules that have important roles in the pathogenesis of HI (i.e., iron, glutamate, calcium, and nitric oxide) are often mediators of crucial developmental processes [9]. The final functional and anatomic neurological pattern that exists after HI depends partly on the gestational age at which the insult occurred [9]. In the immature central nervous system, neurons are selectively vulnerable in a manner that is comparable to cell-type vulnerabilities in adult neurological disorders such as Parkinson's disease and Huntington's disease [10]. This concept of selective vulnerability of different brain structures and neural cells at different stages of development is explained here because it also determines the immature response to HI of the central nervous system [9].

2.4 *Cerebral Hypoxia-Ischemia in the Preterm Neonate*

The most common neurological manifestation of HI in preterm neonates (neonates born at less than 37 weeks gestation) is severe damage to the hemispheric white matter with relative sparing of the gray matter [11]. The most common post-HI diagnoses in preterm neonates are periventricular leukomalacia and periventricular hemorrhagic infarction. Periventricular leukomalacia is hallmarked by a characteristic pattern of white matter injury defined by infarcts, with or without cysts, in periventricular and callosal white matter and susceptible regions of end-artery distribution. Acute and severe damage to the basal ganglia and thalamic nuclei is also occasionally observed [11]. Increasing experimental and clinical data support the notion that the neonatal brain differs fundamentally in function, cellular composition, and connectivity compared to the adult brain [12]. As such, it is believed that periventricular leukomalacia and white matter injuries are most commonly seen in preterm neonates because developing oligodendrocytes and subplate neurons are preferentially vulnerable [10, 13, 14]. Damage to subplate neurons is believed to be involved in the visual loss sometimes associated with HI, as these cells play a critical role in the formation of connections between the thalamus and the visual cortex [15].

In addition to intrinsic properties of cellular populations of the central nervous system, the preterm white matter is rendered particularly vulnerable by the immaturity of penetrator blood vessels into the deep cortex and white matter, as well as lack of autoregulatory control [12, 16]. Thus, when systemic hypotension occurs, there is little compensatory regulation in these regions [12]. The age window of greatest susceptibility to periventricular leukomalacia is between 24 and 32 weeks gestation, when the cerebral white matter is predominantly populated by oligodendrocyte precursors and immature oligodendrocytes [12, 17, 18]. Preoligodendrocytes and oligodendrocyte progenitor cells are more vulnerable to HI than mature oligodendrocytes; this is thought to be largely the result of differences in the capability of these cells to cope with cellular factors such as oxidative stress and excitotoxicity [19]. For example, the immature brain has high concentrations of unsaturated fatty acids and redox-active iron, a high rate of oxygen consumption, and low concentrations of antioxidants [10]. As a result, preterm neonates with many more oligodendrocyte progenitors and preoligodendrocytes are extremely vulnerable to the depletion of antioxidants or exposure to exogenous free radicals. As neonates approach term and the population of mature oligodendrocytes grows, the neonate and its brain cells become more resistant to oxidative stress. This change is thought to occur largely as a result of the increases in antioxidant enzymes that occur with age [10].

Oligodendrocyte precursors are also extremely susceptible to excitotoxicity and owe their preferential vulnerability to differences in α-amino-3-hydroxy-5-methyl-4-isoxazole propionic acid (AMPA) receptor subunit composition [20, 21]. In vivo, glutamate receptor subunit-2 (GluR2)-deficient-AMPA receptors are highly expressed in immature and premyelinating preoligodendrocytes [12]. AMPA-mediated signaling and excitotoxicity depend on the functional properties of the receptor complex, such as calcium permeability, which in turn are dictated by subunit composition [20–23]. AMPA receptors lacking the GluR2 subunit are calcium permeable, and because these are the most abundant receptor subtype on immature oligodendrocytes, it is believed that they are the reason for the described developmental vulnerability to excitotoxicity and white matter damage.

2.5 Cerebral Hypoxia-Ischemia in the Term Neonate

In term neonates (37–42 weeks gestation), a single profound total failure of blood or oxygen supply (i.e., cardiac arrest) typically causes symmetrical damage to the nuclei of the brainstem, including the cranial nerve nuclei, basal ganglia, and thalami [11]. The fact that these areas are predominantly damaged, in contrast to the HI-related cortical injury commonly observed in adults, is because of the high energy demand as well as the excitotoxicity associated with the development of neurotransmitters. Many of these neonates do not survive the insult and those who do show disorders of thermoregulation, sucking, swallowing, and oculomotor paresis [11, 24, 25]. When subacute or partial and more prolonged HI insults occur, the cortical "watershed" regions between major arterial territories can be damaged, such as a parasagittal area for anterior and middle cerebral arteries. The moderate severity of HI insults allows time for fetal adaptive responses, and blood is redistributed to the vital centers of the brainstem and shunted away from the cerebral hemispheres [11]. However, the topography of damage commonly varies, and the two patterns of brain damage in term neonates suffered from HI insults overlap with differential injury to the deep nuclei and cerebral cortex.

2.6 Pathophysiology

When cerebral HI is severe enough to produce irreversible tissue injury, the insult is almost always associated with major perturbations in the energy status of the brain [26, 27]. During the early course of a HI insult, alterations in adenosine triphosphate (ATP), adenosine diphosphate (ADP), adenosine monophosphate

(AMP), and phosphocreatine occur [26, 27]. This loss of cellular ATP during HI severely compromises those metabolic processes that require energy for their completion. Further, oxidative phosphorylation ceases and anaerobic glycolysis becomes the major source of energy. As the neuronal demand for energy is so high, all reserves in the brain are depleted soon after the onset of anoxia (e.g., within 2–3 min in rats), and anaerobic production of ATP cannot sustain brain cells at a functional level [28]. Consequently, ATP-dependent processes such as Na^+/K^+ exchange are curtailed and there is a resultant intracellular accumulation of Na^+, Cl^-, and water [27, 29, 30]. This accumulation results in cytotoxic edema occurring from the net movement of osmotically driven water from the extracellular to the intracellular space [31]. Without the regeneration of ATP, energonic reactions do not resume, intracellular ions and water continue to accumulate, and electrochemical gradients cannot be reestablished. In addition, other factors ultimately influence tissue cellular integrity such as the formation of free radicals and the associated peroxidation of free fatty acids within cellular membranes, and cellular acidosis [32]. As ischemia impairs cellular metabolic rates and depletes the energy required to maintain ionic gradients, neurons and glia lose their membrane potential and depolarize [33]. These depolarizations allow somatodendritic and presynaptic voltage-dependent calcium channels to become activated, and excitatory amino acids are released into the extracellular space [32]. In addition, the presynaptic reuptake of neurotransmitters becomes impaired, and amino acids accumulate in the extracellular synaptic cleft.

A fundamental process responsible for a large proportion of early HI cell death is excitotoxicity, or cell death mediated by excessive stimulation of excitatory amino acid receptors. Normally these receptors mediate the physiological excitatory effects of glutamate, one of the most ubiquitous and versatile neurotransmitters in the brain [6, 7]. However, when excessively stimulated by combinations of elevated synaptic levels of glutamate and membrane depolarization associated with ischemia, the channels associated with these receptors allow a lethal flood of Ca^{2+} and Na^+ to enter neurons [6, 7]. This activation of glutamate receptors, through the attendant failure of ion homeostasis and increases in intracellular calcium, is a major factor involved in the initiation of ischemic cell death [32]. Downstream consequences of this sequence include activation of lipases, endonucleases, proteases, and caspases and degradation of the cytoskeleton and cell membrane.

3 Oxidative Stress in Neonates

Oxidative stress occurs when the production of damaging free radicals and other oxidative molecules exceeds antioxidant defenses. Where babies are concerned, this process typically begins after delivery but can occur in the unborn fetus in the wake of a HI event or in the presence of maternal disease such as preeclampsia or maternal infection [34–36]. After delivery, the immature central nervous system is not only vulnerable to oxidative stress caused by HI injury, it is also vulnerable to oxidative stress precipitated by postdelivery medical interventions and neonatal intensive care practices such as hyperoxia during respiratory support, as well as ischemia-reperfusion and inflammation [36, 37]. The main cellular and molecular sources of reactive oxygen species (ROS) following HI include damaged mitochondria and impaired electron transport systems, cyclooxygenase activity, lipooxygenases, arachidonic acid, enzymatic activity of xanthine oxidase, autooxidation of catecholamines and amino acid neurotransmitters, infiltrating neutrophils and microglia, and nitric oxide synthase [36]. Interestingly, many if not all of these sources are directly or indirectly related to increases in cytosolic calcium. Once ROS are produced they are then capable of initiating apoptotic and necrotic cell death pathways, activating matrix metalloproteinases, nitrosylating proteins, disrupting cell membrane integrity, increasing blood–brain barrier permeability, and causing abnormal arteriolar reactivity [38–43]. The fetal brain is especially prone to lipid peroxidation-mediated injury because of a unique abundance of membrane lipids [44–46]. The aldehydes formed from lipid hydroperoxides are most specifically detected in glia [44, 47], and in the preterm human fetus, elevated lipid peroxidation products have been detected in periventricular white matter [44, 48]. E_2-isoprostranes (peroxidation products) are especially toxic to oligodendrocyte progenitors but not to mature oligodendrocytes, which suggests that specific compounds endogenously formed during oxidative stress might be a potential mechanism for the degeneration of oligodendrocytes in the periventricular white matter injury that is commonly seen following HI [44, 49].

Innate protective antioxidant defenses include enzymes such as superoxide dismutase, catalase, and glutathione peroxidase. Nonenzymatic molecules such as vitamins A, C, E, glutathione, and coenzyme Q also play a role [36]. Other protective antioxidants (i.e., β-carotene, ceruloplasmin) and trace metals such as copper, zinc, iron, and selenium (metal cofactors required for antioxidant synthesis and activity) are deficient in preterm neonates [36, 50, 51]. The neonatal brain may also be more vulnerable to oxidative damage because of a lag in the expression of superoxide dismutase type-1 and type-2, catalase, and glutathione peroxidase, which have been found to be deficient in white matter tissue samples [44, 52].

3.1 Cellular Mechanisms of Oxidative Cell Death

After a HI event, oxidative stress can cause brain cell death through both apoptosis and necrosis; whether a cell will ultimately die through necrosis or apoptosis depends on a variety of factors including the developmental stage, cell type, injury severity, and availability of ATP [53, 54].

Necrotic cell death is often associated with severe insults and involves cell swelling, activation of intracellular proteases, disruption of cellular membranes, spillage of intracellular components into the extracellular space, and a concerted inflammatory response. Necrosis unleashes the uncontrolled function of several proteases and enzymes that disrupt protein assemblies [55]. In contrast, cell death through apoptotic processes is biochemically and genetically regulated, requiring time, energy, and protein synthesis. Nuclear condensation is observed, and cellular and organelle membranes remain intact until end stages. In apoptotic cell death, outer cellular membranes form into blebs that ultimately pinch off into apoptotic bodies which can be engulfed or phagocytosed. Apoptosis can be initiated through three separate pathways, two of which are dependent on activation of intracellular cysteine proteinases and are referred to as intrinsic and extrinsic pathways [56]. The intrinsic pathway involves initiation of apoptosis as a result of disturbances in intracellular homeostasis. This pathway has also been termed the "mitochondrial cell death pathway" as the mitochondria are critical for the execution of cell death. The extrinsic pathway involves the initiation of apoptosis through ligation of plasma membrane death receptors. Although the initiation stage of intrinsic and extrinsic pathways are different, they converge to result ultimately in cellular morphological and biochemical alternations that are characteristic of apoptosis. In both the intrinsic and extrinsic apoptotic pathways, activation of caspases, the final "executioner" proteins, is required [57, 58]. Caspases are intracellular cysteine proteinases that are constitutively expressed as inactive pro-caspases or zymogens [55]. There are two general groups of caspases, the upstream instigators (caspase-8, -9, and -10), which activate other caspases, and the downstream terminators that cleave essential cellular substrates, resulting in cell destruction (caspase-3, -6, and -7). Caspase-1, -2, -4, -5, -11, and -12 can act as both initiators and executioners [55]. The caspases can be activated by external signals such as cytokine binding to cell-surface death receptors (e.g., Fas or tumor necrosis factor receptor (TNFR)) or by activation of the mitochondrial (intrinsic) pathway, which involves Bax translocation to the mitochondrial membrane, leakage of cytochrome *c* into the cytosol, cytochrome *c* binding to "apoptosis protease-activating factor-1" (Apaf-1) and pro-caspase-9, which together activate caspase-3 and cause cell death.

A third apoptotic pathway, a caspase-independent pathway, involves apoptosis-inducing factor (AIF). Under injurious conditions, AIF translocates from the mitochondria to the cell nucleus. Once in the nucleus AIF induces condensation of the chromatin and DNA fragmentation [59]. Cell death associated with AIF appears to contribute to rapid cell death after hypoxic-ischemic injury and is not affected by caspase inhibitors, suggesting this pathway operates independently of caspase activation [60, 61].

3.2 Oxidative Stress and Changes in Neurochemistry Following Cerebral Hypoxia-Ischemia

Free radical production, oxidative stress, and the subsequent cerebral injury induced could be the major target of therapeutic interventions to improve the neurodevelopmental outcome of neonates with asphyxia. Supportive management includes ventilatory care, meticulous fluid management, avoidance of hypotension and hypoglycemia, and treatment of seizures. In addition to supportive care, interventions to reduce ROS generation and the related oxidative stress have been investigated. Among these interventions, therapeutic hypothermia has been shown to be effective, at least for a selected population of neonates with significant asphyxia (reviewed in Chap. 3).

3.2.1 Controlled Reoxygenation

Upon delivery, 5–10% of all neonates require some degree of resuscitation and assistance to begin breathing [8, 62–64]. The aim of resuscitation is to prevent neonatal death and adverse long-term neurodevelopmental sequelae associated with the HI event, in addition to the rapid reversal of fetal hypoxemia and acidosis [65]. The use of 100% oxygen in resuscitation appears to be an intuitive response to maximize the gradient required to drive oxygen into hypoxic cells [66]. However, a building body of evidence derived from animal models has demonstrated that although resuscitation with 100% oxygen restores cerebral perfusion following HI, it may occur at the price of greater biochemical oxidative stress [37, 65]. To avoid hyperoxia-induced oxidative stress, the efficacy of resuscitation with room air instead of 100% oxygen has been extensively studied. To date, the results of these investigations are inconclusive. The investigations conducted by Munkeby et al. suggest that resuscitation of asphyxiated piglets with 100% oxygen is detrimental to the brain; however, in HI neonatal mice, resuscitation with 100% oxygen restores cerebral blood flow significantly faster than with room air and with improved later neurofunctional outcome [67, 68]. Studies performed on asphyxiated human neonates have shown that room air, rather than 100% oxygen, favors clinical recovery [37, 69–73].

The complete neurochemical effects of resuscitating with different oxygen tensions are also not known, especially under long-term survival conditions [37]. In addition to being neurotransmitters, many amino acids are involved in intermediary metabolism, the wiring of neuronal networks, building cytoarchitecture, and protein synthesis [74–76]. Independently, amino acid neurotransmitters are closely linked to the vulnerability of the immature brain to neuronal injury [77–79]. In particular, excitatory amino acids (i.e., glutamate and aspartate) participate in signal transduction and exert trophic influences that affect neuronal survival, growth, and differentiation during restricted developmental periods [76, 80]. In previous investigations, significantly higher levels of excitatory amino acids were found in the striatum of newborn HI piglets reoxygenated with room air compared with piglets reoxygenated with 100% oxygen [68, 81]. It has also been shown that extracellular striatal dopamine level, which can exert neurotoxicity, was twofold higher in piglets resuscitated with 100% oxygen compared with animals resuscitated with room air following systemic hypoxic injury [68, 82]. Investigations performed in our laboratory indicate that certain resuscitation practices can cause significant neurochemical changes that persist for days after HI injury in the neonatal piglet [37]. We found that resuscitation with 100% oxygen significantly increases glutamate and glycine in the dorsolateral cortex contralateral to common carotid artery ligation, compared to resuscitation with room air and that these changes persisted for 4 days following HI [37]. Further, resuscitation with 100% oxygen lead to higher levels of glutamate and aspartate in the basal ganglia, and those piglets resuscitated with 21% oxygen had lower alanine levels as well [37]. These results are especially relevant when it is considered that hypoxia/reoxygenation significantly increases cerebral cortical lactate and alanine levels in the dorsal cortex, lateral cortex, and hypothalamus in neonatal piglets, indicating changes in acid–base balance after injury. Different resuscitation protocols may tip the balance between free radical generation/secondary reperfusion injury and persistent HI damage. For example, hyperoxic resuscitation (with 100% oxygen) clearly precipitates free radical formation, whereas hypoxic resuscitation (with 18% oxygen) may extend the window for HI cell death because oxygen tension remains low throughout the reoxygenation period. Thus, normoxic resuscitation (with room air or an intermediate oxygen concentration) may be more beneficial by balancing these two competing processes and maintaining homeostasis post-HI.

3.2.2 Antioxidants

A number of antioxidants and antiapoptosis agents have been studied for potential clinical use in newborn subjects [83–85]. However, some agents do not have high bioavailability and indeed may cause harm owing to complications such as the increased incidence of sepsis and necrotizing enterocolitis and mortality in vitamin E administration [86, 87]. From the pathophysiological perspective, ROS or a pro-oxidant state may play specific roles in normal physiological conditions ranging from cell differentiation and proliferation in potential target organs such as lung, brain, intestine, and retina to bacterial killing by neutrophils [88]. Transcription factors including the NF-κB and activator protein-1 are sensitive to cellular redox state [89]. Thus, the cellular redox state as governed by the balance between oxidants and antioxidants will modulate the activities of transcription factors, which can affect gene expression and the signaling pathway.

N-Acetylcysteine (NAC) is a thiol-containing compound that has antioxidative properties (restoring cellular redox status) as a free radical scavenger and as a precursor to the main intracellular antioxidant, glutathione [90]. Clinically used as a mucolytic agent and to treat acetaminophen toxicity [91], NAC is able to cross placental and blood–brain barriers [92] and is considered safe during pregnancy [93–97]. Therefore, it may have therapeutic potential in human neonatal HI injury. NAC has been shown to reduce oxidative stress, restore intracellular glutathione levels, scavenge oxygen free radicals, improve cellular redox potential, reduce apoptosis, and reduce inflammation [83]. Previous animal studies of neonatal asphyxia have also suggested that NAC may reduce injuries associated with oxygen-derived free radicals in the heart [98, 99], intestine [100], kidneys [99], lungs, and platelets [101]. The neuroprotection of NAC is related to scavenging ROS, replenishing glutathione that has been depleted by hypoxia-reoxygenation, attenuating the activation of apoptotic proteases [83, 93], and increasing phosphorylated levels of p42 and p44 extracellular signal-regulated kinases [102]. Previously, we have reported that postresuscitation administration of NAC significantly attenuates increases in cortical hydrogen peroxide, but not nitric oxide, during reoxygenation, and that NAC-treated piglets have significantly higher carotid oxygen delivery and lower cerebral lactate levels than those of reoxygenated hypoxic controls [103–105]. The neuroprotective effects of NAC have also been reported in rodent models of hypoxia-reoxygenation and ischemia-reperfusion [93, 106, 107]. Changes in neurochemistry in response to conventional resuscitation practices and NAC have also been investigated [103, 108]. NAC administration maintains cerebral amino acid homeostasis and does not significantly alter amino acid neurochemical profiles post hypoxia and during reoxygenation. These results indicate a potential strength of this therapeutic approach as one would not want to administer a therapy to hypoxic neonates that significantly changed amino acid neurochemistry after injury, which could be detrimental to neurodevelopmental outcome.

The effect of NAC in the clinical arena is controversial. Although its beneficial effect has been evidenced in patients with acetaminophen poisoning [91], the roles of NAC in contrast-induced nephropathy [109] and patients with chronic obstructive pulmonary disease [110] remain to be determined. Indeed, NAC has been used in supplemental parenteral nutrition in premature neonates, who lack the expression of cystathionase in the trans-sulfuration pathway [111], but the effect is controversial with a recent review showing insufficient evidence to support its use in premature neonates [112]. Furthermore, Asfar et al. [113] reported an increased mortality rate in those septic patients when NAC was given more than 24 h after hospital admission. Thus, further clinical and translational investigations are needed to examine the clinical role of NAC in asphyxiated neonates.

4 Summary

Neonatal or perinatal HI is common and is associated with hypoxic-ischemic encephalopathy that may lead to significant neurological, neurodevelopmental, and neurobehavioral morbidities in the survivors. Because of the differences in developmental vulnerability of neurons and in vascular development, the pattern of cerebral HI varies between preterm and term neonates. Thus, periventricular leukomalacia and white matter injuries are most commonly observed in preterm neonates because developing oligodendrocytes and subplate neurons are preferentially vulnerable to damage through free radicals and excitotoxicity. In term neonates, perinatal HI causes damage to the nuclei of the brainstem, basal ganglia, thalamus, hippocampus, and "watershed" regions of the cerebral cortex depending on the severity and acuity of HI. The cerebral tissue injury is secondary to both the hypoxic-ischemic event (leading to energy deprivation) and the reoxygenation-reperfusion event following resuscitation. Although HI events compromise metabolic and ionic homeostatic processes, resulting in cellular dysfunction and cytotoxic edema, reoxygenation-reperfusion phenomena can lead to further tissue damage, with the generation of free radicals causing oxidative stress-induced injury.

The ROS generated may cause abnormal cellular function, disrupt cell membrane integrity, increase blood–brain barrier permeability, and initiate apoptotic and necrotic cell death pathways. The developing brain is especially prone to lipid peroxidation-mediated injury because of its unique abundance of membrane lipids. Resistance to oxidative stress (both the enzymatic and nonenzymatic systems) increases with advancing gestational age.

Controlled reoxygenation can limit the excess oxidative stress incurred during resuscitation while avoiding the prolongation of oxygen debt resulting from HI. In the recent guidelines, although a specific initial inspiratory fraction of oxygen has not been suggested, the possible toxic effects of oxygen and a strict control of oxygenation in the resuscitation of neonatal asphyxia have been recommended. Future studies should examine the role of antioxidants in the alleviation of oxidative stress-induced cerebral injury.

References

1. Sankaran K, Chien LY, Walker R, Seshia M, Ohlsson A, Canadian Neonatal Network. Variations in mortality rates among Canadian neonatal intensive care units. CMAJ. 2002;166:173–8.
2. Carter B, Haverkamp A, Merenstein G. The definition of acute perinatal asphyxia. Clin Perinatol. 1993;20:287–304.
3. Sunshine P. Perinatal asphyxia: an overview. In: Stevenson DK, Benitz WE, Sunshine P, editors. Fetal and neonatal brain injury: mechanisms, management and the risks of practice. Cambridge: Cambridge University Press; 2003. p. 2–29.
4. Adcock L, Papile LA. Perinatal asphyxia. In: Cloherty J, Eichenwald E, Stark A, editors. Manual of neonatal care. Philadelphia: Lippincott Williams & Wilkins; 2008. p. 518–28.
5. Vannucci RC. Hypoxic-ischemic encephalopathy. Am J Perinatol. 2000;17:113–9.
6. Johnston MV, Nakajima W, Hagberg H. Mechanisms of hypoxic neurodegeneration in the developing brain. Neuroscientist. 2002;8:212–20.
7. Johnston MV, Trescher WH, Ishida A, Nakajima W. Neurobiology of hypoxic-ischemic injury in the developing rat brain. Pediatr Res. 2001;49:735–41.
8. Schubert S, Brandl U, Brodhun M, Ulrich C, Spaltmann J, Fiedler N, et al. Neuroprotective effects of topiramate after hypoxia-ischemia in newborn piglets. Brain Res. 2005;1058:129–36.
9. McLean C, Ferriero D. Mechanisms of hypoxic-ischemic injury in the term infant. Semin Perinatol. 2004;28:425–32.
10. Ferriero D. Neonatal brain injury. N Engl J Med. 2004;351:1985–95.
11. Squier W, Cowan FM. The value of autopsy in determining the cause of failure to respond to resuscitation at birth. Semin Neonatol. 2004;9:331–45.
12. Jensen F. Developmental factors regulating susceptibility to perinatal brain injury and seizures. Curr Opinion Pediatr. 2006;18:628–33.
13. McQuillen PS, Ferriero D. Selective vulnerability in the developing central nervous system. Pediatr Neurol. 2004;30:227–35.
14. McQuillen PS, Sheldon RA, Shatz CT, Ferriero D. Selective vulnerability of subplate neurons after early neonatal hypoxia-ischemia. J Neurosci. 2003;23:3308–15.

15. Kanold PO, Kara P, Reid R. Shatz CJ. Role of subplate neurons in functional maturation of visual cortical columns. Science. 2003;301:521–5.
16. Soul JS, Taylor GA, Wypij D, Duplessis AJ, Volpe JJ. Noninvasive detection of changes in cerebral blood flow by near-infrared spectroscopy in a piglet model of hydrocephalus. Pediatr Res. 2000;48:445–9.
17. Back SA, Luo NL, Borenstein NS, Levine JM, Volpe JJ, Kinney HC. Late oligodendrocyte progenitors coincide with the developmental window of vulnerability for human perinatal white matter injury. J Neurosci. 2001;21:1302–12.
18. Kinney HC, Back SA. Human oligodendroglial development: relationship to periventricular leukomalacia. Semin Pediatr Neurol. 1998;5:180–9.
19. Khwaja O, Volpe JJ. Pathogenesis of cerebral white matter injury of prematurity. Arch Dis Child Fetal Neonatal Ed. 2008;93:F153–61.
20. Talos DM, Follett PL, Folkerth RD, Fishman RE, Trachtenberg FL, Volpe JJ, et al. Developmental regulation of alpha-amino-3-hydroxy-5-methyl-4-isoxazole-proprionic acid receptor subunit expression in forebrain and relationship to regional susceptibility to hypoxic/ischemic injury. I. Rodent cerebral white matter and cortex. J Comp Neurol. 2006;497:42–60.
21. Talos DM, Follett PL, Folkerth RD, Fishman RE, Trachtenberg FL, Volpe JJ, et al. Developmental regulation of alpha-amino-3-hydroxy-5-methyl-4-isoxazole-proprionic acid receptor subunit expression in forebrain and relationship to regional susceptibility to hypoxic/ischemic injury. II. Human cerebral white matter and cortex. J Comp Neurol. 2006;497:61–77.
22. Deng W, Rosenberg PA, Volpe JJ, Jensen FE. Calcium-permeable AMPA/kainate receptors mediate toxicity and preconditioning by oxygen-glucose deprivation in oligodendrocyte precursors. Proc Natl Acad Sci U S A. 2003;100:6801–6.
23. Follett PL, Deng W, Dai W, Talos DM, Massillon LJ, Rosenberg PA, et al. Glutamate receptor-mediated oligodendrocyte toxicity in periventricular leukomalacia: a protective role for topiramate. J Neurosci. 2004;24:4412–20.
24. Felderhoff-Mueser U, Rutherford MA, Squier WV, Cox P, Maalouf EF, Counsell SJ, et al. Relationship between MR imaging and histopathologic findings of the brain in extremely sick preterm infants. AJNR Am J Neuroradiol. 1999;20:1349–57.
25. Pasternak JF, Gorey MT. The syndrome of acute near-total intrauterine asphyxia in the term infant. Pediatr Neurol. 1998;18:391–8.
26. Vannucci RC. Experimental biology of cerebral hypoxia-ischemia: relation to perinatal brain damage. Pediatr Res. 1990;27:317–26.
27. Vannucci RC. Mechanisms of perinatal hypoxic-ischemic brain damage. Semin Perinatol. 1993;17:330–7.
28. Nestler EJ, Hyman SE, Malenka RC. Seizures and stroke. In: Nestler EJ, Hyman SE, Malenka RC, editors. Molecular neuropharmacology: a foundation for clinical neuroscience. New York: McGraw Hill; 2001.
29. Vannucci RC. Experimental models of perinatal hypoxic-ischemic brain damage. APMIS Suppl. 1993;40:89–95.
30. Vannucci RC, Christensen MA, Yager JY. Nature, time-course, and extent of cerebral edema in perinatal hypoxic-ischemic brain damage. Pediatr Neurol. 1993;9:29–34.
31. Vexler ZS, Ferriero D. Molecular and biochemical mechanisms of perinatal brain injury. Semin Neonatol. 2001;6:99–108.
32. Dirnagl U, Iadecola C, Moskowitz MA. Pathobiology of ischemic stroke: an integrated view. Trends Neurosci. 1999;22:391–7.
33. Katsura K, Kristian T, Siesjo BK. Energy metabolism, ion homeostasis, and cell damage in the brain. Biochem Soc Trans. 1994;22:991–6.
34. Dani C, Masini E, Bertini G, di Felice AM, Pezzati M, Ciofini S, et al. Role of heme oxygenase and bilirubin in oxidative stress in preterm infants. Pediatr Res. 2004;56:873–7.

35. Buhimschi IA, Buhimschi C, Pupkin M, Weiner CP. Beneficial impact of term labor: nonenzymatic antioxidant reserve in the human fetus. Am J Obstet Gynecol. 2003;189:181–8.
36. Dani C, Cecchi A, Bertini G. Role of oxidative stress as physiopathologic factor in the preterm infant. Minerva Pediatr. 2004;56:381–94.
37. Jantzie LL, Cheung PY, Obaid L, Emara M, Johnson ST, Bigam DL, et al. Persistent neurochemical changes in neonatal piglets after hypoxia-ischemia and resuscitation with 100%, 21% or 18% oxygen. Resuscitation. 2008;77:111–20.
38. Chan PH, Schmidley JW, Fishman RA, Longar SM. Brain injury, edema, and vascular permeability changes induced by oxygen-derived free radicals. Neurology. 1984;34:315–20.
39. Delanty N, Dichter MA. Antioxidant therapy in neurologic disease. Arch Neurol. 2000;57:1265–70.
40. Palmer C, Vannucci RC. Cellular and molecular biology of perinatal hypoxic-ischemic brain damage. In: Stevenson DK, Benitz WE, Sunshine P, editors. Fetal and neonatal brain injury: mechanisms, management and the risks of practice. Cambridge: Cambridge University Press; 2003. p. 58–82.
41. Leffler CW, Busija DW, Armstead WM, Shanklin DR, Mirro R, Thelin O. Activated oxygen and arachidonate effects on newborn cerebral arterioles. Am J Physiol. 1990;259:H1230–8.
42. Schleien CL, Koehler RC, Shaffner DH, Traystman RJ. Blood–brain barrier integrity during cardiopulmonary resuscitation in dogs. Stroke. 1990;21:1185–91.
43. Wei EP, Ellison MD, Kontos HA, Povlishock JT. O_2 radicals in arachidonate-induced increased blood–brain barrier permeability to proteins. Am J Physiol. 1986;251:H693–9.
44. Back SA, Rivkees SA. Emerging concepts in periventricular white matter injury. Semin Perinatol. 2004;28:405–14.
45. Janero DR. Malondialdehyde and thiobarbituric acid-reactivity as diagnostic indices of lipid peroxidation and peroxidative tissue injury. Free Radic Biol Med. 1990;9:515–40.
46. Hasegawa K, Yoshioka H, Sawada T, Nishikawa H. Lipid peroxidation in neonatal mouse brain subjected to two different types of hypoxia. Brain Dev. 1991;13:101–3.
47. Montine KS, Quinn JF, Zhang J, Fessel JP, Roberts LJII, Morrow JD, et al. Isoprostanes and related products of lipid peroxidation in neurodegenerative diseases. Chem Phys Lipids. 2004;128:117–24.
48. Haynes RL, Folkerth RD, Keefe RJ, Sung I, Swzeda LI, Rosenberg PA, et al. Nitrosative and oxidative injury to premyelinating oligodendrocytes in periventricular leukomalacia. J Neuropathol Exp Neurol. 2003;62:441–50.
49. Brault S, Martinez-Bermudez AK, Roberts J II, Cui QL, Fragoso G, Hemdan S, et al. Cytotoxicity of the E(2)-isoprostane 15-E(2t)-IsoP on oligodendrocyte progenitors. Free Radic Biol Med. 2004;37:358–66.
50. Halliwell B, Gutteridge JM. Role of free radicals and catalytic metal ions in human disease: an overview. Methods Enzymol. 1990;186:1–85.
51. Kelly FJ. Free radical disorders of preterm infants. Br Med Bull. 1993;49:668–78.
52. Folkerth RD, Haynes RL, Borenstein NS, Belliveau RA, Trachtenberg F, Rosenberg PA, et al. Developmental lag in superoxide dismutases relative to other antioxidant enzymes in premyelinated human telencephalic white matter. J Neuropathol Exp Neurol. 2004; 63(9):990–9.
53. Fiskum G. Mechanisms of neuronal death and neuroprotection. J Neurosurg Anesthesiol. 2004;16:108–10.
54. Lipton P. Ischemic cell death in brain neurons. Physiol Rev. 1999;79:1431–568.
55. Ferrer I, Planas AM. Signaling of cell death and cell survival following focal cerebral ischemia: life and death struggle in the penumbra. J Neuropathol Exp Neurol. 2003;62:329–39.
56. Ashe PC, Berry MD. Apoptotic signaling cascades. Prog Neuropsychopharmacol Biol Psychiatry. 2003;27:199–214.

57. Eldadah BA, Faden AI. Caspase pathways, neuronal apoptosis, and CNS injury. J Neurotrauma. 2000;17:811–29.
58. MacManus JP, Linnik MD. Gene expression induced by cerebral ischemia: an apoptotic perspective. J Cereb Blood Flow Metab. 1997;17:815–32.
59. Daugas E, Nochy D, Ravagnan L, Loeffler M, Susin SA, Zamzami N, et al. Apoptosis-inducing factor (AIF): a ubiquitous mitochondrial oxidoreductase involved in apoptosis. FEBS Lett. 2000;476:118–23.
60. Zhu C, Qiu L, Wang X, Hallin U, Candé C, Kroemer G, et al. Involvement of apoptosis-inducing factor in neuronal death after hypoxia-ischemia in the neonatal rat brain. J Neurochem. 2003;86:306–17.
61. Zhu C, Wang X, Qiu L, Peeters-Scholte C, Hagberg H, Blomgren K. Nitrosylation precedes caspase-3 activation and translocation of apoptosis-inducing factor in neonatal rat cerebral hypoxia-ischaemia. J Neurochem. 2004;90:462–71.
62. American Heart Association, American Academy of Pediatrics. 2005 American Heart Association (AHA) guidelines for cardiopulmonary resuscitation (CPR) and emergency cardiovascular care (ECC) of pediatric and neonatal patients: neonatal resuscitation guidelines. Pediatrics. 2006;117:e1029–38.
63. Davis PG, Tan A, O'Donnell C, Schulze A. Resuscitation of newborn infants with 100% oxygen or air: a systematic review and meta-analysis. Lancet. 2004;364:1329–33.
64. Tan A, Schulze A, O'Donnell C, Davis PG. Air versus oxygen for resuscitation of infants at birth. Cochrane Database Syst Rev. 2005;CD002273.
65. Martin RJ, Walsh MC, Carlo WA. Reevaluating neonatal resuscitation with 100% oxygen. Am J Respir Crit Care Med. 2005;172:1360.
66. Corff KE, McCann DL. Room air resuscitation versus oxygen resuscitation in the delivery room. J Perinat Neonatal Nurs. 2005;19:379–90.
67. Munkeby BH, Børke WB, Bjørnland K, Sikkeland LI, Borge GI, Halvorsen B, et al. Resuscitation with 100% O_2 increases cerebral injury in hypoxemic piglets. Pediatr Res. 2004;56:783–90.
68. Presti AL, Kishkurno SV, Slinko SK, Randis TM, Ratner VI, Polin RA, et al. Reoxygenation with 100% oxygen versus room air: late neuroanatomical and neurofunctional outcome in neonatal mice with hypoxic-ischemic brain injury. Pediatr Res. 2006;60:55–9.
69. Saugstad OD. Room air resuscitation: two decades of neonatal research. Early Hum Dev. 2005;81:111–6.
70. Saugstad OD, Ramji S, Vento M. Resuscitation of depressed newborn infants with ambient air or pure oxygen: a meta analysis. Biol Neonate. 2005;87:27–34.
71. Vento M, Asensi M, Sastre J, García-Sala F, Pallardó FV, Viña J. Resuscitation with room air instead of 100% oxygen prevents oxidative stress in moderately asphyxiated term infants. Pediatrics. 2001;107:642–7.
72. Vento M, Asensi M, Sastre J, García-Sala F, Viña J. Six years experience in the use of room air for the resuscitation of asphyxiated newly born term infants. Biol Neonate. 2001;79:261–7.
73. Vento M, Sastre J, Asensi M, Vina J. Room-air resuscitation causes less damage to heart and kidney than 100% oxygen. Am J Respir Crit Care Med. 2005;172:1393–8.
74. Ben-Ari Y, Khazipov R, Leinekugel X, Caillard O, Gaiarsa JL. GABA-A, NMDA, and AMPA receptors: a developmentally regulated "menage a trios." Trends Neurosci. 1997;20:523–9.
75. Cooper JR, Bloom FE, Roth RH. The biochemical basis of neuropharmacology. 8th ed. New York: Oxford University Press; 2003.
76. Herlenius E, Lagercrantz H. Neurotransmitters and neuromodulators during early human development. Early Hum Dev. 2001;65:21–37.
77. Ikonomidou C, Bittigau P, Koch C, Genz K, Hoerster F, Felderhoff-Mueser U, et al. Neurotransmitters and apoptosis in the developing brain. Biochem Pharmacol. 2001;62:401–5.

78. Johnston MV. Neurotransmitters and vulnerability of the developing brain. Brain Dev. 1995;17:301–6.
79. Johnston MV, Ishiwa S. Ischemia and excitotoxins in development. Mental Retard Dev Disabil Res Rev. 1995;1:193–200.
80. McDonald JW, Johnston MV. Physiological and pathological roles of excitatory amino acids during central nervous system development. Brain Res Rev. 1990;15:41–70.
81. Solås AB, Kutzsche S, Vinje M, Saugstad OD. Cerebral hypoxemia-ischemia and reoxygenation with 21% or 100% oxygen in newborn piglets: effects on extracellular levels of excitatory amino acids and microcirculation. Pediatr Crit Care Med. 2001;2:340–5.
82. Huang CC, Yonetani M, Lajevardi N, Delivoria-Papadopoulos M, Wilson DF, Pastuszko A. Comparison of postasphyxial resuscitation with 100% and 21% oxygen on cortical oxygen pressure and striatal dopamine metabolism in newborn piglets. J Neurochem. 1995;64:292–8.
83. Jatana M, Singh I, Singh A, Jenkins D. Combination of systemic hypothermia and *N*-acetylcysteine attenuates hypoxic-ischemic brain injury in neonatal rats. Pediatr Res. 2006;59:684–9.
84. Payton KS, Sheldon RA, Mack DW, Zhu C, Blomgren K, Ferriero DM, et al. Antioxidant status alters levels of Fas-associated death domain-like IL-1B-converting enzyme inhibitory protein following neonatal hypoxia-ischemia. Dev Neurosci. 2007;29:403–11.
85. Yesilirmak DC, Kumral A, Tugyan K, Cilaker S, Baskin H, Yilmaz O, et al. Effects of activated protein C on neonatal hypoxic ischemic brain injury. Brain Res. 2008;1210:56–62.
86 Johnson L, Bowen FW Jr, Abbasi S, Herrmann N, Weston M, Sacks L, et al. Relationship of prolonged pharmacologic serum levels of vitamin E to incidence of sepsis and necrotizing enterocolitis in infants with birth weight 1,500 grams or less. Pediatrics. 1985;75:619–38.
87. Lorch V, Murphy D, Hoersten LR, Harris E, Fitzgerald J, Sinha SN. Unusual syndrome among premature infants: association with a new intravenous vitamin E product. Pediatrics. 1985;75:598–602.
88. Jankov RP, Negus A, Tanswell AK. Antioxidants as therapy in the newborn: some words of caution. Pediatr Res. 2001;50:681–7.
89. Allen RG, Tresini M. Oxidative stress and gene regulation. Free Radic Biol Med. 2000;28:463–99.
90. Zafarullah M, Li WQ, Sylvester J, Ahmad M. Molecular mechanisms of *N*-acetylcysteine actions. Cell Mol Life Sci. 2003;60:6–20.
91. Buhimschi I, Buhimschi C, Weiner C. Protective effect of *N*-acetylcysteine against fetal death and preterm labor induced maternal inflammation. Am J Obstet Gynecol. 2003;188:203–8.
92. Farr SA, Poon HF, Dogrukol-Ak D, Drake J, Banks WA, Eyerman E, et al. The antioxidants alpha-lipoic acid and *N*-acetylcysteine reverse memory impairment and brain oxidative stress in aged SAMP8 mice. J Neurochem. 2003;84:1173–83.
93. Wang X, Svedin P, Nie C, Lapatto R, Zhu C, Gustavsson M, et al. *N*-Acetylcysteine reduces lipopolysaccharide-sensitized hypoxic-ischemic brain injury. Ann Neurol. 2007;61:263–71.
94. Beloosesky R, Gayle DA, Amidi F, Nunez SE, Babu J, Desai M, et al. *N*-Acetyl-cysteine suppresses amniotic fluid and placenta inflammatory cytokine responses to lipopolysaccharide in rats. Am J Obstet Gynecol. 2006;194:268–73.
95. Beloosesky R, Gayle DA, Ross MG. Maternal *N*-acetylcysteine suppresses fetal inflammatory cytokine responses to maternal lipopolysaccharide. Am J Obstet Gynecol. 2006;195:1053–7.
96 Horowitz RS, Dart RC, Jarvie DR, Bearer CF, Gupta U. Placental transfer of *N*-acetylcysteine following human maternal acetaminophen toxicity. J Toxicol Clin Toxicol. 1997;35:447–51.
97. Riggs BS, Bronstein AC, Kulig K, Archer PG, Rumack BH. Acute acetaminophen overdose during pregnancy. Obstet Gynecol. 1989;74:247–53.
98. Cheung PY, Danial H, Jong J, Schulz R. Thiols protect the inhibition of myocardial aconitase by peroxynitrite. Arch Biochem Biophys. 1998;350:104–8.

99. Johnson ST, Bigam DL, Emara M, Obaid L, Slack G, Korbutt G, et al. *N*-Acetylcysteine improves the hemodynamics and oxidative stress in hypoxic newborn pigs reoxygenated with 100% oxygen. Shock. 2007;28:484–90.
100. Ferrer JV, Ariceta J, Guerrero D, Gomis T, Larrea MM, Balén E, et al. Allopurinol and *N*-acetylcysteine avoid 60% of intestinal necrosis in an ischemia-reperfusion experimental model. Transplant Proc. 2672;1998;30.
101. Chan E, Obaid L, Johnson ST, Bigam DL, Cheung PY. *N*-Acetylcysteine administration improves platelet aggregation in hypoxia-reoxygenation injury. Proc West Pharmacol Soc. 2007;50:53–7.
102. Lin CH, Kuo SC, Huang LJ, Gean PW. Neuroprotective effect of *N*-acetylcysteine on neuronal apoptosis induced by a synthetic gingerdione compound: involvement of ERK and p38 phosphorylation. J Neurosci Res. 2006;84:1485–94.
103. Lee TF, Jantzie LL, Todd KG, Cheung PY. Postresuscitation *N*-acetylcysteine treatment reduces cerebral hydrogen peroxide in the hypoxic piglet brain. Intensive Care Med. 2008;34:190–7.
104. Lee TF, Tymafichuk CN, Bigam DL, Cheung PY. Effects of post-resuscitation *N*-acetylcysteine on cerebral free radical production and perfusion during reoxygenation of hypoxic newborn piglets. Pediatr Res. 2008;64:256–61.
105. Ytrebø LM, Korvald C, Nedredal GI, Elvenes OP, Nielsen Grymyr OJ, Revhaug A. *N*-Acetylcysteine increases cerebral perfusion pressure in pigs with fulminant hepatic failure. Crit Care Med. 2001;29:1989–95.
106. Khan M, Sekhon B, Jatana M, Giri S, Gilg AG, Sekhon C, et al. Administration of *N*-acetylcysteine after focal cerebral ischemia protects brain and reduces inflammation in rat model of experimental stroke. J Neurosci Res. 2004;76:519–27.
107. Yip L, Dart RC, Hurlbut KM. Intravenous administration of oral *N*-acetylcysteine. Crit Care Med. 1998;26:40–3.
108. Jantzie LL, Todd KG, Johnson ST, Bigam DL, Cheung P-Y. The use of *N*-acetylcysteine in the reoxygenation cerebral injury of asphyxiated newborn piglets (abstract). EPAS. 2007;617932.20.
109. Walker PD, Brokering KL, Theobald JC. Fenoldopam and *N*-acetylcysteine for the prevention of radiographic contrast material-induced nephropathy: a review. Pharmacotherapy. 2003;23:1617–26.
110. Sadowska AM, Manuel-Y-Keenoy B, De Backer WA. Antioxidant and anti-inflammatory efficacy of NAC in the treatment of COPD: discordant in vitro and in vivo dose-effects: a review. Pulm Pharmacol Ther. 2007;20:9–22.
111. Viña J, Vento M, García-Sala F, Puertes IR, Gascó E, Sastre J, et al. l-Cysteine and glutathione metabolism are impaired in premature infants due to cystathionase deficiency. Am J Clin Nutr. 1995;61:1067–9.
112. Soghier LM, Brion LP. Cysteine, cystine or *N*-acetylcysteine supplementation in parenterally fed neonates. Cochrane Database Syst Rev. 2006;4:CD004869.
113. Asfar P, De Backer D, Meier-Hellmann A, Radermacher P, Sakka SG. Clinical review: influence of vasoactive and other therapies on intestinal and hepatic circulations in patients with septic shock. Crit Care. 2004;8:170–9.

Chapter 5
Oxidative Stress in Multiple Sclerosis Pathology and Therapeutic Potential of Nrf2 Activation

Helga E. de Vries, Gerty Schreibelt, and Jack van Horssen

Abstract Reactive oxygen species contribute to the formation and persistence of multiple sclerosis (MS) lesions by acting on distinct pathological processes. To counteract the detrimental effects of reactive oxygen species, the central nervous system is endowed with a protective mechanism consisting of enzymatic and nonenzymatic antioxidants. Expression of most antioxidant enzymes is regulated through the transcription factor nuclear factor-E_2-related factor (Nrf2), and antioxidant response elements (ARE) in the genes encoding enzymatic antioxidants and are induced by oxidative stress. In brain tissue of MS patients, enhanced expression of Nrf2/ARE-regulated antioxidants suggests the occurrence of oxidative stress in these lesions. Antioxidant therapy may therefore represent an attractive treatment of MS. Several studies have shown that antioxidant therapy is beneficial in vitro and in vivo in animal models for MS. However, the use of exogenous antioxidants for MS treatment has drawbacks, as large amounts of antioxidants are required to achieve functional antioxidant levels in the central nervous system. Therefore, the induction of endogenous antioxidant enzymes by activators of the Nrf2/ARE pathway may be an interesting approach to obtain sufficient levels of antioxidants to interfere with pathological processes underlying MS lesion formation. Here we discuss and summarize the biological role, regulation, and potential therapeutic effects of endogenous antioxidant enzymes in MS.

Keywords Multiple sclerosis · Neuroinflammation · Reactive oxygen species · Oxidative stress · Antioxidants · Antioxidant enzymes

---

J. van Horssen (✉)
Department of Molecular Cell Biology and Immunology,
VU University Medical Center, Amsterdam, The Netherlands
e-mail: j.vanhorssen@vumc.nl

N. Gadoth and H.H. Göbel (eds.), *Oxidative Stress and Free Radical Damage in Neurology*, Oxidative Stress in Applied Basic Research and Clinical Practice,
DOI 10.1007/978-1-60327-514-9_5,

1 Introduction

Multiple sclerosis (MS) is a chronic inflammatory and demyelinating disease of the central nervous system (CNS) particularly affecting young adults [1, 2]. MS patients may suffer from a wide variety of clinical symptoms, including changes in sensation, visual problems, muscle weakness, and difficulties with coordination and speech. Neuropathological examination of brain tissue of individuals with MS demonstrates multiple lesions in the white matter [3]. To date, evidence is accumulating that extensive demyelination also occurs in the cerebral cortex of chronic MS patients [4–6]. Common pathological features of MS white matter plaques include blood–brain barrier (BBB) leakage, destruction of myelin sheaths, oligodendrocyte damage and cell death, axonal damage and axonal loss, glial scar formation, and the presence of inflammatory infiltrates that generally consist of lymphocytes and macrophages [3]. In particular, monocyte-derived macrophages contribute to MS lesion formation [7, 8] as they phagocytose myelin, which ultimately leads to damage of myelin sheaths and oligodendrocyte cell death. Importantly, during inflammation, macrophages secrete various inflammatory mediators, including cytokines, chemokines, nitric oxide, and reactive oxygen species (ROS), which all contribute to the progression of the disease [9, 10]. In spite of tremendous efforts, the mechanisms underlying the pathogenesis of MS remain enigmatic. Although MS is traditionally considered as an inflammatory autoimmune demyelinating disease, recent findings have challenged this concept, and evidence is emerging that primary degeneration of oligodendrocytes may be involved in the initial phase of the disease [11].

2 ROS Are Intimately Involved in MS Pathogenesis

To date, evidence is accumulating that ROS-induced oxidative stress and damage contribute to the pathogenesis of MS lesions. In the initial phase of MS lesion formation, locally produced ROS provoke BBB breakdown and enhance transendothelial monocyte migration [12, 13]. In the brain parenchyma, ROS mediate lesion persistence by inducing myelin phagocytosis, oligodendrocyte injury, and axonal damage [14–17]. Both activated microglia and infiltrated leukocytes produce vast amounts of ROS and reactive nitrogen species (RNS). ROS and RNS generated by macrophages have been implicated as mediators of demyelination and axonal injury in MS and experimental autoimmune encephalomyelitis (EAE), a widely used animal model for MS. Additionally, noninflammatory mechanisms, such as mitochondrial dysfunction, contribute to the formation of ROS [18–20]. Mitochondrial alterations have been implicated in MS lesion development and persistence and may occur as a response to demyelination because this pathological process enhances the energy demand in axons and thereby affects mitochondrial activity [21, 22]. Together, these events might lead to accumulating free radical damage, compromised mitochondrial functioning, and subsequent axonal loss. In EAE, nitration of mitochondrial proteins even preceded infiltration of leukocytes and induced major mitochondrial changes and apoptotic cell death [23]. Furthermore, reduced concentrations of antioxidants have been detected in serum of MS patients and in MS plaques [24, 25], suggesting that increased levels of ROS may have resulted in the depletion of cellular antioxidants.

At physiological concentrations, ROS mediate numerous regulatory processes [26]. However, when the rate of free radical generation overwhelms the capacity of antioxidant defense, oxidative stress occurs. Enhanced levels of ROS cause damage to biological macromolecules, such as polyunsaturated fatty acids in membrane lipids, essential proteins, and nucleotides in RNA and DNA [24]. Increased levels of indicators of oxidative stress and/or decreased levels of antioxidant enzymes and antioxidants are observed in blood and cerebrospinal fluid during the active phases of MS [27–31]. Additionally, enhanced expression of markers of oxidative damage in MS brain tissue has been reported, indicating the occurrence of widespread oxidative damage in demyelinating MS plaques. Immunohistochemical detection of nitrotyrosine, a putative footprint for peroxynitrite formation, demonstrated enhanced nitrotyrosine staining in acute but not in chronic MS lesions [32]. Nitrotyrosine was detected on astrocytes, macrophages, and oligodendrocytes in MS lesions [33, 34]. Additionally, markers for lipid peroxidation, such as malondialdehyde and 4-hydroxynonenal, and oxidative damage to nucleotides are present in MS lesions [35]. Interestingly, increased protein carbonylation was detected in normal-appearing white and gray matter of patients with MS [36]. Oligodendrocytes are particularly prone to oxidative damage as they contain high amounts of iron [37] and reduced levels of superoxide dismutase-2, glutathione, and glutathione peroxidase [38]. Oligodendrocytes produce large myelin sheets, which mainly consist of polyunsaturated fatty acids that can react with ROS, thereby triggering lipid peroxidation. ROS is able to induce both necrotic cell death as well as apoptosis; however, the mechanisms that trigger oligodendrocyte cell death under oxidative stress condition remain poorly understood.

The CNS is equipped with an endogenous antioxidant defense mechanism, which consists of antioxidant enzymes and nonenzymatic antioxidants. Nonenzymatic antioxidants can be divided into endogenous molecules, such as glutathione, uric acid, and nicotinamide adenine dinucleotide phosphate (NADPH), and exogenous antioxidants, such as ascorbic acid (vitamin C), α-tocopherol (vitamin E), α-lipoic acid, and flavonoids. Protective effects of exogenous antioxidants in the development and progression of EAE have been reported by several research groups [24, 25, 39, 40], indicating that antioxidant therapy may be an attractive therapeutic approach for MS treatment. However, despite promising results in the treatment of EAE, data on successful antioxidant therapy in MS patients are still limited, emphasizing the need for further epidemiological and clinical studies on antioxidant strategies. Importantly, there are a number of drawbacks to the use of exogenous antioxidants for MS treatment, as most antioxidants do not efficiently cross the BBB. Consequently, high doses are generally required to achieve protective effects in EAE animals. Alternative strategies aimed at increasing the production of endogenous antioxidants might limit the harmful effects of ROS [41].

3 Endogenous Antioxidant Enzymes

Oxidative stress activates an adaptive mechanism aimed to protect cells against ROS-induced injury and to control tissue redox balance. This stress response is regulated at the transcriptional level by the transcription factor Nrf2. Under physiological conditions, Nrf2 is linked to the actin-bound Kelch-like ECH-associated protein 1 (Keap1) and located in the cytoplasm. Upon exposure to ROS or electrophiles Nrf2 dissociates from Keap1 and translocates to the nucleus, where it binds to antioxidant response elements (ARE) in genes encoding antioxidant enzymes [42]. To date, more than 200 Nrf2-ARE-driven genes involved in detoxification and antioxidant defense have been identified, including catalase [43], superoxide dismutases (SODs) [44], glutathione peroxidases [45], peroxiredoxins (Prxs) [43, 46], heme oxygenases [47, 48], NAD(P)H:quinone oxidoreductase 1 (NQO1), and NHR: quinone oxidoreductase 2 (NQO2) [49–51].

3.1 Superoxide Dismutases

Superoxide is one of the main radicals in the cell, and as such SODs play a key role in the detoxification of superoxide. SODs are metal-containing enzymes that promote dismutation of superoxide anion into molecular oxygen and hydrogen peroxide [52]. Three distinct forms of SOD have been identified: copper- and zinc-containing cytosolic SOD1 (Cu/ZnSOD) [53], manganese-containing mitochondrial SOD2 (MnSOD) [54], and extracellular Cu/ZnSOD (SOD3) [55]. SOD1 and SOD2 are abundantly expressed throughout the brain. SOD1 is mainly expressed in astrocytes whereas SOD2 is generally observed in neurons. In contrast, the basal activity of SOD1 and SOD2 is relatively low in microglia and oligodendrocytes [56]. Several groups have demonstrated the involvement of SOD in neurodegenerative and neuroinflammatory disorders, and increased expression of SOD has been detected in Alzheimer's disease [57, 58] and stroke brain tissue [59, 60]. Enhanced gene expression of SOD1 has been observed in active demyelinating MS lesions [61]. Increased immunostaining of SOD2, but not SOD1, was found in guinea pigs suffering from EAE [62]. Remarkably, SOD administration did not influence the development of EAE in a rat model [9] and only had minor effects on EAE and experimental allergic optic neuritis in guinea pigs [63, 64]. Compounds that mimic the active site of metalloenzymes, such as SOD, can cross the BBB and were effective in animal models of ischemia and Parkinson's disease [65, 66]. This finding emphasizes the need for future research to study the putative protective effects of SOD mimetics in EAE models.

3.2 *Catalase*

Conversion of superoxide by SODs results in the formation of hydrogen peroxide, which is able to diffuse across cellular membranes. In the presence of transition metals, such as iron and copper, hydrogen peroxide can be converted to highly reactive hydroxyl radicals. To remove hydrogen peroxide, cells are equipped with several hydrogen peroxide-removing enzymes, such as catalase, glutathione peroxidases, and Prxs. Catalase is ubiquitously expressed in the CNS and is mainly located in peroxisomes where it catalyzes the conversion of hydrogen peroxide into water and molecular oxygen. Impaired peroxisomal function has been demonstrated in CNS homogenates of rats suffering from EAE, which was accompanied by reduced catalase gene expression and activity [67]. Catalase treatment of guinea pigs suffering from EAE significantly reduced demyelination of the optic nerves, increased BBB integrity, and ameliorated neurological symptoms [63, 68]. Administration of catalase reduced the severity of clinical signs in a rat model for EAE [9]. Furthermore, upregulation of catalase expression via viral vectors ameliorates EAE, implicating its positive effect in neuroinflammatory disorders [69, 70].

3.3 *Glutathione Peroxidases*

Glutathione peroxidases constitute a family of six different selenium-containing enzymes that detoxify cellular organic peroxides and hydrogen peroxide. The activity of glutathione peroxidase-1 is generally higher than that of catalase [71], suggesting that in the CNS glutathione peroxidase may even be more important than catalase in removing hydrogen peroxide. Tajouri and coworkers demonstrated that glutathione peroxidase gene expression is significantly increased in inflammatory MS lesions [61], and Guy and colleagues showed that treatment with glutathioneperoxidase reduces loss of BBB integrity in chronic EAE [63], suggesting that glutathione peroxidase plays a protective role in neuroinflammation.

3.4 *Peroxiredoxins*

The family of Prxs consists of six distinct groups of antioxidant proteins that are involved in the enzymatic degradation of hydrogen peroxide, organic hydroperoxides and peroxynitrite. Peroxiredoxins are abundant in the cytosol [72, 73], mitochondria, peroxisomes, plasma, and nuclei and are associated with membranes [74]. Peroxiredoxin-1 is primarily expressed in astrocytes while peroxiredoxin-2 is predominantly localized to neurons [75]. Krapfenbauer and colleagues examined the distribution of Prxs in various neurodegenerative diseases that are associated with oxidative stress, including Down syndrome, Alzheimer's disease and Pick's disease.

They showed that peroxiredoxin-1 expression was not different from control brains, whereas expression of peroxiredoxin-2 and -6 was significantly increased. In contrast, peroxiredoxin-3 expression was reduced, indicating that expression of Prxs is differentially regulated in neurodegenerative disorders [76]. Recently, we reported that peroxiredoxin-1 overexpression protected brain endothelial cells from ROS-induced cell death, reduced adhesion and subsequent transendothelial migration of monocytes, and enhanced the integrity of the brain endothelial cell layer [77]. Furthermore, we observed a striking increase of vascular peroxiredoxin-1 immunostaining in inflammatory lesions of experimental EAE animals and inflammatory demyelinating MS lesions. Enhanced vascular peroxiredoxin-1 expression may reflect ongoing oxidative stress in EAE and MS, or it may function as a protective mechanism to limit ROS-mediated damage and leukocyte infiltration into the CNS [77].

3.5 Heme Oxygenase

Heme oxygenase-2 is constitutively expressed by a wide variety of cell types, whereas heme oxygenase-1 is inducible by a variety of stimuli, including its substrate heme and oxidative stress [48]. Heme oxygenases catalyze the rate-limiting step in the catabolism of heme and break down the porphyrin ring into biliverdin, free iron, and carbon monoxide. Various research groups have demonstrated protective effects of heme oxygenase-1 in vitro and in vivo. Induction of EAE in heme oxygenase-1 knockout mice led to enhanced CNS demyelination, paralysis, and mortality, as compared with wild-type mice [78]. Upregulation of heme oxygenase-1 by either heminin [79] or cobalt protoporphyrin IX administration [78] reduced the clinical severity of EAE, whereas tin mesoporphyrin, a well-known inhibitor of heme oxygenase-1, markedly exacerbated EAE. In contrast, treatment with the heme oxygenase-1 inhibitor tin-protoporphyrin attenuated clinical scores in murine EAE [80]. In the active MS and EAE lesions, heme oxygenase-1-immunoreactivity was strikingly upregulated in activated microglia, infiltrated macrophages, and hypertrophic astrocytes [81–83].

Heme oxygenase degrades heme into biliverdin, which is subsequently converted by biliverdin reductase into bilirubin, while the pro-oxidant iron is sequestered and inactivated by coinduced ferritin. Bilirubin exerts potent antioxidant activity, and both suppressed ongoing EAE and halted EAE progression when administered after disease onset, likely by reducing BBB permeability and oxidative damage [84]. Interestingly, bilirubin protected rat oligodendrocytes against hydrogen peroxide-mediated cell death. Treatment with biliverdin reductase, as with bilirubin, reduced oxidative injury in EAE lesions and significantly suppressed clinical symptoms of EAE [85]. Together, these studies show that heme oxygenase-1 or downstream products of heme metabolism may be potential therapeutics for MS treatment.

3.6 *Quinone Oxidoreductases*

NQO1 and NQO2 are cytosolic flavoproteins that have broad-spectrum antioxidant properties [86–89]. Besides their function as antioxidants, NQO1 and NQO2 function to maintain both α-tocopherol and coenzyme Q10 in their reduced antioxidant state. NQO1 is expressed in tissues that require high levels of antioxidant protection, such as lung respiratory epithelium and the CNS [90, 91]. NQO1 is predominantly found in astrocytes and brain endothelial cells of healthy human brain tissue [92, 93]. In Alzheimer's disease and Parkinson's disease brains, astrocytes are highly NQO1-immunoreactive in areas of ongoing disease activity, indicating the occurrence of oxidative stress [92, 94]. We recently showed that NQO1 is markedly upregulated in inflammatory MS lesions, particularly in hypertrophic astrocytes and myelin-laden macrophages [93]. In human tissue, NQO2 gene expression has been observed in kidney, skeletal muscle, liver, lung, and heart, whereas very low expression was found in the brain [95]. To date, little is known about the regulation of NQO2 expression in neurodegenerative or neuroinflammatory diseases.

4 Therapeutic Potential of Antioxidant Enzyme Upregulation

Enhanced antioxidant enzyme expression might act as an endogenous compensatory system to counteract ROS-induced damage. Indeed, Nrf2/ARE-regulated antioxidant enzymes are upregulated in CNS tissue of EAE animals and MS patients. Despite this cellular protective mechanism, oxidative damage to essential biological macromolecules is abundant in inflammatory MS lesions, suggesting that although Nrf2-mediated antioxidant enzymes are induced this response may be insufficient to counteract ROS-induced cellular damage. We propose that additional activation of the Nrf2/ARE system via monofunctional inducers might represent a novel therapeutic approach to limit oxidative stress in neuroinflammatory diseases such as MS. Monofunctional inducers, such as tert-butylhydroquinone (tBHQ), dimethylfumarate, sulforaphane, and 3-hydroxycoumarin, are well tolerated, have the ability to cross the BBB, and promote transcription of endogenous antioxidant enzymes [96, 97]. Several reports have highlighted the protective effects of Nrf2 activation in reducing oxidative stress in both in vitro and in vivo models of neurodegenerative disorders. Induction of Nrf2-driven antioxidants by dietary tBHQ demonstrated beneficial effects in animal models for neurodegeneration and cerebral ischemia [98, 99]. Similarly, sulforaphane reduces the severity of traumatic brain injury and cerebral ischemia in vivo [100, 101].

In addition to the coordination of antioxidant enzyme production, Nrf2 is also involved in the regulation of monocyte chemoattractant protein-1 (MCP-1) and vascular cell adhesion molecule-1 (VCAM-1) [102]. Nrf2 suppressed tumor necrosis factor-alpha (TNF-α)-induced MCP-1 and VCAM-1 mRNA and protein expression and limited TNF-α-induced monocyte adhesion to endothelial cells. Hence, we hypothesize that Nrf2/ARE activation might play a protective role in the pathogenesis of MS by operating on distinct levels: (1) antioxidant enzymes can directly scavenge ROS, thereby restoring BBB integrity; (2) Nrf2 activation might reduce leukocyte adhesion and subsequent transendothelial leukocyte migration; (3) increased levels of antioxidant enzymes might limit myelin phagocytosis and breakdown; and (4) upregulation of antioxidant enzymes might prevent ROS-induced oligodendrocyte and neuroaxonal damage.

In conclusion, ROS play a central role in various pathological processes underlying MS pathology. Enhanced expression of Nrf2/ARE-regulated antioxidants in EAE and MS tissue suggests the occurrence of ongoing oxidative stress. Antioxidant therapy may therefore represent an attractive treatment of MS. Induction of endogenous antioxidant enzymes by activators of the Nrf2/ARE pathway may be an interesting approach to obtain sufficient levels of antioxidants to reduce leukocyte migration into the CNS, demyelination, and oxidative damage. Future studies are warranted to gain more insight into the putative protective effects of Nrf2/ARE induction in the treatment of neuroinflammatory diseases.

5 Summary

ROS contribute to the formation and persistence of MS lesions by acting on distinct pathological processes. In the initial phase of leukocyte infiltration ROS are produced, which induce BBB disruption, thereby facilitating transendothelial migration of inflammatory cells. Furthermore, infiltrated leukocytes produce significant amounts of ROS, which induce oxidative damage to vulnerable brain cells. To counteract the detrimental effects of ROS, the CNS is endowed with a protective mechanism consisting of enzymatic and nonenzymatic antioxidants. Expression of most antioxidant enzymes is regulated through the transcription factor Nrf2, and ARE in the genes encoding enzymatic antioxidants and are induced by oxidative stress. In brain tissue of MS patients, enhanced expression of Nrf2/ARE-regulated antioxidants suggests the occurrence of oxidative stress in these lesions. Antioxidant therapy may therefore represent an attractive treatment of MS. Several studies have shown that antioxidant therapy is beneficial in vitro and in vivo in animal models for MS. However, the use of exogenous antioxidants for MS treatment has drawbacks, as large amounts of antioxidants are required to achieve functional antioxidant levels in the CNS. Therefore, the induction of endogenous antioxidant enzymes by activators of the Nrf2/ARE pathway may be an interesting approach to obtain sufficient levels of antioxidants to interfere with pathological processes underlying MS lesion formation.

Acknowledgments This work was supported by grants from "The Dutch Foundation of MS Research," The Netherlands project numbers 05–567 MS and 05–358c MS (J. van Horssen) and MS 02–358 (H.E. de Vries), and the Institute for Clinical and Experimental Neurosciences, VU University Medical Center, Amsterdam, The Netherlands (G. Schreibelt).

References

1. Hafler DA. Multiple sclerosis. J Clin Invest. 2004;113(6):788–94.
2. Keegan BM, Noseworthy JH. Multiple sclerosis. Annu Rev Med. 2002;53:285–302.
3. Frohman EM, Racke MK, Raine CS. Multiple sclerosis: the plaque and its pathogenesis. N Engl J Med. 2006;354(9):942–55.
4. Geurts JJ, Bo L, Pouwels PJ, Castelijns JA, Polman CH, Barkhof F. Cortical lesions in multiple sclerosis: combined postmortem MR imaging and histopathology. AJNR Am J Neuroradiol. 2005;26(3):572–7.
5. Bo L, Geurts JJ, Mork SJ, van der Valk P. Grey matter pathology in multiple sclerosis. Acta Neurol Scand Suppl. 2006;183:48–50.
6. Vercellino M, Plano F, Votta B, Mutani R, Giordana MT, Cavalla P. Grey matter pathology in multiple sclerosis. J Neuropathol Exp Neurol. 2005;64(12):1101–7.
7. Bruck W, Stadelmann C. The spectrum of multiple sclerosis: new lessons from pathology. Curr Opin Neurol. 2005;18(3):221–4.
8. Hendriks JJ, Teunissen CE, de Vries HE, Dijkstra CD. Macrophages and neurodegeneration. Brain Res Brain Res Rev. 2005;48(2):185–95.
9. Ruuls SR, Bauer J, Sontrop K, Huitinga I, 't Hart BA, Dijkstra CD. Reactive oxygen species are involved in the pathogenesis of experimental allergic encephalomyelitis in Lewis rats. J Neuroimmunol. 1995;56(2):207–17.
10. Benveniste EN. Cytokine actions in the central nervous system. Cytokine Growth Factor Rev. 1998;9(3–4):259–75.
11. Barnett MH, Prineas JW. Relapsing and remitting multiple sclerosis: pathology of the newly forming lesion. Ann Neurol. 2004;55(4):458–68.
12. Schreibelt G, Musters RJ, Reijerkerk A, de Groot LR, Van Der Pol SM, Hendrikx EM, et al. Lipoic acid affects cellular migration into the central nervous system and stabilizes blood-brain barrier integrity. J Immunol. 2006;177(4):2630–7.
13. van der Goes A, Wouters D, Van Der Pol SM, Huizinga R, Ronken E, Adamson P, et al. Reactive oxygen species enhance the migration of monocytes across the blood-brain barrier in vitro. FASEB J. 2001;15(10):1852–4.
14. Smith KJ, Kapoor R, Felts PA. Demyelination: the role of reactive oxygen and nitrogen species. Brain Pathol. 1999;9(1):69–92.
15. Smith KJ, Kapoor R, Hall SM, Davies M. Electrically active axons degenerate when exposed to nitric oxide. Ann Neurol. 2001;49(4):470–6.
16. van der Goes A, Brouwer J, Hoekstra K, Roos D, Van den Berg TK, Dijkstra CD. Reactive oxygen species are required for the phagocytosis of myelin by macrophages. J Neuroimmunol. 1998;92(1–2):67–75.
17. van Meeteren ME, Hendriks JJ, Dijkstra CD, van Tol EA. Dietary compounds prevent oxidative damage and nitric oxide production by cells involved in demyelinating disease. Biochem Pharmacol. 2004;67(5):967–75.
18. Kalman B, Laitinen K, Komoly S. The involvement of mitochondria in the pathogenesis of multiple sclerosis. J Neuroimmunol. 2007;188(1–2):1–12.
19. Kalman B, Leist TP. A mitochondrial component of neurodegeneration in multiple sclerosis. Neuromol Med. 2003;3(3):147–58.
20. Kalman B. Role of mitochondria in multiple sclerosis. Curr Neurol Neurosci Rep. 2006;6 (3):244–52.
21. Lu F, Selak M, O'Connor J, Croul S, Lorenzana C, Butunoi C, et al. Oxidative damage to mitochondrial DNA and activity of mitochondrial enzymes in chronic active lesions of multiple sclerosis. J Neurol Sci. 2000;177(2):95–103.
22. Mahad D, Ziabreva I, Lassmann H, Turnbull D. Mitochondrial defects in acute multiple sclerosis lesions. Brain. 2008;131(7):1722–35.

23. Qi X, Lewin AS, Sun L, Hauswirth WW, Guy J. Mitochondrial protein nitration primes neurodegeneration in experimental autoimmune encephalomyelitis. J Biol Chem. 2006; 281(42):31950–62.
24. Gilgun-Sherki Y, Melamed E, Offen D. The role of oxidative stress in the pathogenesis of multiple sclerosis: the need for effective antioxidant therapy. J Neurol. 2004;251(3):261–8.
25. van Meeteren ME, Teunissen CE, Dijkstra CD, van Tol EA. Antioxidants and polyunsaturated fatty acids in multiple sclerosis. Eur J Clin Nutr. 2005;59(12):1347–61.
26. Fialkow L, Wang Y, Downey GP. Reactive oxygen and nitrogen species as signaling molecules regulating neutrophil function. Free Radic Biol Med. 2007;42(2):153–64.
27. Karg E, Klivenyi P, Nemeth I, Bencsik K, Pinter S, Vecsei L. Nonenzymatic antioxidants of blood in multiple sclerosis. J Neurol. 1999;246(7):533–9.
28. Greco A, Minghetti L, Sette G, Fieschi C, Levi G. Cerebrospinal fluid isoprostane shows oxidative stress in patients with multiple sclerosis. Neurology. 1999;53(8):1876–9.
29. Calabrese V, Raffaele R, Cosentino E, Rizza V. Changes in cerebrospinal fluid levels of malondialdehyde and glutathione reductase activity in multiple sclerosis. Int J Clin Pharmacol Res. 1994;14(4):119–23.
30. Ferretti G, Bacchetti T, Principi F, Di Ludovico F, Viti B, Angeleri VA, et al. Increased levels of lipid hydroperoxides in plasma of patients with multiple sclerosis: a relationship with paraoxonase activity. Mult Scler. 2005;11(6):677–82.
31 Koch M, Ramsaransing GS, Arutjunyan AV, Stepanov M, Teelken A, Heersema DJ, et al. Oxidative stress in serum and peripheral blood leukocytes in patients with different disease courses of multiple sclerosis. J Neurol. 2006;253(4):483–7.
32. Liu JS, Zhao ML, Brosnan CF, Lee SC. Expression of inducible nitric oxide synthase and nitrotyrosine in multiple sclerosis lesions. Am J Pathol. 2001;158(6):2057–66.
33. Diaz-Sanchez M, Williams K, DeLuca GC, Esiri MM. Protein co-expression with axonal injury in multiple sclerosis plaques. Acta Neuropathol. 2006;111(4):289–99.
34. Jack C, Antel J, Bruck W, Kuhlmann T. Contrasting potential of nitric oxide and peroxynitrite to mediate oligodendrocyte injury in multiple sclerosis. Glia. 2007;55(9):926–34.
35. Vladimirova O, O'Connor J, Cahill A, Alder H, Butunoi C, Kalman B. Oxidative damage to DNA in plaques of MS brains. Mult Scler. 1998;4(5):413–8.
36. Bizzozero OA, DeJesus G, Callahan K, Pastuszyn A. Elevated protein carbonylation in the brain white matter and gray matter of patients with multiple sclerosis. J Neurosci Res. 2005;81 (5):687–95.
37. Connor JR. Iron regulation in the brain at the cell and molecular level. Adv Exp Med Biol. 1994;356:229–38.
38. Juurlink BH, Thorburne SK, Hertz L. Peroxide-scavenging deficit underlies oligodendrocyte susceptibility to oxidative stress. Glia. 1998;22(4):371–8.
39. Kean RB, Spitsin SV, Mikheeva T, Scott GS, Hooper DC. The peroxynitrite scavenger uric acid prevents inflammatory cell invasion into the central nervous system in experimental allergic encephalomyelitis through maintenance of blood-central nervous system barrier integrity. J Immunol. 2000;165(11):6511–8.
40. Hendriks JJ, Alblas J, Van Der Pol SM, van Tol EA, Dijkstra CD, de Vries HE. Flavonoids influence monocytic GTPase activity and are protective in experimental allergic encephalitis. J Exp Med. 2004;200(12):1667–72.
41. Carlson NG, Rose JW. Antioxidants in multiple sclerosis: do they have a role in therapy? CNS Drugs. 2006;20(6):433–41.
42. Itoh K, Wakabayashi N, Katoh Y, Ishii T, O'Connor T, Yamamoto M. Keap1 regulates both cytoplasmic-nuclear shuttling and degradation of Nrf2 in response to electrophiles. Genes Cells. 2003;8(4):379–91.
43. Dringen R, Pawlowski PG, Hirrlinger J. Peroxide detoxification by brain cells. J Neurosci Res. 2005;79(1-2):157–65.
44. McCord JM, Edeas MA. SOD, oxidative stress and human pathologies: a brief history and a future vision. Biomed Pharmacother. 2005;59(4):139–42.

45. Thimmulappa RK, Mai KH, Srisuma S, Kensler TW, Yamamoto M, Biswal S. Identification of Nrf2-regulated genes induced by the chemopreventive agent sulforaphane by oligonucleotide microarray. Cancer Res. 2002;62(18):5196–203.
46. Kim YJ, Ahn JY, Liang P, Ip C, Zhang Y, Park YM. Human prx1 gene is a target of Nrf2 and is up-regulated by hypoxia/reoxygenation: implication to tumor biology. Cancer Res. 2007; 67(2):546–54.
47. Ishii T, Itoh K, Takahashi S, Sato H, Yanagawa T, Katoh Y, et al. Transcription factor Nrf2 coordinately regulates a group of oxidative stress-inducible genes in macrophages. J Biol Chem. 2000;275(21):16023–9.
48. Wagener FA, Volk HD, Willis D, Abraham NG, Soares MP, Adema GJ, et al. Different faces of the heme–heme oxygenase system in inflammation. Pharmacol Rev. 2003;55(3):551–71.
49. Li Y, Jaiswal AK. Regulation of human NAD(P)H:quinone oxidoreductase gene. Role of AP1 binding site contained within human antioxidant response element. J Biol Chem. 1992;267 (21):15097–104.
50. Jaiswal AK. Regulation of genes encoding NAD(P)H:quinone oxidoreductases. Free Radic Biol Med. 2000;29(3–4):254–62.
51. Iskander K, Li J, Han S, Zheng B, Jaiswal AK. NQO1 and NQO2 regulation of humoral immunity and autoimmunity. J Biol Chem. 2006;281(41):30917–24.
52. Johnson F, Giulivi C. Superoxide dismutases and their impact upon human health. Mol Aspects Med. 2005;26(4–5):340–52.
53. McCord JM, Fridovich I. Superoxide dismutase. An enzymic function for erythrocuprein (hemocuprein). J Biol Chem. 1969;244(22):6049–55.
54. McCord JM. Iron- and manganese-containing superoxide dismutases: structure, distribution, and evolutionary relationships. Adv Exp Med Biol. 1976;74:540–50.
55. Marklund SL. Human copper-containing superoxide dismutase of high molecular weight. Proc Natl Acad Sci USA. 1982;79(24):7634–8.
56. Maier CM, Chan PH. Role of superoxide dismutases in oxidative damage and neurodegenerative disorders. Neuroscientist. 2002;8(4):323–34.
57. Delacourte A, Defossez A, Ceballos I, Nicole A, Sinet PM. Preferential localization of copper zinc superoxide dismutase in the vulnerable cortical neurons in Alzheimer's disease. Neurosci Lett. 1988;92(3):247–53.
58. Ceballos I, Javoy-Agid F, Delacourte A, Defossez A, Lafon M, Hirsch E, et al. Neuronal localization of copper-zinc superoxide dismutase protein and mRNA within the human hippocampus from control and Alzheimer's disease brains. Free Radic Res Commun. 1991;12–13(pt 2):571–80.
59. Liu XH, Kato H, Nakata N, Kogure K, Kato K. An immunohistochemical study of copper/zinc superoxide dismutase and manganese superoxide dismutase in rat hippocampus after transient cerebral ischemia. Brain Res. 1993;625(1):29–37.
60. Matsuyama T, Michishita H, Nakamura H, Tsuchiyama M, Shimizu S, Watanabe K, et al. Induction of copper-zinc superoxide dismutase in gerbil hippocampus after ischemia. J Cereb Blood Flow Metab. 1993;13(1):135–44.
61. Tajouri L, Mellick AS, Ashton KJ, Tannenberg AE, Nagra RM, Tourtellotte WW, et al. Quantitative and qualitative changes in gene expression patterns characterize the activity of plaques in multiple sclerosis. Brain Res Mol Brain Res. 2003;119(2):170–83.
62. Qi X, Guy J, Nick H, Valentine J, Rao N. Increase of manganese superoxide dismutase, but not of Cu/Zn-SOD, in experimental optic neuritis. Invest Ophthalmol Vis Sci. 1997;38 (6):1203–12.
63. Guy J, Ellis EA, Hope GM, Rao NA. Antioxidant enzymes reduce loss of blood-brain barrier integrity in experimental optic neuritis. Arch Ophthalmol. 1989;107(9):1359–63.
64. Guy J, Ellis EA, Hope GM, Rao NA. Influence of antioxidant enzymes in reduction of optic disc edema in experimental optic neuritis. J Free Radic Biol Med. 1986;2(5–6):349–57.

65. Shimizu K, Rajapakse N, Horiguchi T, Payne RM, Busija DW. Neuroprotection against hypoxia-ischemia in neonatal rat brain by novel superoxide dismutase mimetics. Neurosci Lett. 2003;346(1–2):41–4.
66. Pong K. Oxidative stress in neurodegenerative diseases: therapeutic implications for superoxide dismutase mimetics. Expert Opin Biol Ther. 2003;3(1):127–39.
67. Singh I, Paintlia AS, Khan M, Stanislaus R, Paintlia MK, Haq E, et al. Impaired peroxisomal function in the central nervous system with inflammatory disease of experimental autoimmune encephalomyelitis animals and protection by lovastatin treatment. Brain Res. 2004;1022 (1–2):1–11.
68. Guy J, Ellis EA, Hope GM, Rao NA. Antioxidant enzyme suppression of demyelination in experimental optic neuritis. Curr Eye Res. 1989;8(5):467–77.
69. Guy J, Qi X, Hauswirth WW. Adeno-associated viral-mediated catalase expression suppresses optic neuritis in experimental allergic encephalomyelitis. Proc Natl Acad Sci USA. 1998;95 (23):13847–52.
70. Guy J, Qi X, Wang H, Hauswirth WW. Adenoviral gene therapy with catalase suppresses experimental optic neuritis. Arch Ophthalmol. 1999;117(11):1533–9.
71. Marklund SL. Human copper-containing superoxide dismutase of high molecular weight. Proc Natl Acad Sci USA. 1982;79(24):7634–8.
72. Kang SW, Chae HZ, Seo MS, Kim K, Baines IC, Rhee SG. Mammalian peroxiredoxin isoforms can reduce hydrogen peroxide generated in response to growth factors and tumor necrosis factor-alpha. J Biol Chem. 1998;273(11):6297–302.
73. Seo MS, Kang SW, Kim K, Baines IC, Lee TH, Rhee SG. Identification of a new type of mammalian peroxiredoxin that forms an intramolecular disulfide as a reaction intermediate. J Biol Chem. 2000;275(27):20346–54.
74. Hofmann B, Hecht HJ, Flohe L. Peroxiredoxins. Biol Chem. 2002;383(3-4):347–64.
75. Sarafian TA, Verity MA, Vinters HV, Shih CC, Shi L, Ji XD, et al. Differential expression of peroxiredoxin subtypes in human brain cell types. J Neurosci Res. 1999;56(2):206–12.
76. Krapfenbauer K, Engidawork E, Cairns N, Fountoulakis M, Lubec G. Aberrant expression of peroxiredoxin subtypes in neurodegenerative disorders. Brain Res. 2003;967(1–2):152–60.
77. Schreibelt G, van Horssen J, Haseloff RF, Reijerkerk A, Van Der Pol SM, Nieuwenhuizen O, et al. Protective effects of peroxiredoxin-1 at the injured blood-brain barrier. Free Radic Biol Med. 2008;45(3):256–64.
78. Chora AA, Fontoura P, Cunha A, Pais TF, Cardoso S, Ho PP, et al. Heme oxygenase-1 and carbon monoxide suppress autoimmune neuroinflammation. J Clin Invest. 2007;117 (2):438–47.
79. Liu Y, Zhu B, Luo L, Li P, Paty DW, Cynader MS. Heme oxygenase-1 plays an important protective role in experimental autoimmune encephalomyelitis. Neuroreport. 2001;12 (9):1841–5.
80. Chakrabarty A, Emerson MR, LeVine SM. Heme oxygenase-1 in SJL mice with experimental allergic encephalomyelitis. Mult Scler. 2003;9(4):372–81.
81. Stahnke T, Stadelmann C, Netzler A, Bruck W, Richter-Landsberg C. Differential upregulation of heme oxygenase-1 (HSP32) in glial cells after oxidative stress and in demyelinating disorders. J Mol Neurosci. 2007;32(1):25–37.
82. Mehindate K, Sahlas DJ, Frankel D, Mawal Y, Liberman A, Corcos J, et al. Proinflammatory cytokines promote glial heme oxygenase-1 expression and mitochondrial iron deposition: implications for multiple sclerosis. J Neurochem. 2001;77(5):1386–95.
83. Emerson MR, LeVine SM. Heme oxygenase-1 and NADPH cytochrome P450 reductase expression in experimental allergic encephalomyelitis: an expanded view of the stress response. J Neurochem. 2000;75(6):2555–62.
84. Liu Y, Zhu B, Wang X, Luo L, Li P, Paty DW, et al. Bilirubin as a potent antioxidant suppresses experimental autoimmune encephalomyelitis: implications for the role of oxidative stress in the development of multiple sclerosis. J Neuroimmunol. 2003;139(1–2):27–35.

85. Liu Y, Liu J, Tetzlaff W, Paty DW, Cynader MS. Biliverdin reductase, a major physiologic cytoprotectant, suppresses experimental autoimmune encephalomyelitis. Free Radic Biol Med. 2006;40(6):960–7.
86. Beyer RE, Segura-Aguilar J, di Bernardo S, Cavazzoni M, Fato R, Fiorentini D, et al. The two-electron quinone reductase DT-diaphorase generates and maintains the antioxidant (reduced) form of coenzyme Q in membranes. Mol Aspects Med. 1997;18(Suppl.1): S15–S23.
87. Cadenas E. Antioxidant and prooxidant functions of DT-diaphorase in quinone metabolism. Biochem Pharmacol. 1995;49(2):127–40.
88. Siegel D, Gustafson DL, Dehn DL, Han JY, Boonchoong P, Berliner LJ, et al. NAD(P)H: quinone oxidoreductase 1: role as a superoxide scavenger. Mol Pharmacol. 2004;65(5): 1238–47.
89. Ross D, Kepa JK, Winski SL, Beall HD, Anwar A, Siegel D. NAD(P)H:quinone oxidoreductase 1 (NQO1): chemoprotection, bioactivation, gene regulation and genetic polymorphisms. Chem Biol Interact. 2000;129(1-2):77–97.
90. Siegel D, Ross D. Immunodetection of NAD(P)H:quinone oxidoreductase 1 (NQO1) in human tissues. Free Radic Biol Med. 2000;29(3-4):246–53.
91. Schultzberg M, Segura-Aguilar J, Lind C. Distribution of DT diaphorase in the rat brain: biochemical and immunohistochemical studies. Neuroscience. 1988;27(3):763–76.
92. van Muiswinkel FL, de Vos RA, Bol JG, Andringa G, Jansen Steur EN, Ross D, et al. Expression of NAD(P)H:quinone oxidoreductase in the normal and Parkinsonian substantia nigra. Neurobiol Aging. 2004;25(9):1253–62.
93. van Horssen J, Schreibelt G, Bo L, Montagne L, Drukarch B, van Muiswinkel FL, et al. NAD (P)H:quinone oxidoreductase 1 expression in multiple sclerosis lesions. Free Radic Biol Med. 2006;41(2):311–7.
94. SantaCruz KS, Yazlovitskaya E, Collins J, Johnson J, DeCarli C. Regional NAD(P)H: quinone oxidoreductase activity in Alzheimer's disease. Neurobiol Aging. 2004;25(1):63–9.
95. Long DJ, Jaiswal AK. NRH:quinone oxidoreductase 2 (NQO2). Chem Biol Interact. 2000;129(1-2):99–112.
96. Wierinckx A, Breve J, Mercier D, Schultzberg M, Drukarch B, Van Dam AM. Detoxication enzyme inducers modify cytokine production in rat mixed glial cells. J Neuroimmunol. 2005;166(1-2):132–43.
97. Nguyen T, Sherratt PJ, Pickett CB. Regulatory mechanisms controlling gene expression mediated by the antioxidant response element. Annu Rev Pharmacol Toxicol. 2003;43:233–60.
98. Shih AY, Imbeault S, Barakauskas V, Erb H, Jiang L, Li P, et al. Induction of the Nrf2-driven antioxidant response confers neuroprotection during mitochondrial stress in vivo. J Biol Chem. 2005;280(24):22925–36.
99. Shih AY, Li P, Murphy TH. A small-molecule-inducible Nrf2-mediated antioxidant response provides effective prophylaxis against cerebral ischemia in vivo. J Neurosci. 2005;25(44):10321–35.
100. Zhao J, Moore AN, Redell JB, Dash PK. Enhancing expression of Nrf2-driven genes protects the blood–brain barrier after brain injury. J Neurosci. 2007;27(38):10240–8.
101. Zhao J, Kobori N, Aronowski J, Dash PK. Sulforaphane reduces infarct volume following focal cerebral ischemia in rodents. Neurosci Lett. 2006;393(2–3):108–12.
102. Chen XL, Kunsch C. Induction of cytoprotective genes through Nrf2/antioxidant response element pathway: a new therapeutic approach for the treatment of inflammatory diseases. Curr Pharm Des. 2004;10(8):879–91.

Chapter 6
Free Radicals in Central Nervous System Inflammation

D. Craig Hooper, Marzena J. Fabis, and Anirban Roy

Abstract Free radical production is central to the development and pathogenesis of central nervous system (CNS) inflammation, the vascular and cellular response that results in the accumulation of circulating cells and factors in CNS tissues. Oxygen- and nitrogen-based radicals produced by invading neutrophils and monocytes have been implicated in the CNS tissue damage that occurs in multiple sclerosis, viral encephalitis, and other neuroinflammatory diseases, secondary inflammation following CNS injury, and a number of neurodegenerative conditions with inflammatory elements, such as Parkinson's and Alzheimer's diseases. However in addition to cytotoxic effects, there is accumulating evidence that radical activity contributes to the increased blood–brain barrier (BBB) permeability associated with the invasion of leukocytes into the tissues in both pathological and protective CNS immune responses. Radical production during different types of CNS inflammatory conditions is reviewed from the perspective that the sources and nature of the radicals produced at least partly dictate the extent of CNS tissue pathology. Therapeutic strategies to control the production or activity of free radicals in the CNS are also discussed.

Keywords CNS inflammation · Free radicals · Autoimmunity · Encephalitis · Antioxidants

D.C. Hooper (✉)
WHO Center for Neurovirology, Department of Cancer Biology,
Kimmel Cancer Center, Philadelphia, PA, USA
and
Department of Neurological Surgery, Thomas Jefferson University, Philadelphia, PA, 19107-6731, USA
e-mail: douglas.hooper@jefferson.edu

N. Gadoth and H.H. Göbel (eds.), *Oxidative Stress and Free Radical Damage in Neurology*, Oxidative Stress in Applied Basic Research and Clinical Practice,
DOI 10.1007/978-1-60327-514-9_6, © Springer Science+Business Media, LLC 2011

1 Introduction

Inflammation is a complex vascular and cellular response to an injury, infection, or autoimmune stimulus that results in the accumulation and activity of circulating cells and factors in target tissues. The vasculature of the central nervous system (CNS) is less permeable than the peripheral vasculature and normally prevents circulating cells and factors from infiltrating CNS tissues. This specialization of the neurovasculature, known as the blood–brain barrier (BBB) places an additional constraint on the generation of an inflammatory response in the CNS: the induction of elevated permeability to allow contact between potentially invasive cells in the circulation and chemoattractive factors produced by CNS resident cells. Historically, CNS inflammation is defined histologically, as the accumulation of leukocytes and edema in CNS tissues. However, there are clear differences in the nature of the CNS inflammatory response between diverse stimuli. For example, CNS inflammation may be predominantly mediated by polymorphonuclear leukocytes (PMN) and monocytes, as seen in the response to injury, or can involve the invasion of T and B cells, as is evident when there is an autoimmune or infectious etiology. Moreover, inflammatory or proinflammatory changes can occur in the CNS tissues without cell invasion, for example, in neurodegenerative diseases such as Parkinson's and Alzheimer's. Free radicals have been implicated in the induction of the BBB permeability changes associated with cell infiltration into CNS tissues as well as the pathology that is often associated with this process.

Initial interest in radical production during inflammatory responses centered on the "respiratory burst," the production of large amounts of oxygen-based radical species by phagocytic inflammatory cells such as PMN and monocytes in response to pathogens and other triggers. In general, the beneficial contributions of these radicals to the control of an infection outweigh their nonspecific, destructive reactivities with surrounding normal tissues. However, because of the limited capacity of neurons to regenerate, radical production in the CNS presents a particular risk, and CNS inflammation is most often associated with neuropathology. For example, the formation of the CNS lesions that are characteristic of multiple sclerosis (MS) is associated with the activity of monocytes expressing inducible nitric oxide synthase (NOS-2), an enzyme that catalyzes the production of high levels of nitric oxide ($^{\bullet}$NO) [1, 2]. There is also extensive evidence that radicals contribute to the pathogenesis of a variety of neurodegenerative diseases with inflammatory components. Although the destructive contributions of free radicals to inflammatory pathologies are incontrovertible, it is also clear that these radicals are produced in a tightly regulated fashion by a wide variety of cells and mediate a number of normal physiological functions [3]. Recent findings indicate that this is also the case during a CNS inflammatory response. For example, there is evidence that certain radicals contribute to the vascular changes that facilitate immune/inflammatory cell invasion into target tissues in both pathological and therapeutic CNS inflammatory/immune responses [4]. It is important to consider that many of the radical species that have been implicated in the induction or pathogenesis of CNS inflammation have important roles in other systems, the activities of •NO in vascular homeostasis and neural transmission being a good example. For the purposes of this chapter, we focus on radicals that have been implicated in the generation of CNS inflammation and the inflammatory functions of the cells involved, with an emphasis on the differences in radical activity during CNS autoimmunity, infection, and trauma.

2 Radical Production in CNS Inflammatory Responses

The free radicals and their reactive products that are formed during an inflammatory response are generally grouped into reactive oxygen species (ROS) and reactive nitrogen species (RNS) and the enzymes catalyzing production of the different radicals generally provide useful identifiers of the cell types involved. Many of these have important roles in normal physiology but are overproduced in inflammation. In this section, features of the radicals likely to be associated with different aspects of diverse CNS inflammatory conditions are reviewed from the perspective of radicals released by cells during CNS inflammation. Although it is not always known why a particular stimulus causes the release of radicals from an inflammatory cell, these radicals likely bear primary responsibility for bystander cell pathology and other important aspects of CNS inflammation. However, it should be noted that the activity of radicals produced within cells, for example, the destruction of engulfed viruses, bacteria, or parasites by phagocytes, is an important facet of the CNS inflammatory response to many pathogens. In addition, the contributions of radical activity to cell signaling are now being recognized [5] and may contribute to cell function during inflammation.

2.1 Radical Production by Infiltrating Cells

Much of the focus upon radical-producing cells in CNS inflammation has historically been on infiltrating cells as these are readily associated with the production of high levels of ROS and RNS and, consequently, readily identifiable CNS tissue damage. Depending on the stimulus of cell infiltration, which will be discussed further next, different cell types capable of producing free radicals invade CNS tissues. The predominant radical-elaborating cell types associated with CNS inflammation are PMN, primarily neutrophils, and monocytes. These cells, often referred to collectively as "inflammatory cells," express high levels of three key enzymes associated with radical production: (1) the Nox2 isoform of NADPH oxidase that catalyzes the formation of superoxide ($^{\bullet}O_2^-$) [6]; (2) NOS-2, the inducible isoform of nitric oxide synthase that is responsible for high levels of •NO production; and (3) myeloperoxidase (MPO), an enzyme that catalyzes the formation of hypochlorous acid (HOCl) from hydrogen peroxide (H_2O_2). While superoxide can also be produced by cells expressing other Nox isoforms or dual oxidase (Duox) enzymes and is a by-product of aerobic respiration in all metabolically active cells, Nox2 expression is associated with the high-level production of hydroxyl radical ($^{\bullet}OH$), H_2O_2, and oxygen-centered organic radicals that is seen in CNS inflammation. The expression of both Nox and NOS-2 by an inflammatory cell results in the formation of peroxynitrite ($ONOO^-$), the product of the reaction of superoxide and nitric oxide. Under physiological conditions, $ONOO^-$ rapidly reacts with carbon dioxide (CO_2) to produce the highly reactive radicals nitrogen dioxide ($^{\bullet}NO_2$) and $^{\bullet}CO_3^-$ and can also decompose to form $^{\bullet}OH$ [7]. As discussed further below, these radicals are now believed to make key contributions to the pathogenesis of CNS inflammation.

The nitration of tyrosine residues on proteins has proven to be a useful marker of the reactivity of these radicals and is often detected in association with NOS-2-positive cells in CNS inflammatory responses [1, 2, 8]. Tyrosine nitration can also be catalyzed by MPO through the oxidation of nitrite to $^{\bullet}NO_2$ [9]. Analysis of cytokine expression can also provide insight into radical production during an inflammatory response. The principal inflammatory cytokine that contributes to the induction of both NOS-2 and Nox2 in monocytes and PMN neutrophils in most CNS inflammatory responses is interferon-gamma (IFN-γ), which is largely produced by activated CD4 Th1 cells, CD8 T cells, and natural killer (NK) cells. Interleukin (IL)-17, produced by CD4 Th17 cells, is also known to induce NOS-2 expression by various cell types [10]. The proinflammatory cytokines interleukin-1β (IL-1β), tumor necrosis factor-alpha (TNF-α), and interleukin 6 (IL-6), which can be produced by a variety of cells including CNS resident cells, also make important contributions to the modulation of the expression and activity of these enzymes [11]. Despite producing cytokines that induce radical production, T cells are not known to elaborate radicals. On the other hand, subsets of NK cells, in addition to expressing the important inflammatory cytokine IFN-γ, can also express NOS-2 [12].

2.2 *Radical Production by CNS Resident Cells During CNS Inflammation*

It is now clear that most CNS resident cell populations can be induced to produce ROS and RNS under appropriate conditions. Foremost among these are microglia, bone marrow-derived cells of the monocyte lineage that can express high levels of Nox2 [13] and NOS-2 [14]. These cells are therefore capable of making significant amounts of $^{\bullet}NO$, $^{\bullet}O_2^-$, and their radical products. However, because of the difficulty in differentiating between the cells it is often unclear whether cells accumulating in an inflammatory lesion are invading monocytes or activated microglia. Astrocytes can express Nox1, -2, -4, and Duox1, -2 [15] and NOS-2 [16], whereas neurons constitutively express NOS-1. The activity of neurovascular endothelial cells is central to BBB function. These cells constitutively express NOS-3, and

Table 1 Key radical production during central nervous system (CNS) inflammation

Source enzyme(s)[a]	Primary radical[b]	Secondary radicals	Cell expression[c]
Nox2	O_2^-	OH H_2O_2	Monocytes, polymorphonuclear leukocytes (PMN), neurovascular endothelial cells
NOS-2 (NOS-1, NOS-3)	$^{\bullet}NO$	$ONOO^- \rightarrow {}^{\bullet}NO_2$, $^{\bullet}CO_3^-$	
MPO	HOCl		PMN, monocytes

[a] Multiple enzymes and pathways are responsible for the production of different radicals and only the primary enzyme associated with CNS inflammation is shown, except for nitric oxide synthase (NOS)-1 and NOS-3, which have been implicated in certain responses
[b] The principal product of the enzyme is shown, with certain of the secondary radicals that result from their reactivity
[c] These are the principal cells expressing the enzymes during a CNS inflammatory response but other cell types, such as microglia and astrocytes, can also express radical-producing enzymes and are likely to contribute to radical production

there is evidence that they can also express Nox1, -2, and -4 as well as NOS-2 [17]. Mast cells likely capable of producing $^{\bullet}NO$ are associated with CNS blood vessels [18]. As discussed next, the production of RNS by neurovascular endothelial cells, and possibly other associated cell types, makes important contributions to the modulation of BBB integrity during certain classes of CNS inflammatory response. Table 1 lists the central radicals produced at high levels during conventional CNS inflammatory responses.

3 CNS Inflammatory Processes

The nature of a CNS inflammatory response largely depends on its trigger and the types of cells that infiltrate CNS tissues. We can define different classes of CNS inflammation based on these two criteria: (1) antigen-driven infiltration of T cells, neutrophils, and monocytes as seen in experimental allergic encephalomyelitis (EAE), an animal model of MS [4]; (2) antigen-specific infiltration of T and B cells with limited monocyte or neutrophil invasion as associated with the clearance of attenuated rabies virus from the CNS [4]; (3) antigen-independent infiltration of primarily neutrophils and monocytes as occurs following CNS trauma [19]; and (4) innate inflammatory or proinflammatory changes in the CNS tissues in the absence of detectable immune/inflammatory cell invasion, a hallmark of a number of neurodegenerative diseases such as Parkinson's and Alzheimer's diseases [20, 21]. Although the latter diseases do not fit the classical description of CNS inflammation with respect to infiltrating cell accumulation, we have included these because their inflammatory features include free radical activity. In the following discussion, we have grouped CNS inflammatory responses based on their etiology. It should be noted, however, that CNS inflammation with similar characteristics can be induced by diverse stimuli.

3.1 Autoimmunity

Much of our understanding of the contribution of free radicals to CNS inflammatory pathology comes from studies of EAE and MS. Accumulations of lymphocytes and monocytes, associated with astrogliosis, local demyelination, and neuronal destruction, are common to both diseases. The development of these lesions can be thought of as occurring in three steps: (1) the extravasation of immune/inflammatory cells, which is associated with the loss of BBB integrity; (2) the invasion of the cells into the CNS parenchyma; and (3) CNS tissue pathology. Free radical activity has been implicated at each of these steps. The inflammatory cell types that infiltrate and are likely responsible for much of the radical-mediated CNS pathology in MS and EAE are primarily macrophages, but there is evidence from animal models that PMNs may also play an important initiating role [22]. Both these cell types are notable for their ability to produce high levels of ROS and RNS for release into phagolysosomes following phagocytosis of a pathogen. However, in a pathological CNS inflammatory response the free radicals involved are released from the cells to damage surrounding cells and tissues. Although there is no doubt that both oxygen- and nitrogen-centered radicals can cause tissue pathology during CNS inflammation, the contributions of different radicals to the pathogenesis of CNS inflammation are not always clear. For example, MPO has been detected in MS lesions [23], suggesting that HOCl may contribute to the pathogenesis of the disease but MPO-knockout mice are more susceptible to EAE [24]. A somewhat similar paradox exists for the relationship between NOS-2, $^{\bullet}$NO, and the pathogenesis of MS and EAE. Cells expressing NOS-2 are present in the CNS lesions of these diseases, leading to considerable interest in the possibility that $^{\bullet}$NO contributes to their pathogenesis. Further support for this concept comes from studies in conventional EAE models in which inhibition of NOS-2 has been found to have therapeutic effects [25]. Paradoxically, mice lacking NOS-2 exhibit increased EAE severity compared to normal controls [26]. Although there are a number of possible explanations for these and other similar findings, together they suggest that $^{\bullet}$NO may not directly cause inflammatory pathology in the CNS. In this regard it is notable that nitrotyrosine residues are also found in active MS and EAE lesions, suggesting that a product of $^{\bullet}$NO, $ONOO^-$, may be involved [1, 2, 8]. Proteins, lipids, and nucleic acids can all be modified by $ONOO^-$-dependent reactions, which are highly toxic to neurons and other CNS resident cells in vitro, likely through several pathways [27, 28]. Moreover, inactivation of $ONOO^-$ or $ONOO^-$-dependent radicals is therapeutic in EAE [1, 29].

There is accumulating evidence that, in addition to mediating pathological changes in CNS tissues, radicals originating from NOS-2-positive cells contribute to the loss of BBB integrity necessary to provide circulating cells and factors access to CNS tissues. This concept has been advanced by our studies to establish whether $^{\bullet}$NO or an $^{\bullet}$NO product such as $ONOO^-$ makes a more important contribution to the pathogenesis of CNS inflammation as modeled by EAE. We chose urate in an attempt to scavenge $ONOO^-$ because there was some evidence, now confirmed under more physiological conditions, that it may be more selective for this molecule than another physiological antioxidant, ascorbate [30]. Raising the serum urate levels of mice immunized to induce the development of EAE prevented not only the induction of disease, but also immune/inflammatory cell accumulation in CNS tissues [31]. Further studies have shown that urate treatment prevents the loss of BBB integrity in conventional EAE models as well as in certain neurotrophic virus infections and bacterial meningitis [31, 32, 33].

Further studies have established that urate is an efficient scavenger not of $ONOO^-$, but of $^{\bullet}NO_2$ and $^{\bullet}CO_3^-$ radicals formed by the reaction of CO_2 with $ONOO^-$ [34]. Because inactivation of these radical species by urate maintains BBB integrity, innate inflammatory changes in CNS tissues triggered by the loss of BBB integrity as well as radical production by invading inflammatory cells is prevented [35]. To gain insight into the role of urate-sensitive radicals in CNS tissue damage, we allowed animals to develop EAE and CNS lesions before administering urate. The animals recovered from clinical signs of EAE within 4–5 days of urate treatment [31]. Lesions containing NOS-2-positive cells were present in their CNS tissues, but nitrotyrosine was less evident and BBB integrity was restored. These findings suggest that $^{\bullet}NO$ and other radicals produced by the accumulated inflammatory cells have less impact on the acute clinical signs of EAE than do the radicals targeted by urate. More recently we have established a model of CNS inflammation where elevated BBB permeability occurs in the presence of urate treatment but NOS-2-positive monocytes do not enter CNS tissues [4]. This finding suggests that $ONOO^-$-dependent radicals are involved in the invasion of NOS-2-positive cells through the CNS parenchyma. The fact that $ONOO^-$ can activate matrix metalloproteinase-2 [36], an enzyme involved in degradation of the extracellular matrix, may be relevant to this finding. The concept that $ONOO^-$-dependent radicals make an important contribution to the generation and pathogenesis of autoimmune CNS inflammatory responses is also supported by a variety of studies in which reagents that interfere with the production or reactivity of $^{\bullet}NO$, $^{\bullet}O_2^-$, and their radical products have been found to have therapeutic effects in models of CNS inflammation [1, 29].

3.2 Infection

The production of $^{\bullet}NO$ and its reactive products in the CNS has been demonstrated to be a key antiviral mechanism during infections of the CNS with a variety of viruses including herpes simplex virus-1 [37], ectromelia virus [38], and sindbis virus [39]. Free radical production can be induced during the innate response to an invading pathogen that triggers pattern recognition receptors such as the Toll-like receptors (TLRs). For example, interactions with TLR-2, -4, -7, and -9 in glial cells can lead to elevated $^{\bullet}NO$ production in the CNS [40]. These mechanisms, which may be related to the role of radicals in phagocytosis, generally induce low levels of radical production, but the amount of radicals elaborated by cells activated by T cell-derived cytokines during the adaptive, antigen-specific inflammatory response is likely to be orders of magnitude higher. •NO production during viral encephalitis can be associated with NOS-2, NOS-1, or NOS-3 activity, and this may be relevant to the nature of the infection. For example, both NOS-2 and NOS-1 are involved in vesicular stomatitis virus (VSV) infection [41], raising the possibility that •NO production by NOS-1 is more important for viruses that directly infect neurons.

Accumulating evidence suggests that •NO and its reaction products $^{\bullet}NO_2$ and $^{\bullet}CO_3^-$ play a major role in the delivery of immune effectors across the BBB into CNS tissues, as is the case for autoimmune CNS inflammation. The vasodilatory effects of •NO in the neurovasculature are likely to be important in facilitating immune/inflammatory cell adherence to the neurovasculature. However, there is also evidence that $ONOO^-$-dependent radicals are an important trigger of the changes in the barrier properties of the neurovasculature that allow circulating cells to enter CNS tissues [31]. The elevated BBB permeability coupled with immune effector invasion into the CNS tissues of animals clearing attenuated rabies virus is evidently IFNγ dependent and associated with nitrotyrosine formation. NOS-3 and NOS-2, expressed by cells of the neurovascular unit as opposed to invading inflammatory cells, are both evidently associated with this process, which occurs in the absence of radical pathology in the CNS tissues [42].

The radical-induced mechanisms that permit the infiltration of immune effectors across the BBB into CNS tissues and radical-mediated damage to neurotrophic pathogens may be necessary to clear an infectious agent from the CNS. For instance, an •NO-dependent process involved in the immune response in bacterial meningitis that interferes with the radical-dependent inflammatory changes which allow immune cells to reach the infection has lethal consequences [43]. On the other hand, inflammation is often responsible for significant CNS tissue damage. For example, the pathogenesis of certain viral infections of the CNS is a consequence of the inflammatory response to viral antigens rather than the replication of the virus in CNS tissues. In some cases the inflammatory response to the virus involves radical-driven mechanisms that are indistinguishable from CNS autoimmunity. Borna disease in rats is an example. Rats normally succumb to Borna disease with the neurotrophic Borna disease virus (BDV) within 30 days of infection; however, the animals can survive with only relatively subtle signs of the infection despite the presence of high levels of BDV in the CNS if an immune/inflammatory response to the virus is avoided. The characteristics of the CNS inflammatory response in BDV-infected rats resembles those of EAE with invasion of T cells and NOS-2-positive cells as well as nitrotyrosine formation in the brain parenchyma. Similar to EAE, urate treatment has a therapeutic effect in Borna disease by maintaining BBB integrity and preventing cell invasion [32]. The expression of NOS-2 by glial cells and invading macrophages has also been associated with CNS tissue damage during Theiler's murine EAE virus infection [44]. In VSV infection, CNS pathology with extensive BBB failure is associated with high viral titers in CNS tissues [45]. However, BBB integrity is maintained, pathology lessened, and the CNS virus load reduced in VSV-infected mice lacking NOS-3 [45]. These findings underline the importance of •NO and its products in the modulation of BBB integrity and the pathogenesis of CNS inflammation.

Although acute viral infections of the CNS are often associated with an inflammatory response with high levels of ROS and RNS production and tissue damage, there are situations where pathological inflammatory processes do not appear to be associated with immune-mediated virus clearance from the CNS. An example is the clearance of laboratory-attenuated rabies viruses from the CNS where the BBB permeability changes required to provide immune effector cells access to the CNS are evidently mediated by the expression of NOS-2 and $ONOO^-$-dependent radicals in the neurovasculature [42]. Thus, $^{\bullet}NO_2$ and possibly $^{\bullet}CO_3^-$ appear to be the primary trigger for BBB permeability changes in both pathological CNS inflammation and protective CNS immune responses. However, the BBB permeability changes associated with this process in rabies virus infection are limited to relatively small molecules whereas those seen in EAE are more extensive, permitting molecules of up to at least 150 kDa to infiltrate CNS tissues from the circulation [4]. Based on our comparison of EAE and virus clearance, we speculate that inflammatory pathology is avoided during the latter by limiting the extent of BBB permeability and the accumulation of NOS-2-positive cells in the CNS parenchyma. A comparison of the patterns of accumulation of NOS-2-positive cells and nitrotyrosine in permeable areas of the CNS tissues in mice with EAE versus infected with attenuated rabies virus that illustrates the differences is shown in Fig. 1.

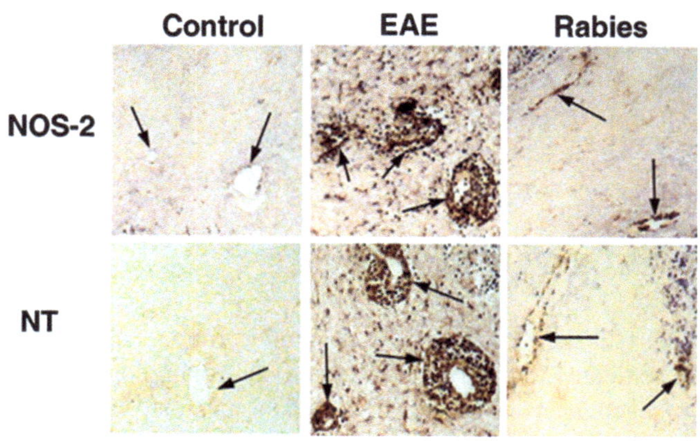

Fig. 1 Differences in the patterns of NOS-2 and nitrotyrosine (NT) expression between autoimmune central nervous system (CNS) inflammation (EAE) and therapeutic virus clearance (rabies). Sections from the cerebella of normal (Control), myelin basic protein (MBP)-immunized (EAE), and rabies CVS-F3-infected (Rabies) PLSJL mice were stained for NOS-2 (*upper panels*) and nitrotyrosine (*lower panels*) when peak blood–brain barrier (BBB) permeability occurs. CNS blood vessels are *arrowed*. Immunization, infection, and immunohistochemical methodology is detailed elsewhere [4]

3.3 Injury

Trauma and ischemia/reperfusion injury to the CNS tissues both can cause secondary inflammation, which can exacerbate the initial damage, interfere with recovery, and contribute to the development of a chronic inflammatory response [46]. In both cases the genesis of the inflammatory response is a consequence of the effects of ROS and RNS on the cells of the neurovasculature. Glial cells produce radicals following trauma [47], whereas ischemia can cause radical production by a variety of cells [48] and high levels are produced by vascular endothelial cells during reperfusion following circulatory disruption [49]. This sequence can lead to the elevated expression of adhesion molecules on vascular endothelial cells, increased vascular permeability, and inflammatory cell invasion into the CNS tissues. For example, the accumulation in CNS tissues of neutrophils expressing MPO and NOS-2 that is associated with tissue pathology occurring approximately 24 h after a spinal cord injury can be prevented by treatment with urate [50]. Similar processes have been associated with CNS tissue pathology following ischemia/reperfusion injury [51]. Several cytotoxic pathways are likely to be involved, including the $^{\bullet}NO$-dependent elevated expression of the proapoptotic protein Bax [52] and $ONOO^{-}$-mediated poly(ADP)-ribose polymerase-1 (PARP-1) activation [27], which both lead to DNA fragmentation. The possibility that free radical-mediated processes relevant to inflammation may have neuroprotective effects after CNS injury should also be noted. For example, T-cell function has been implicated in recovery following spinal cord injury [53], and it is possible that the invasion of these cells into spinal cord tissue is radical dependent [50]. Other neuroprotective contributions of free radical production in the CNS that are likely to be mediated by specific signal transduction mechanisms resulting in the activity of cytokines and growth factors have also been described [54].

3.4 Neurodegenerative Diseases

It is now well established that free radicals produced by mechanisms that may be considered to be related to innate inflammation make pathological contributions to chronic neurodegenerative diseases. Free radical-mediated damage occurs at an early stage during the development of Alzheimer's disease [21], and oxidative and nitrative modifications of proteins relevant to Parkinson's disease have been demonstrated in the substantia nigra [20]. Parkinson's-like disease induced in animal models by administration of the neurotoxin 1-methyl-4-phenyl-1,2,3,6-tetrahydropyridine (MPTP) is associated with the elevated production of $^{\bullet}O_2^{-}$ and $^{\bullet}NO$ [55]. Moreover, a low serum level of the $ONOO^{-}$-dependent radical scavenger urate has been associated with the incidence of both diseases [56, 57]. Further support for the hypothesis that inflammatory free radicals play a role in the pathogenesis of neurodegenerative diseases is provided by associations between disease incidence and anomalies in genes encoding enzymes that regulate radical production. For example, the MPO gene polymorphism associated with an increased risk of MS is also associated with Alzheimer's disease [58], and a superoxide dismutase (SOD) mutation has been implicated in certain forms of amyotrophic lateral sclerosis (ALS) [59]. This latter finding illustrates the significance of cellular antioxidants in neuroprotection. Further evidence of the importance of cellular antioxidants comes from studies of neurological disorders induced by exposure to metals such as methylmercury, where damage has been associated with a reduction in the activity of the glutathione antioxidant system [60]; uranium, where toxic levels are correlated with changes in the levels of a number of cellular antioxidants [61]; and cadmium, where elevated expression of SOD or glutathione peroxidase protects exposed astrocytes [62].

4 Therapeutic Intervention

Free radicals are highly toxic and can cause significant cell toxicity, which is particularly important in the case of largely irreplaceable neurons. Nevertheless, free radical activity is associated with cell respiration as well as the response to infection. Under normal circumstances the contribution of radicals to beneficial functions is balanced by a reducing physiological environment, cellular and serum antioxidants, and tightly controlled production. The overproduction of free radicals in CNS tissues from diverse sources is associated with a number of acute and chronic neurodegenerative diseases, and it is in this context that therapeutic intervention is generally viewed. There are two general approaches toward therapeutically controlling free radical activity: (1) inhibition of free radical production by acting on the redox enzymes that catalyze their formation or the processes causing their activity; and (2) inhibition by chemically inactivating free radicals with scavengers, endogenous or synthetic antioxidants, or other sacrificial reactants. In this section, we briefly survey these approaches as well as comment on situations where radical activity in the CNS may be desired.

4.1 Inhibition of Radical Generation

Therapeutic intervention against the generation of radicals has been approached through interfering with the activity of the involved enzymes as well as inhibition of pathways that lead to their expression or activation. Because of the importance of NOS in the formation of highly neurotoxic nitrogen-centered radicals, studies of reagents targeted at these enzymes are more common and a variety of inhibitors selective for the different NOS isoforms have been identified [63]. Most promising are reagents that selectively prevent $^{\bullet}NO$ production by NOS-2, such as aminoguanidine and L-*N*-iminoethyl-lysine, which have been shown to possess protective effects in diverse CNS inflammatory conditions including EAE [25], Murray Valley viral encephalitis [64], and traumatic brain injury [65]. However, the fact that mice lacking NOS-2 are more susceptible to autoimmune CNS inflammation [26] raises concerns about this approach.

There are many classes of reagents that inhibit the generation of free radicals through a variety of indirect pathways. For example, nonsteroidal antiinflammatory drugs (NSAIDS) exert their effects through inhibition of cyclooxygenase-2, which catalyzes the production of prostaglandins and thromboxanes, lipid compounds with inflammatory properties. On the other hand, the antiinflammatory effects of endogenous, plant, and synthetic cannabinoids are evidently mediated through suppression of the activities of cells of the immune system and microglia that express CB2 cannabinoid receptors [66]. Bioavailability to CNS tissues and the broad range of action of many candidate therapeutic reagents limits their utility in CNS inflammation. Moreover, most of the radical species implicated in the pathogenesis of CNS inflammation have important roles in normal physiology. Consequently, general inhibition of their formation is likely to have undesired as well as protective effects.

4.2 *Endogenous Antioxidants and Radical Scavengers*

The physiological production of radicals by respiring cells necessitates a reducing environment that is maintained by the collective activity of a number of antioxidant systems naturally present in cells. Referred to as cellular antioxidants, these include enzymes and reducing agents that inactivate radicals in cells. Many also offer protection against exogenous radicals produced in a CNS inflammatory response, particularly if they are present in the extracellular milieu or the radicals can diffuse across the cell membrane. Important cellular antioxidant enzymes known to impact radicals involved in CNS inflammation include SOD, which catalyzes the dismutation of $^{\bullet}O_2^-$ into oxygen and H_2O_2, catalase, which catalyzes the decomposition of H_2O_2 to water and oxygen, glutathione reductase, which reduces oxidized glutathione, and glutathione peroxidases, which reduce lipid hydroperoxides to alcohols and H_2O_2. There are also a wide variety of cellular antioxidants that normally protect cells against radicals produced by metabolic processes and likely contribute to protection against free radicals produced by CNS inflammation; these include nicotinamide adenine dinucleotide phosphate (NADP), an important reducing agent in a variety of biosynthetic reactions, glutathione, an antioxidant that reduces thiol groups, and lipoic acid, a BBB permeant enzyme cofactor, transition metal ion chelator, and scavenger of ROS and RNS [67].

Several important general antioxidants are present at high levels in humans: α-tocopherol (vitamin E), a lipid-soluble antioxidant with high bioavailability, and ascorbate (vitamin C), a reducing agent and enzyme cofactor [68], are essential for

Table 2 Endogenous antioxidants tested in neurological diseases with an inflammatory component

Endogenous antioxidants	CNS disease	Species[a]
Vitamin C (ascorbic acid)	Alzheimer's disease	Human [72]
	Parkinson's disease	Human [73]
	Experimental allergic encephalomyelitis (EAE)	Rodent [74]
	Ischemia and stroke	Human [75]
Vitamin E (α-tocopherol)	Alzheimer's disease	Human [76]
	Parkinson's disease	Human [73]
	Ischemia and stroke	Human [75]
Coenzyme Q_{10}	Parkinson's and Huntington's diseases	Human [77]
Uric acid	Ischemic stroke	Human [78]
	Multiple sclerosis (MS)	Human [79]
	Borna disease viral encephalitis	Rodent [32]
	Spinal cord injury	Rodent [50]
Melatonin	Cerebral ischemia	Rodent [68]
	Venezuelan equine encephalomyelitis virus infection	Rodent [80]
α-Lipoic acid	Ischemia-reperfusion injury	Rodent [68]
	Rickettsial infection	Rodent [81]

[a] Reviews are referenced where available

the maintenance of normal physiological function but derived from dietary sources. Urate, a scavenger with selectivity for $^{\bullet}NO_2$ and $^{\bullet}CO_3^-$ radicals, is the end product of purine metabolism in humans and derived from both endogenous and dietary purines. The distribution of these and other endogenous antioxidants likely is a major determinant of the vulnerability of a particular tissue to radical-mediated damage. For example, ascorbate, despite being water soluble and therefore unlikely to passively diffuse across the BBB, is found at higher concentrations in the CNS than in peripheral organs [69]. In contrast, urate levels in the CNS tissues are considerably lower than those found in the circulation, and the CNS tissues may therefore be more susceptible to damage mediated by $^{\bullet}NO_2$ and $^{\bullet}CO_3^-$. On the other hand, α-tocopherol is lipid soluble and likely to be particularly important in protecting cell membranes from oxidation by inactivating lipid radicals. Nevertheless, the concentration of α-tocopherol is low in the cerebellum compared to other organs and to the rest of the CNS tissues [70]. This difference could contribute to the fact that the cerebellum is a more common target for $ONOO^-$-mediated tyrosine nitration during EAE and rabies virus infection than the cortex [4].

Among the natural radical scavengers, urate holds particular interest for the treatment of CNS inflammatory diseases such as MS because of its effects on CNS inflammation in animal models [4], its naturally high levels in humans, and the inverse relationship between serum levels and MS [71]. Examples of endogenous antioxidants that have been assessed for therapeutic utility in neurological diseases with an inflammatory component are presented in Table 2.

Table 3 Example synthetic antioxidants and radical scavengers

Antioxidant[a]	Mode of action[b]
Methyl prednisolone, 21-aminosteroids such as lazaroids	Glucocorticoid analogues [82]
Pegorgotein, nitroxides	Superoxide dismutase activity [83]
Ebselen	Glutathione peroxidase, mimetic that inhibits a number of enzymes involved in inflammation [84]
N-acetylcysteine, *N,N′*-propylenedinicotinamide, nicaraven	Radical scavenging-modified amino compounds [85]
α-Phenyl-*t*-butyl-nitrone (PBN)	Spin-trap scavenging agents [86]
Deferoxamine, desferral	Metal chelators [87, 88]
FeTPPS (5,10,15,20-tetrakis-[4-sulfonatophenyl] porphyrinato iron(III) chloride)	$ONOO^-$ decomposition catalyst [89]

[a] This list represents only limited examples of different classes of antioxidants
[b] Not all the activities of different antioxidants are fully understood

4.3 *Pharmacologic Radical Scavengers and Antioxidants*

A wide variety of physiological, plant-derived, and synthetic compounds have been tested in cell culture, in animal models, and in humans for their antioxidant and radical scavenging properties. It is probably not an overestimate that hundreds of compounds have exhibited antioxidant properties in vitro: few have proven suitable for use in CNS inflammation. The capacity to cross the BBB is believed to be a major issue; therefore, lipid-soluble reagents are thought to be better therapeutic candidates. However, CNS inflammation is generally associated with the loss of BBB integrity, which may provide a reagent with access to CNS tissues. Although a number of endogenous antioxidants have proven to have therapeutic activity in animal models of CNS inflammation, they often have broad specificity, additional functions, metabolic properties, or other concerns that preclude their use at sufficient concentrations without impacting important physiological processes. Consequently, a wide variety of synthetic compounds, some based on natural analogues, have been developed with diverse antioxidant properties. Some examples that have been tested in models of CNS inflammation are presented in Table 3. The results of different studies assessing the therapeutic utility of these agents in different oxidative neurological diseases including CNS inflammatory diseases, which have been reviewed previously [68], have been variable but generally support the concept that targeting free radicals can be protective.

Studies of traditional or alternative medicines have revealed that a number of herbal remedies for neurological disorders have potent antioxidant properties that may explain their historical usage: these include extracts of green tea, *Ginkgo biloba*, blueberry, grape, and a variety of other plant components. Many of the compounds responsible for these natural antioxidant activities have been identified and are being investigated for potential therapeutic uses in CNS inflammatory processes [90, 91, 92, 93]. Plant polyphenols such as the flavonoids (e.g., catechin, epicatechin, quercetin) are examples.

4.4 *Induction of Radical Activity*

As already noted, •NO plays a key modulatory role in the induction of CNS inflammation, particularly at the level of the neurovasculature. There are pathways through which raising •NO levels may have therapeutic effects in a CNS inflammatory response, as illustrated by the fact that mice lacking NOS-2 develop more severe EAE than congenic controls [26]. Any expected benefit would likely depend upon whether the formation of more toxic radicals can be avoided. However, there are certain situations where the radical-dependent induction of a CNS inflammatory response can have a beneficial outcome regardless of collateral damage. The lethality of virulent rabies viruses is in part the result of the maintenance of BBB integrity, which prevents antiviral immune effectors capable of clearing the virus from entering CNS tissues. In this case, the use of a CNS autoimmune inflammatory reaction, EAE, to open the BBB and allow immune effectors to clear the virus permits some animals to survive this otherwise lethal infection despite the pathological consequences of EAE [94]. Better approaches to modulate the radical-mediated induction of BBB permeability changes would clearly benefit the numerous situations where reagent delivery to the CNS is desired.

5 Conclusions

Regardless of its etiology, CNS inflammation is associated with the production of a variety of free radicals and, in certain cases, radical activity triggers the elevated BBB permeability associated with the infiltration of immune and inflammatory cells into CNS tissues. A variety of factors determine whether radical production will lead to CNS pathology, including the type of radicals produced, the location, and accessible targets including antioxidants and cell proteins, lipids, and nucleic acids as well as extracellular components. For instance, when activated neutrophils and monocytes expressing NOS-2 and Nox2 enter CNS tissues, their radical production can overwhelm existing antioxidant defenses and cause considerable damage to CNS resident cells, an example being the MS plaque. However, high levels of production of similar radicals in the neurovasculature do not appear to result in significant CNS pathology [4]. Strategies to modulate radical production and activity in neurological diseases with an inflammatory component are likely to have substantial therapeutic impact, providing that the contributions of the targeted radicals to normal physiology are minimal or can be preserved.

Acknowledgments We thank Rhonda Kean for assistance with the manuscript. Our studies in CNS immunity and inflammation are supported in part by National Institutes of Health Grants AI-077033, AI-060005, and AT-003432, and by a grant to the Biotechnology Foundation Inc. from the Commonwealth of Pennsylvania.

References

1. Hooper DC, Bagasra O, Marini JC, et al. Prevention of experimental allergic encephalomyelitis by targeting nitric oxide and peroxynitrite: implications for the treatment of multiple sclerosis. Proc Natl Acad Sci USA. 1997;94:2528–33.
2. Liu JS, Zhao ML, Brosnan CF, Lee SC. Expression of inducible nitric oxide synthase and nitrotyrosine in multiple sclerosis lesions. Am J Pathol. 2001;158:2057–66.
3. Calabrese V, Guagliano E, Sapienza M, Mancuso C, Butterfield DA, Stella AM. Redox regulation of cellular stress response in neurodegenerative disorders. Ital J Biochem. 2006;55:263–82.
4. Fabis MJ, Phares TW, Kean RB, Koprowski H, Hooper DC. Blood-brain barrier changes and cell invasion differ between therapeutic immune clearance of neurotrophic virus and CNS autoimmunity. Proc Natl Acad Sci USA. 2008;105:15511–6.
5. Riccio A Alvania RS, Lonze BE, et al. A nitric oxide signaling pathway controls CREB-mediated gene expression in neurons. Mol Cell. 2006;21:283–294.
6. Lambeth JD, Kawahara T, Diebold B. Regulation of Nox and Duox enzymatic activity and expression. Free Radic Biol Med. 2007;43:319–31.
7. Denicola A, Freeman BA, Trujillo M, Radi R. Peroxynitrite reaction with carbon dioxide/bicarbonate: kinetics and influence on peroxynitrite-mediated oxidations. Arch Biochem Biophys. 1996;333:49–58.
8. Cross AH, Manning PT, Stern MK, Misko TP. Evidence for the production of peroxynitrite in inflammatory CNS demyelination. J Neuroimmunol. 1997;80:121–30.
9. Baldus S, Eiserich JP, Mani A, et al. Endothelial transcytosis of myeloperoxidase confers specificity to vascular ECM proteins as targets of tyrosine nitration. J Clin Invest. 2001;108:1759–70.
10. Miljkovic D, Trajkovic V. Inducible nitric oxide synthase activation by interleukin-17. Cytokine Growth Factor Rev. 2004;15:21–32.
11. Brown GC. Mechanisms of inflammatory neurodegeneration: iNOS and NADPH oxidase. Biochem Soc Trans. 2007;35:1119–21.
12. Cifone MG, D'Alò S, Parroni R, et al. Interleukin-2-activated rat natural killer cells express inducible nitric oxide synthase that contributes to cytotoxic function and interferon-gamma production. Blood. 1999;93:3876–84.
13. McCann SK, Dusting GJ, Roulston CL. Early increase of Nox4 NADPH oxidase and superoxide generation following endothelin-1-induced stroke in conscious rats. J Neurosci Res. 2008;86:2524–34.
14. Marques CP, Cheeran MC, Palmquist JM, Hu S, Lokensgard JR. Microglia are the major cellular source of inducible nitric oxide synthase during experimental herpes encephalitis. J Neurovirol. 2008;14:229–38.
15. Reinehr R, Görg B, Becker S, et al. Hypoosmotic swelling and ammonia increase oxidative stress by NADPH oxidase in cultured astrocytes and vital brain slices. Glia. 2007;55:758–71.
16. Hewett SJ, Corbett JA, McDaniel ML, Choi DW. Interferon-gamma and interleukin-1 beta induce nitric oxide formation from primary mouse astrocytes. Neurosci Lett. 1993;164:229–32.
17. Miller AA, Drummond GR, Sobey CG. Novel isoforms of NADPH-oxidase in cerebral vascular control. Pharmacol Ther. 2006;111:928–48.

18. Strbian D, Karjalainen-Lindsberg ML, Tatlisumak T, Lindsberg PJ. Cerebral mast cells regulate early ischemic brain swelling and neutrophil accumulation. J Cereb Blood Flow Metab. 2006;26:605–12.
19. Blight AR. Macrophages and inflammatory damage in spinal cord injury. J Neurotrauma. 1992;9 Suppl 1:S83–91.
20. Danielson SR, Andersen JK. Oxidative and nitrative protein modifications in Parkinson's disease. Free Radic Biol Med. 2008;44:1787–94.
21. Sonnen JA, Breitner JC, Lovell MA, Markesbery WR, Quinn JF, Montine TJ. Free radical-mediated damage to brain in Alzheimer's disease and its transgenic mouse models. Free Radic Biol Med. 2008;45:219–30.
22. McColl SR, Staykova MA, Wozniak A, Fordham S, Bruce J, Willenborg DO. Treatment with anti-granulocyte antibodies inhibits the effector phase of experimental autoimmune encephalomyelitis. J Immunol. 1998;161:6421–6.
23. Nagra RM, Becher B, Tourtellotte WW, et al. Immunohistochemical and genetic evidence of myeloperoxidase involvement in multiple sclerosis. J Neuroimmunol. 1997;78:97–107.
24. Brennan M, Gaur A, Pahuja A, Lusis AJ, Reynolds WF. Mice lacking myeloperoxidase are more susceptible to experimental autoimmune encephalomyelitis. J Neuroimmunol. 2001;112:97–105.
25. Zhao W, Tilton RG, Corbett JA, et al. Experimental allergic encephalomyelitis in the rat is inhibited by aminoguanidine, an inhibitor of nitric oxide synthase. J Neuroimmunol. 1996;64:123–33.
26. Fenyk-Melody JE, Garrison AE, Brunnert SR, et al. Experimental autoimmune encephalomyelitis is exacerbated in mice lacking the NOS2 gene. J Immunol. 1998;160:2940–6.
27. Szabó C. Multiple pathways of peroxynitrite cytotoxicity. Toxicol Lett. 2003; 140–141:105–12.
28. Pacher P, Beckman JS, Liaudet L. Nitric oxide and peroxynitrite in health and disease. Physiol Rev. 2007;87:315–424.
29. Cross AH, San M, Stern MK, Keeling RM, Salvemini D, Misko TP. A catalyst of peroxynitrite decomposition inhibits murine experimental autoimmune encephalomyelitis. J Neuroimmunol. 2000;107:21–8.
30. Whiteman M, Ketsawatsakul U, Halliwell B. A reassessment of the peroxynitrite scavenging activity of uric acid. Ann N Y Acad Sci. 2002;962:242–59.
31. Hooper DC, Scott GS, Zborek A, et al. Uric acid, a peroxynitrite scavenger, inhibits CNS inflammation, blood-CNS barrier permeability changes, and tissue damage in a mouse model of multiple sclerosis. FASEB J. 2000;14:691–8.
32. Hooper DC, Kean RB, Scott GS, et al. The central nervous system inflammatory response to neurotropic virus infection is peroxynitrite dependent. J Immunol. 2001;167:3470–7.
33. Kastenbauer S, Koedel U, Pfister HW. Role of peroxynitrite as a mediator of pathophysiological alterations in experimental pneumococcal meningitis. J Infect Dis. 1999;180:1164–70.
34. Squadrito GL, Cueto R, Splenser AE, et al. Reaction of uric acid with peroxynitrite and implications for the mechanism of neuroprotection by uric acid. Arch. Biochem Biophys. 2000;376 2:333–337.
35. Scott GS, Kean RB, Fabis MJ, et al. ICAM-1 upregulation in the spinal cords of PLSJL mice with experimental allergic encephalomyelitis is dependent upon TNF-alpha production triggered by the loss of blood-brain barrier integrity. J Neuroimmunol. 2004;155:32–42.
36. Rajagopalan S, Meng XP, Ramasamy S, Harrison DG, Galis ZS. Reactive oxygen species produced by macrophage-derived foam cells regulate the activity of vascular matrix metalloproteinases in vitro. Implications for atherosclerotic plaque stability. J Clin Invest. 1996;98:2572–9.
37. Croen KD. Evidence for an antiviral effect of nitric oxide: inhibition of herpes simplex virus type 1 replication. J Clin Invest. 1993;91:2446–52.
38. Karupiah G, Xie QW, Buller RM, Nathan C, Duarte C, MacMicking JD. Inhibition of viral replication by interferon-gamma-induced nitric oxide synthase. Science. 1993;261:1445–8.

39. Tucker PC, Griffin DE, Choi S, Bui N, Wesselingh S. Inhibition of nitric oxide synthesis increases mortality in Sindbis virus encephalitis. J Virol. 1996;70:3972–7.
40. Ebert S, Gerber J, Bader S, et al. Dose-dependent activation of microglial cells by Toll-like receptor agonists alone and in combination. J Neuroimmunol. 2005;159:87–96.
41. Komatsu T, Ireland DD, Chen N, Reiss CS. Neuronal expression of NOS-1 is required for host recovery from viral encephalitis. Virology. 1999;258:389–95.
42. Phares TW, Fabis MJ, Brimer CM, Kean RB, Hooper DC. A peroxynitrite-dependent pathway is responsible for blood-brain barrier permeability changes during a central nervous system inflammatory response: TNF-alpha is neither necessary nor sufficient. J Immunol. 2007;178:7334–43.
43. Koedel U, Paul R, Winkler F, Kastenbauer S, Huang PL, Pfister HW. Lack of endothelial nitric oxide synthase aggravates murine pneumococcal meningitis. J Neuropathol Exp Neurol. 2001;60:1041–50.
44. Rose JW, Hill KE, Wada Y, et al. Nitric oxide synthase inhibitor, aminoguanidine, reduces inflammation and demyelination produced by Theiler's virus infection. J Neuroimmunol. 1998;81:82–9.
45. Komatsu T, Ireland DD, Chung N, Doré A, Yoder M, Reiss CS. Regulation of the BBB during viral encephalitis: roles of IL-12 and NOS. Nitric Oxide. 1999;3:327–39.
46. Szocs K. Endothelial dysfunction and reactive oxygen species production in ischemia/reperfusion and nitrate tolerance. Gen Physiol Biophys. 2004;23:265–95.
47. Fawcett JW, Asher RA. The glial scar and central nervous system repair. Brain Res Bull. 1999;49:377–91.
48. Vanden Hoek TL, Becker LB, Shao Z, Li C, Schumacker PT. Reactive oxygen species released from mitochondria during brief hypoxia induce preconditioning in cardiomyocytes. J Biol Chem. 1998;273:18092–8.
49. Zweier JL, Kuppusamy P, Thompson-Gorman S, Klunk D, Lutty GA. Measurement and characterization of free radical generation in reoxygenated human endothelial cells. Am J Physiol. 1994;266(3 pt 1):C700–8.
50. Scott GS, Cuzzocrea S, Genovese T, Koprowski H, Hooper DC. Uric acid protects against secondary damage after spinal cord injury. Proc Natl Acad Sci USA. 2005;102:3483–8.
51. Coeroli L, Renolleau S, Arnaud S, et al. Nitric oxide production and perivascular tyrosine nitration following focal ischemia in neonatal rat. J Neurochem. 1998;70:2516–25.
52. Mishra OP, Zubrow AB, Ashraf QM, Delivoria-Papadopoulos M. Effect of nitric oxide synthase inhibition during post-hypoxic reoxygenation on Bax and Bcl-2 protein expression and DNA fragmentation in neuronal nuclei of newborn piglets. Brain Res. 2006;1101:20–8.
53. Ziv Y, Avidan H, Pluchino S, Martino G, Schwartz M. Synergy between immune cells and adult neural stem/progenitor cells promotes functional recovery from spinal cord injury. Proc Natl Acad Sci USA. 2006;103:13174–9.
54. Hammarberg H, Lidman O, Lundberg C, et al. Neuroprotection by encephalomyelitis: rescue of mechanically injured neurons and neurotrophin production by CNS-infiltrating T and natural killer cells. J Neurosci. 2000;20:5283–91.
55. Tieu K, Ischiropoulos H, Przedborski S. Nitric oxide and reactive oxygen species in Parkinson's disease. IUBMB Life. 2003;55:329–35.
56. Church WH, Ward VL. Uric acid is reduced in the substantia nigra in Parkinson's disease: effect on dopamine oxidation. Brain Res Bull. 1994;33:419–25.
57. Maesaka JK, Wolf-Klein G, Piccione JM, Ma CM. Hypouricemia, abnormal renal tubular urate transport, and plasma natriuretic factor(s) in patients with Alzheimer's disease. J Am Geriatr Soc. 1993;41:501–6.
58. Reynolds WF, Rhees J, Maciejewski D, et al. Myeloperoxidase polymorphism is associated with gender specific risk for Alzheimer's disease. Exp Neurol. 1999;155:31–41.
59. Orrell R, de Belleroche J, Marklund S, Bowe F, Hallewell R. A novel SOD mutant and ALS. Nature. 1995;374:504–5.

60. Stringari J, Nunes AK, Franco JL, et al. Prenatal methylmercury exposure hampers glutathione antioxidant system ontogenesis and causes long-lasting oxidative stress in the mouse brain. Toxicol Appl Pharmacol. 2008;227:147–54.
61. Linares V, Sánchez DJ, Bellés M, Albina L, Gómez M, Domingo JL. Pro-oxidant effects in the brain of rats concurrently exposed to uranium and stress. Toxicology. 2007;236:82–91.
62. Yang CS, Tzou BC, Liu YP, Tsai MJ, Shyue SK, Tzeng SF. Inhibition of cadmium-induced oxidative injury in rat primary astrocytes by the addition of antioxidants and the reduction of intracellular calcium. J Cell Biochem. 2008;103:825–34.
63. Southan GJ, Szabó C. Selective pharmacological inhibition of distinct nitric oxide synthase isoforms. Biochem Pharmacol. 1996;51:383–94.
64. Andrews DM, Matthews VB, Sammels LM, Carrello AC, McMinn PC. The severity of Murray valley encephalitis in mice is linked to neutrophil infiltration and inducible nitric oxide synthase activity in the central nervous system. J Virol. 1999;73:8781–90.
65. Gahm C, Holmin S, Wiklund PN, Brundin L, Mathiesen T. Neuroprotection by selective inhibition of inducible nitric oxide synthase after experimental brain contusion. J Neurotrauma. 2006;23:1343–54.
66. Ortega-Gutiérrez S, Molina-Holgado E, Guaza C. Effect of anandamide uptake inhibition in the production of nitric oxide and in the release of cytokines in astrocyte cultures. Glia. 2005;52:163–8.
67. Salinthone S, Yadav V, Bourdette DN, Carr DW. Lipoic acid: a novel therapeutic approach for multiple sclerosis and other chronic inflammatory diseases of the CNS. Endocr Metab Immune Disord Drug Targets. 2008;8:132–42.
68. Gilgun-Sherki Y, Rosenbaum Z, Melamed E, Offen D. Antioxidant therapy in acute central nervous system injury: current state. Pharmacol Rev. 2002;54:271–84.
69. Schreiber M, Trojan S. Ascorbic acid in the brain. Physiol Res. 1991;40:413–8.
70. Vatassery GT, Angerhofer CK, Knox CA, Deshmukh DS. Concentrations of vitamin E in various neuroanatomical regions and subcellular fractions, and the uptake of vitamin E by specific areas, of rat brain. Biochim Biophys Acta. 1984;792:118–22.
71. Hooper DC, Spitsin S, Kean RB, et al. Uric acid, a natural scavenger of peroxynitrite, in experimental allergic encephalomyelitis and multiple sclerosis. Proc Natl Acad Sci USA. 1998;95:675–80.
72. Quinn J, Suh J, Moore MM, Kaye J, Frei B. Antioxidants in Alzheimer's disease: vitamin C delivery to a demanding brain. J Alzheimers Dis. 2003;5:309–13.
73. Fahn S. A pilot trial of high-dose alpha-tocopherol and ascorbate in early Parkinson's disease. Ann Neurol. 1992;32 Suppl:S128–32.
74. Spitsin SV, Scott GS, Mikheeva T, et al. Comparison of uric acid and ascorbic acid in protection against EAE. Free Radic Biol Med. 2002;33:1363–71.
75. Ullegaddi R, Powers HJ, Gariballa SE. Antioxidant supplementation enhances antioxidant capacity and mitigates oxidative damage following acute ischaemic stroke. Eur J Clin Nutr. 2005;59:1367–73.
76. Sano M, Ernesto C, Thomas RG, et al. A controlled trial of selegiline, alpha-tocopherol, or both as treatment for Alzheimer's disease. The Alzheimer's Disease Cooperative Study. N Engl J Med. 1997;336:1216–1222.
77. Young AJ, Johnson S, Steffens DC, Doraiswamy PM. Coenzyme Q10: a review of its promise as a neuroprotectant. CNS Spectr. 2007;12:62–8.
78. Amaro S, Planas AM, Chamorro A. Uric acid administration in patients with acute stroke: a novel approach to neuroprotection. Expert Rev Neurother. 2008;8:259–70.
79. Spitsin S, Hooper DC, Leist T, Streletz LJ, Mikheeva T, Koprowskil H. Inactivation of peroxynitrite in multiple sclerosis patients after oral administration of inosine may suggest possible approaches to therapy of the disease. Mult Scler. 2001;7:313–9.
80. Valero N, MarinaEspina L, Bonilla E, Mosquera J. Melatonin decreases nitric oxide production and lipid peroxidation and increases interleukin-1 beta in the brain of mice infected by the Venezuelan equine encephalomyelitis virus. J Pineal Res. 2007;42:107–12.

81. Rydkina E, Sahni SK, Santucci LA, Turpin LC, Baggs RB, Silverman DJ. Selective modulation of antioxidant enzyme activities in host tissues during *Rickettsia conorii* infection. Microb Pathog. 2004;36:293–301.
82. Hall ED, McCall JM. Antioxidant action of lazaroids. Methods Enzymol. 1994;234:548–55.
83. Young B, Runge JW, Waxman KS, et al. Effects of pegorgotein on neurologic outcome of patients with severe head injury. A multicenter, randomized controlled trial. JAMA. 1996;276:538–43.
84. Schewe T. Molecular actions of ebselen: an antiinflammatory antioxidant. Gen Pharmacol. 1995;26:1153–69.
85. Yamamoto S, Teng W, Nishizawa S, Kakiuchi T, Tsukada H. Improvement in cerebral blood flow and metabolism following subarachnoid hemorrhage in response to prophylactic administration of the hydroxyl radical scavenger, AVS, (+/–)-*N*,*N*′-propylenedinicotinamide: a positron emission tomography study in rats. J Neurosurg. 2000;92:1009–15.
86. Endoh H, Kato N, Fujii S, et al. Spin trapping agent, phenyl *N-tert*-butylnitrone, reduces nitric oxide production in the rat brain during experimental meningitis. Free Radic Res. 2001;35:583–91.
87. Palmer C, Roberts RL, Bero C. Deferoxamine posttreatment reduces ischemic brain injury in neonatal rats. Stroke. 1994;25:1039–45.
88. Zhang R, Shohami E, Beit-Yannai E, Bass R, Trembovler V, Samuni A. Mechanism of brain protection by nitroxide radicals in experimental model of closed-head injury. Free Radic Biol Med. 1998;24:332–40.
89. Yeo JF, Ling SF, Tang N, Ong WY. Antinociceptive effect of CNS peroxynitrite scavenger in a mouse model of orofacial pain. Exp Brain Res. 2008;184:435–8.
90. Rezai-Zadeh K, Arendash GW, Hou H, et al. Green tea epigallocatechin-3-gallate (EGCG) reduces beta-amyloid mediated cognitive impairment and modulates tau pathology in Alzheimer transgenic mice. Brain Res. 2008;1214:177–87.
91. Bastianetto S, Zheng WH, Quirion R. The *Ginkgo biloba* extract (EGb 761) protects and rescues hippocampal cells against nitric oxide-induced toxicity: involvement of its flavonoid constituents and protein kinase C. J Neurochem. 2000;74:2268–77.
92. Lau FC, Bielinski DF, Joseph JA. Inhibitory effects of blueberry extract on the production of inflammatory mediators in lipopolysaccharide-activated BV2 microglia. J Neurosci Res. 2007;85:1010–7.
93. Sun AY, Simonyi A, Sun GY. The "French Paradox" and beyond: neuroprotective effects of polyphenols. Free Radic Biol Med. 2002;32:314–8.
94. Roy A, Hooper DC. Lethal silver-haired bat rabies virus infection can be prevented by opening the blood-brain barrier. J Virol. 2007;81:7993–8.

Chapter 7
The Role of Reactive Oxygen Species in the Pathogenesis of Traumatic Brain Injury

Esther Shohami and Ron Kohen

Abstract Traumatic brain injury (TBI) is a major cause of death in the young age group and leads to persisting neurological impairment in many of its victims. TBI involves a primary mechanical impact followed by the development of vasogenic and cytotoxic edema and impairment of energy metabolism and ionic homeostasis. Primary injury sets in motion a cascade of events that activate molecular and cellular responses. The relatively rapid process of primary cell death is followed by secondary degeneration of adjacent neurons having escaped the initial insult. The primary death of brain cells concomitantly also causes accumulation of harmful physiological substances such as glutamate, reactive oxygen species (ROS), and pro-inflammatory cytokines, creating a toxic environment for neighboring neurons and resulting in functional deficits. The mammalian brain is vulnerable to oxidative stress because of the high oxygen consumption needed for maintaining neuronal ion homoeostasis during the propagation of action potentials. Interruption of mitochondrial function involves oxidative stress and leads to impaired energy production, followed by rapidly developing brain damage. For nearly three decades, ROS have been the focus of interest as possible candidates for the elicitation of deleterious responses in the pathogenesis of ischemia and TBI; however, despite numerous clinical trials, no antioxidants have made their way into clinical practice. This chapter focuses on the role of oxidative stress and tissue antioxidant capacity in the pathogenesis of TBI. Oxidative stress in the brain and its biological targets are discussed along with the tissue's intrinsic defense mechanisms, including antioxidant enzymes and low molecular weight antioxidants (LMWA). Post-TBI oxidative stress-induced damage is described, highlighting its major hallmarks, namely mitochondrial damage, lipid peroxidation, antioxidant enzymes, and LMWA. Finally, several therapeutic agents

E. Shohami (✉)
Department of Pharmacology, The Hebrew University School of Pharmacy, Jerusalem, Israel
e-mail: esty@cc.huji.ac.il

N. Gadoth and H.H. Göbel (eds.), *Oxidative Stress and Free Radical Damage in Neurology*, Oxidative Stress in Applied Basic Research and Clinical Practice,
DOI 10.1007/978-1-60327-514-9_7,

harboring antioxidant properties are presented at the end of the chapter for their implications in both experimental and clinical settings.

Keywords Antioxidant enzymes · Low molecular weight antioxidants · Lipid peroxidation · Mitochondria · Oxidative stress

1 Introduction

The brain comprises the tissue most vulnerable to oxidative damage, a feature mainly the result of the high rate of oxidative metabolic activity and relatively low antioxidant capacity of the tissue as well as several additional tissue-specific characteristics. Among the factors further contributing to the brain's enhanced susceptibility are intensive production of reactive oxygen metabolites and low repair mechanism activity combined with the nonreplicating nature of neuronal cells and high membrane surface-to-cytoplasm ratio [1]. Additionally, the brain contains high levels of redox-active transition metals such as iron, which are capable of catalyzing the production of highly toxic radicals via the metal-mediated Haber–Weiss reaction [2].

To contend with the continuous exposure to reactive oxygen species (ROS), the living cell has developed several lines of defense that assist it in coping directly with ROS and preventing oxidative damage. This chapter focuses on the role of oxidative stress and tissue antioxidant capacity in the pathogenesis of traumatic brain injury (TBI). Cellular stress responses and their modulation by endogenous antioxidant defense mechanisms are highlighted.

1.1 TBI Damage: Detrimental Effects of ROS and Possible Protective Mechanisms

Traumatic injury of the central nervous system (CNS) is a major health problem in developed countries. Annual incidence rates are estimated to be about two million for TBI and 11,000 for spinal cord injury in the United States, with persisting neurological impairment in many trauma-affected patients [3–6]. These injuries involve a primary mechanical impact that leads to disruption of the blood–brain barrier (BBB), development of vasogenic and cytotoxic edema, and impairment of energy metabolism and ionic homeostasis [7, 8]. Primary injury sets in motion a cascade of events that activate molecular and cellular responses, leading to a secondary injury. The evolution of secondary damage is an active process in which many biochemical pathways are involved (for a review, see reference [9]).

Most of these pathways are detrimental, yet a large body of literature indicates the concomitant activation of endogenous, self-protective mechanisms.

The outcome of CNS injury depends not only on the extent of the primary impact but also on the amount of secondary degeneration that subsequently develops [10, 11]. Neurons that are directly damaged will inevitably die, regardless of the location (white or gray matter) of the lesion. The relatively rapid process of primary death is followed by secondary degeneration of adjacent neurons that have escaped the initial insult. The primary death of brain cells causes accumulation of harmful physiological substances such as glutamate, reactive oxygen, and nitrogen species with pro-inflammatory cytokines, creating a toxic environment for neighboring neurons [12–16]. These events have been shown to be the cause of the majority of nerve cell loss and disruption of neuronal circuitry, resulting in functional deficits. The mammalian brain is considered to be especially vulnerable to oxidative stress [17]. The brain's high oxygen consumption is needed consequent to the vast amounts of ATP used to maintain neuronal ion homoeostasis during the propagation of action potentials. Interruption of mitochondrial function as a result of oxidative stress leads to impaired energy production, followed by rapidly developing brain damage. For nearly three decades, ROS have been the focus of interest as possible candidates for eliciting detrimental responses in the pathogenesis of ischemia and TBI; however, despite numerous clinical trials none of the examined antioxidants has made its way into clinical use. Unraveling the cellular/molecular signaling leading to these events may provide a link to better targeting, design, and clinical application of novel neuroprotective therapeutic agents.

1.2 Reactive Oxygen Species, Oxidative Stress in the Brain, and Biological Targets

The brain consumes about 20% of the inspired oxygen even though it is a relatively small organ, amounting to only 2% of the whole human body [1, 13, 18, 19]. The oxygen molecule itself is an oxidant [20] as it contains two parallel unpaired electrons and can therefore be considered as a bi-radical. However, because of the spin limitation, it is a weak radical. Importantly, oxygen can be transformed into a variety of compounds including powerful oxidants that might cause damage to biological targets. In general, these metabolites are referred to as ROS or reactive nitrogen species (RNS) and can be classified into two groups, radicals and non-radicals, both of which are deleterious. The radical group contains compounds such as superoxide (O_2^-) radicals, hydroxyl radicals, peroxyl radicals, alkoxyl radicals, nitric oxide (NO) radicals, and one form of singlet oxygen. The non-radicals group contains compounds such as hydrogen peroxide, lipid hydroperoxide, singlet oxygen, peroxynitrite ($ONOO^-$), and hypochlorous acid. Although most of the oxygen molecules are consumed for the production of energy in the form of ATP molecules, a small proportion (about 5%) is used in alternative pathways leading to the production of reactive oxygen metabolites. Although these reactive species have many beneficial roles such as activation of genes and second messengers or certain components of the immune system, they may also be deleterious [13]. The brain is rich with mitochondria, which serve as the major source for endogenous ROS production. The small but continuous efflux of ROS from the mitochondria can potentially damage essential targets in the brain and lead to deleterious consequences. Other endogenous sources of ROS in the brain include enzymatic processes in which ROS are generated directly or indirectly (e.g., monoamine oxidase, nitric oxide synthase [NOS], and cyclooxygenase) and various cells such as activated neutrophils. Excitatory amino acids also generate high levels of ROS subsequent to their massive release and receptor activation following injury [21]. Additionally, there are exogenous factors that can initiate and enhance ROS production in the brain: these include ionizing irradiation, air pollutants, drugs, xenobiotics, and TBI (see following).

Radicals are short-lived species. They are highly reactive and upon production can interact with biological targets in their immediate surroundings such as enzymes [22, 23], lipidic membranes [22, 24], and DNA [25]. The continuous efflux of ROS from endogenous and exogenous sources results in sustained and accumulative oxidative damage to cellular components and alters many cellular functions [2]. Nonradical oxygen metabolites may exist for a longer period of time as compared to radicals, but this does not mean that they are less deleterious. On the contrary, compounds such as hydrogen peroxide can accumulate and diffuse from the site of their origin to other parts of the brain, possibly more vulnerable to oxidative damage, or they might serve as a source for more reactive metabolites. For example, peroxynitrite (not a radical) might degrade to highly reactive species such as hydroxyl radicals, carbonate radicals, and NO radicals [26, 27].

The brain is rich with targets for oxidative damage. Potentially damage-susceptible targets include proteins, DNA, and polyunsaturated fatty acids, highly abundant within the brain. These targets may sustain direct or chain reaction-type damage. For example, lipid peroxidation causes damage to biological membranes and nerves and leads to the production of secondary reactive metabolites such as aldehydes.

1.3 Defense Mechanisms

In light of its particular sensitivity and continuous exposure to reactive metabolites, brain tissue could not survive without extremely efficient defense mechanisms against oxidative stress. To cope with the continuous exposure of brain targets to reactive species, efficient direct and indirect defense mechanisms have evolved in this tissue. The antioxidant system is extremely important for its ability to directly remove pro-oxidants and its variety of compounds acting as antioxidants to ensure maximum protection of biological sites. This defense system contains two major groups: antioxidant enzymes and low molecular weight antioxidants (LMWA) [19].

1.3.1 Antioxidant Enzymes

The enzyme-containing group is composed of direct-acting proteins and supporting enzymes. The direct-acting group contains proteins such as superoxide dismutase (SOD; different types exist, such as Cu-Zn SOD, MnSOD, EC-SOD, and Fe-SOD in plants), catalase, and peroxidase, which can remove hydrogen peroxide, the product of the dismutation reaction. It is noteworthy that the more the tissue is exposed to oxidative stress the more efficient and complex is its antioxidant defense system. Hence, the brain, which is particularly vulnerable to oxidative damage, possesses an extremely efficient antioxidant activity. Brain tissue contains various types of antioxidants; some of which are unique to it. The levels of the antioxidant enzymes vary in different brain regions and among various species. Although the activity of SOD and peroxidase in the brain is quite high, the activity of catalase, acting to degrade hydrogen peroxide to water and oxygen, is relatively low [28, 29]. Rat brain contains high levels of glutathione peroxidase, which removes hydrogen peroxide.

The role of the endogenous antioxidant enzymes in acute CNS injury was shown by Chan et al. in genetically manipulated mice. Transgenic mice that overexpress CuZn-SOD and antiapoptotic Bcl-2 in the brain were found to be protected in a model of brain ischemia/reperfusion injury, whereas knockout mutants deficient of SOD were more vulnerable to the same type of injury [30].

1.3.2 Major Low Molecular Weight Antioxidants in the Brain

LMWA compounds that react directly with radicals (e.g., scavengers and chain-breaking antioxidants) are extremely important in combating oxidative stress. This group contains several hundred compounds originating from endogenous and exogenous sources. Some of the most widely known LMWA are listed next.

Glutathione. This tripeptide (glu-cys-gly) is the most abundant non-protein thiol found in the brain. Glutathione acts as an antioxidant via its capacity to serve as a substrate for the enzyme glutathione peroxidase and seems to mainly be found in astrocytes [31].

Tocopherols (*vitamin E*). This diet-derived (in humans) vitamin comprises a lipid-soluble antioxidant, usually accumulated in the cell membrane. Vitamin E acts as a scavenger by donating an electron to peroxyl radicals, thus preventing propagation of lipid peroxidation [29].

Ascorbic acid. Vitamin C is considered a powerful hydrophilic scavenger in biological fluids and tissues [32]. Humans and other primates have lost the ability to synthesize ascorbic acid and rely on dietary sources of this compound. Vitamin C is distributed throughout the brain, and its concentration in the cerebrospinal fluid is about tenfold higher than in the plasma. The vitamin serves as a strong reducing agent by donating electron(s), thus directly neutralizing ROS. In addition, it also acts to recycle the tocopherol radical to its active reduced form. However, ascorbate may also acquire a strong pro-oxidant activity by reacting with iron ions in the so-called Udenfriend reaction to produce highly toxic hydroxyl radicals [33]. This reaction might occur in the brain after injury following mobilization of iron ions and their interaction with ascorbic acid [22, 29, 34].

Histidine-related compounds. Carnosine (*D*-alanyl-L-histidine) was discovered at the beginning of the twentieth century in skeletal muscle. Since then, carnosine and related compounds anserine (8-alanyl-3-methyl-L-histidine) and homocarnosine (γ-aminobutyryl-L-histidine) have been reported to be present in brain tissue, in concentrations ranging from 0.3 to 5 mM. Together, these compounds have been shown to dominate the nonprotein nitrogenous compound pool of the human brain [18, 35, 36]. The distribution of these entities in the brain is not homogeneous, with the highest concentration located in the olfactory bulbs (2–5 mM). Homocarnosine is present in cerebrospinal fluid and brain (2–50 μM) and anserine is present in the brain [18]. Carnosine is under metabolic control and is produced by the enzyme carnosine synthase [37, 38]. The compound and its derivatives act as neuroprotective antioxidant effectors in the brain or peripheral tissues [18, 36, 37, 39–41]. Carnosine, homocamosine, and anserine are biologically significant and possess many antioxidant functions. They act as chelating agents capable of binding transition metals and preventing the involvement of the latter in the metal-mediated Haber–Weiss reaction. The dipeptides scavenge peroxyl and hydroxyl radicals in addition to binding hydrogen peroxide and removing it from targets susceptible to oxidative damage and can also interact with singlet oxygen, mimic SOD, and show peroxidase-like activity [42].

Melatonin. *N*-Acetyl-5-methoxytryptamine is produced in the pineal gland and readily crosses the BBB to enter neurons and glial cells [29]. It is a potent scavenger of peroxyl and hydroxyl radicals, shown to prevent both initiation and propagation of lipid peroxidation and stimulate brain glutathione peroxidase. Melatonin acts as an antioxidant in both lipophilic and hydrophilic environments because of its solubility properties. Recently it was shown that the compound inhibits NOS, thus preventing the toxic effect obtained after its interaction with superoxide radicals [29, 43].

Uric acid. Uric acid, a waste product of the living cell produced by xanthine oxidase (XO), is widely distributed in relatively high concentrations in the body. Urate contributes up to 60% of the total antioxidant activity of plasma in healthy subjects [44, 45], acting as a preventive antioxidant by interacting with 10–15% of the hydroxyl radicals produced daily. The compound is an efficient scavenger of both peroxyl radicals and singlet oxygen [46] and is also capable of binding iron [47] or acting indirectly by stabilizing plasma ascorbate [48]. Using a model of ischemic preconditioning in rats, Glantz et al. [49] proposed that at least part of the neuroprotection afforded is mediated via high levels of uric, but not ascorbic acid. In contrast, Benzie and Strain [45] recently hypothesized that urate acts as a pro-oxidant at high concentrations and suggested that hyperuricemia is a risk factor for oxidative stress-associated disorders.

Lipoic acid. The lipid-soluble dithiol α-lipoate is absorbed from the diet, crosses the BBB, and is taken up by cells. It possesses a wide spectrum of activities as a metabolic antioxidant, and its reduced form, dihydrolipoate, exits the cells to act as an extracellular antioxidant. α-Lipoate scavenges a large variety of species including hydroxyl radicals, hypochlorous acid, NO, peroxynitrite, hydrogen peroxide, and singlet oxygen. It is also an effective chelator of transition metals and may regenerate various antioxidants such as ubisemiquinone, dehydroascorbate radical and glutathione disulfide to their active form. Thus, α-lipoate and its reduced product act as potent antioxidants in both intra- and extracellular environments (see Packer et al. for review: [50]) and were recently reported to show neuroprotective properties in a variety of brain and neuronal tissue pathologies [51].

The aforementioned LMWA are considered as reducing agents capable of donating electrons to ROS and neutralizing them. These agents, also known as scavengers, are responsible in part for creating a reducing environment within the brain. Although there are numerous compounds possessing reducing properties, not all of them significantly affect the overall measurable reducing power within the cell. Thus, compounds present at low concentrations (below ten nanomolar) are not considered here as responsible for the overall cellular reducing power, although they may play a role in specific microcellular environments.

2 Production of ROS and Oxidative Stress Damage After TBI

2.1 *Mitochondrial Damage*

Oxygen is required by eukaryotic cells for energy production via the electron transport chain in the mitochondria. The latter are located in the cytoplasm, play a crucial role in cell survival and function, have a large capacity for calcium uptake, and are an important site of calcium storage in cells. As mitochondrial calcium flux depends on its inner membrane potential [52], intracellular calcium must be tightly regulated to maintain their homeostasis.

The failure of mitochondrial function is one of the hallmarks of oxidative stress and is involved in both necrotic and apoptotic cell death. It is an important mechanism that precedes the onset of neuronal loss after TBI. Drugs that target mitochondria, such as cyclosporine A (CsA), have been shown to be neuroprotective [53, 54]. Oxidative stress occurs within minutes after TBI [55, 56] and may account for the observation of early (30 min) mitochondrial dysfunction following injury. Recent studies on the time course of mitochondrial failure during the first 72 h after TBI revealed that the mitochondrial respiratory dysfunction, as a result of changes in state III and IV respiratory rates, occurred simultaneously with the appearance of oxidative markers such as 4-hydroxynonenal and protein nitration and was sustained at least up to 72 h [54]. Singh et al. [54] also demonstrated that, after TBI, ipsilateral cortical mitochondria displayed progressive (3–12 h) structural damage consisting of swelling and breakage of the outer membranes. These authors also provide evidence that implicates the RNS $ONOO^-$, the product of the reaction between NO radical and superoxide radical [57], in posttraumatic mitochondrial dysfunction.

2.2 Lipid Peroxidation

Among the secondary post-TBI neurotoxic events that are mediated via excessive ROS formation, lipid peroxidation is well documented as having highly damaging effects [58]. Numerous experimental studies have reported that TBI results in an increase in glutamate release, which, via the activation of the *N*-methyl D-aspartate (NMDA) receptor, leads to cytosolic calcium overload. As a result, the activation of phospholipase enzymes along with a direct ROS attack on membrane fatty acid side chains generates carbon-centered radicals. Subsequently, lipoperoxyl radicals are formed, facilitating an attack on adjacent fatty acids and propagating further damage [58]. Accumulation of lipid hydroperoxides alters membrane permeability and fluidity and oxidizes membrane proteins, leading to alterations in ion transport, notably the intracellular flux of Ca^{2+} [59].

2.3 Nitric Oxide Production

Rises in Ca^{2+} can also activate neuronal nitric oxide synthase (nNOS). NO is produced in a reaction that converts arginine to citrulline under control of NOS enzymes. There are three NOS isoforms, namely, the constitutive, neuronal, and endothelial variants (nNOS and eNOS), and an additional inducible form (iNOS). NO itself is a multipotent messenger whose functions include control of cerebral blood flow, interneuronal communications, synaptic plasticity, memory formation, receptor functions, intracellular signal transmission, and release of neurotransmitters. Similar to many other mediators in the brain (e.g., glutamate, cytokines, calcium ions), it may have both beneficial and detrimental effects, depending upon the exact postinjury period examined [60].

NO accumulates in the brain immediately after injury, probably from the increased activity of eNOS and nNOS. A later peak in NO, occurring several days after TBI, is probably caused by activation of iNOS. Inhibition of iNOS was also shown to convey neuroprotective effects in most experimental models of TBI [61]. Interestingly, the cerebrospinal fluid levels of nitrotyrosine, produced via a chain reaction between reactive species, may be an in vivo marker of oxidative NO-mediated damage, as shown by Darwish et al. [62].

3 Defense Mechanisms Against ROS After TBI

3.1 Evaluation of Low Molecular Weight Antioxidants (LMWA)

Most of the methods used to evaluate antioxidant activity are related to specific scavengers or particular reactive species involved. As such, these methods are insufficient to indicate the overall reducing (antioxidant) status of the tissue.

A different approach to LMWA activity measurements in tissues and fluids was developed by Kohen and Nyska, who demonstrated that the antioxidant activity of biological tissues and fluids correlates with their biological oxidation potential and level, that is, anodic current [19]. Thus, LMWA activity of a biological sample can be evaluated by measuring its reducing power (expressed by both parameters: biological oxidation potential and anodic current). Consequently, a convenient method of measuring the overall reducing power of biological fluids or tissue homogenate is cyclic voltammetry (CV) [63]. In particular, CV tracing records the biological oxidation potential specific to the type of scavenger(s) present and the anodic current that depends on their concentration(s).

3.1.1 Antioxidant Capacity (Levels of LMWA) Predicts Functional Outcome After TBI

Changes in the levels of LMWA reflect the consumption and/or recruitment of these antioxidant molecules, acting as defense mechanism to neutralizing ROS. We used this approach to examine the temporal changes in brain LMWA in response to experimental TBI. In addition, the ability of an animal to cope with oxidative stress after TBI was evaluated by measuring brain levels of LMWA. Indeed, we showed that animals that were able to mobilize higher levels of antioxidants within minutes after TBI or ischemia, whether because of differences in physiological or pathological status [13, 49, 64, 65] or specific genetic manipulations [66, 67], displayed greater recovery of function after injury. Thus, the lower antioxidative postinjury response in apoE-deficient mice suggests that this phenomenon may play a role in their impaired ability to recover after TBI [66, 67] and possibly contributes to rendering their brain susceptible to the development of neurodegenerative disorders.

Melatonin, the main pineal hormone possessing antioxidant properties [68], was given to mice 1 h after TBI and was found to facilitate recovery and decrease infarct volume. This treatment was associated with sustained elevation of LMWA levels inhibition of activation of the "redox-sensitive" transcription factors NF-κB and AP-1. Our data support the view that melatonin does not only act as a radical scavenger: it also induces the activities of transcription factors toward pro-survival signals, rather than death-promoting ones, during the secondary phase after TBI [69].

It is interesting to note that, after isolated TBI, there is a whole-body response of oxidative stress, shown by a transient decrease in LMWA levels in the heart, kidney, lung, and liver 1 h after TBI. These decreases suggest the consumption of LMWA, probably from interaction with locally produced ROS [70].

3.1.2 Diabetes as a Physiological Model for Low Antioxidant Capacity

Diabetes is a metabolic disorder associated with structural and functional changes in the peripheral and CNS. Accumulating evidence suggests that oxidative stress is implicated in the pathogenesis of diabetic neuropathy by inducing neurovascular defects that result in endoneural hypoxia and subsequent nerve dysfunction [71]. Studies on antioxidant treatment in experimental models of diabetes point to the involvement of ROS in the reduced nutritive blood flow that causes peripheral nerve hypoxia and dysfunction [72]. Diabetic rats subjected to TBI exhibited lower concentrations of LMWA and of vitamins C and E, suggesting chronic oxidative stress [65]. After TBI, these diabetic rats displayed a slower rate of recovery and greater and more sustained edema as compared with their controls. At all times diabetic rats had higher levels of thiobarbituric acid-reactive substances and conjugated dienes, supporting the hypothesis that diabetic rats suffer greater neurological dysfunction associated with oxidative stress and lipid peroxidation after TBI.

3.1.3 Aging: A Progressive Decline in Antioxidant Capacity

Age is an important risk factor for the outcome of stroke [73] and TBI [74–76] as well as for the decline in motor and cognitive abilities and development of neurodegenerative diseases such as Alzheimer's and Parkinson's. Theoretically, a number of causes could underlie this phenomenon, including a decrease in neuron reserves and neuronal plasticity, reduced capacity of repair and regeneration mechanisms, and a decline in the capacity of the body to cope with physiological stress in general. Oxidative stress in the brain is emerging as a potential causal factor in aging and age-related neurodegenerative disorders [29, 77–79].

Decreased ability of managing oxidative stress is also a common hallmark of neurodegenerative diseases as well as normal aging. The notion that the capacity of brain tissue to neutralize harmful oxidative agents declines in old age is derived from the free radical theory of aging [80, 81] and implies that the vulnerability of the brain to oxidative stress may well increase with age. Indeed, we have demonstrated that in old rats the basal levels of ascorbate in the extracellular fluid is significantly lower than in adults [82]. Moreover, fewer antioxidants were found to accumulate in the extracellular space in response to TBI in the aged brain when compared with younger controls. Thus, lower levels of antioxidants may imply less effective combat of ROS and greater oxidative damage. This finding is further corroborated by our observation that old rats displayed increased neuronal and axonal damage [82].

3.2 *Antioxidant Enzymes in TBI*

3.2.1 Superoxide Dismutase (SOD)

The stepwise activity of the enzymatic antioxidant network consists of dismutation of superoxide anions (O_2^-) by SOD to form hydrogen peroxide (H_2O_2) [83]. This product, in turn, is converted to water through the activities of mitochondrial and cytosolic glutathion peroxidase (GpX) or paroxisomal catalase. In addition, glutathione reductase and 6-phosphate dehydrogenase act to support an integrated antioxidant system [2]. Three isoforms of SOD were identified on the basis of their subcellular localization and metal ion requirements. SOD1 is a dimeric cytosolic enzyme, requiring copper and zinc ions as cofactors, SOD2 is a tetrameric mitochondrial enzyme requiring manganese, and SOD3 is an extracellular, high molecular weight, copper-containing enzyme. Interestingly, although SOD1 is constitutively present in all cells, SOD2 is inducible upon oxidative or inflammatory stimuli, suggesting distinct roles for the two isoforms [84].

The particular role of SOD1 in neuroprotection has been widely investigated but remains controversial. Several groups demonstrated that overexpression of SOD1 is associated with a dose-related decrease in brain edema, tissue infarction, and BBB disruption in models of cold and contusion brain injury [85–87] and following transient cortical ischemia [88–90]. In contrast, SOD1 deficiency did not affect brain edema, infarct volume, or early release of mitochondrial cytochrome *cs* following permanent focal cerebral ischemia [91]. Furthermore, increased SOD1 activity caused pronounced effects on other antioxidant enzymes [92], aggravated oxidative injury [93], and impaired cognitive function in SOD1 transgenic mice [94].

Beni et al. [95] have reported that TBI-induced NF-κB activation was absent in $SOD1^{-/-}$, mice suggesting that SOD1 deficiency impaired H_2O_2-mediated activation of this redox-sensitive transcription factor, decreasing death-promoting signals, and leading to better late outcome. However, in the same study we demonstrated that SOD1 overexpression failed to enhance neurobehavioral or cognitive recovery after TBI [95]. Such duality in the effect of SOD after TBI may lie in the fact that although overexpression of SOD1 facilitates O_2^- removal, it also enhances H_2O_2 production, which in turn favors generation of hydroxyl radicals via Fenton reactions. Therefore, efficient detoxification of O_2^- requires an integrated, fine-tuned activity of SOD together with GpX or catalase, such that accumulation of H_2O_2 and further perturbation of cellular redox-homeostasis are prevented [96, 97].

3.2.2 Glutathione Peroxidase (GpX)

Using both transgenic and knockout mice, Xiong et al. [98] provided evidence that cellular glutathione peroxidase (GpX1) is essential for maintaining brain mitochondrial function after TBI. Mice were subjected to TBI, and the efficiency of oxidative phosphorylation in brain mitochondria was determined. They demonstrated that the susceptibility of mouse brain mitochondria to TBI-induced decrease in NAD-linked bioenergetic capacity is inversely correlated with the expression level of Gpx1. Their results suggest a role for GpX1 in antioxidant defense mechanisms against TBI, and implicate hydrogen peroxide, alkyl hydroperoxides, or both as toxic ROS in secondary brain injury after TBI.

3.2.3 Heme Oxygenase (HO)

The ubiquitous molecule heme, found in mammalian tissues, plays a key role in the delivery of molecular oxygen and affects oxidative reactions associated with cellular metabolism [99], whereas free heme is deposited only under pathological conditions. Following TBI, heme is released from red blood cells and from injured neurons and glia and is also extravasated via the BBB. The heme oxygenase isozymes (HO-1 and HO-2) catalyze the degradation of heme to produce carbon monoxide (CO), ferrous iron (Fe^{2+}), and biliverdin, which is converted to bilirubin [100]. HO-2 is the predominant, constitutively expressed isozyme found in the adult rodent brain and testes and is regulated only by adrenal glucocorticoids [101, 102]. Although HO-2 is immunolocalized in neurons before and after TBI, HO-1 is highly upregulated in glia only after TBI. To study the role of HO after TBI, Chang et al. [103] determined HO activity in brain sonicates from injured HO-2-knockout (KO) mice. They found it to be significantly lower than that of wild types, despite the induction of HO-1 expression after TBI. They also found significantly greater cell loss in KO mice in the cortex, CA3 region of hippocampus, and lateral dorsal thalamus. Furthermore, recovery of motor function was impaired in the KO mice. Interestingly, brain tissue from injured HO-2 KO mice exhibited decreased ability to reduce oxidative stress and higher levels of end products of lipid peroxidation as compared with the wild type. These findings demonstrate that HO-2 expression protects neurons against TBI by reducing lipid peroxidation via the catabolism of free heme.

3.2.4 Xanthine Oxidase (XO)

XO catalyzes the oxidative hydroxylation of hypoxanthine to xanthine and also the subsequent hydroxylation of xanthine to uric acid, the final two steps of purine metabolism in mammals. Hillered et al. [104] reported that the level of hypoxanthine is increased in the extracellular fluid of the striatum early after middle cerebral artery occlusion in rats. However, as this change was not accompanied by appearance of respective degradation products, their work indicated that this compound is not a useful marker for evaluation of energetic homeostasis disruption.

Importantly, xanthine degradation by XO leads to the formation of hydrogen peroxide and hydroxyl radicals. Hence, after TBI or cerebral ischemia the XO pathway may contribute to the generation of ROS and lipid peroxidation. Recently, Solaroglu et al. reported that increased activity of XO was found as early as 2 h after TBI and was sustained up to 24 h. Concomitantly, a time-dependent increase in the levels of thiobarbituric acid reactive substances, a marker of oxidative stress, was noted up to 24 h. These results indicate that the immediate rise in XO activity following TBI only partially explains the increase in ROS [105].

4 Therapeutic Agents with Antioxidant Properties After TBI

4.1 *Tempol*

The spin-trap agent tempol (4-hydroxy 2,2,6,6,-tetramethylpiperidine-1-oxyl) is a commercially available nitroxide antioxidant with no significant side effects [106] that is used as a contrast agent in imaging procedures [107]. In contrast to other antioxidants that are generally active against only one species of radicals, tempol possesses SOD-like activity and was shown to modulate NO levels as well as reduce hypervalent metals. The compound can also react with peroxyl and carbon-centered radicals and dismutate oxygen radicals. Importantly, tempol has a catalytic mode of action and replenishes its antioxidant capacities, in contrast to other agents which finally exhaust their antioxidant capacity [108–110]. Furthermore, tempol is capable of penetrating cells and exerting important intracellular antioxidant actions, whereas antioxidants used in previous experiments did not penetrate cell membranes effectively and thus had only limited intracellular antioxidant effects. Tempol has also been shown to be a potent neuroprotectant in TBI [111] and to protect against myocardial ischemia [112], radiation damage [113], malignancy [114], hemorrhagic shock [115], and focal ischemia [116].

4.2 *Aminosteroids*

The lazaroids are unique compounds possessing the membrane-stabilizing effects of glucocorticoids without their receptor-dependent side effects. The lazaroids exert their antilipid peroxidation action through two mechanisms, free radical scavenging and membrane stabilization. Early animal studies of traumatic and ischemic injury [117–119] suggested that the lazaroids inhibit membrane lipid peroxidation by scavenging peroxyl radicals, a mechanism similar to that of vitamin E [120]. One of these compounds, tirilazad mesylate (U-74006F), has been selected for clinical development as a parenteral neuroprotective agent. In experimental head injury models, tirilazad decreased hydroxyl radical production in the brain upon concussive head injury. Additional observations in gerbils with bilateral carotid occlusion indicated that it may also scavenge hydroxyl radicals [119]. Tirilazad mesylate is a 21-aminosteroid that has been shown to inhibit lipid peroxidation in experimental animals [121]. It penetrates the intact BBB poorly but has a strong affinity for the vascular endothelium [122]. The effect of tiralazad on TBI patients was examined in a randomized clinical study, using mortality and Glasgow outcome score (GOS) at 6 months post TBI as the primary outcome measures. Although no significant difference was found between the drug- and placebo-treated groups, subgroup analysis suggested that treatment reduced mortality rates in males suffering from severe TBI with accompanying traumatic subarachnoid hemorrhage [123].

4.3 Cyclosporine A

As already mentioned, CsA inhibits mitochondrial dysfunction in the CNS. By preventing mitochondrial calcium efflux, the compound presumably interrupts the cascade of events leading to apoptosis and/or necrosis [53–55]. Results of a phase II clinical trial using CsA have been published recently [124]. In this prospective, randomized, placebo-controlled clinical trial, dose escalation was performed as to obtain data regarding CsA effects in TBI patients and concomitantly identify optimal dosage in 30 patients with severe TBI (Glasgow Coma Score, 4–8). It was shown that patients with severe TBI demonstrate a rapid clearance and larger distribution volume of CsA, an effect that will need to be accounted for in future safety and efficacy studies.

Although proven effective in many preclinical studies, the use of antioxidants in the management of TBI has failed thus far [123]. Because the half-lives of most ROS are very short, and given their high reactivity toward many biological targets, chances are scarce that daily-administered antioxidants given during the first 7 days following injury will prevent damage. Future studies should be more focused on the appropriate timing of drug delivery and at facilitating high concentrations of therapeutic agents before or at the time of the ROS formation.

Further elucidation the multifactorial mechanisms and complex network governing the role of ROS in progressive post-TBI damage and identification of endogenous neuroprotective mechanisms that combat ROS are needed to develop novel therapeutics for TBI].

Acknowledgments This study was partially supported the Israel Science Foundation, (grant no. 241/04). E.S. and R.K. are affiliated with the David R. Bloom Center of Pharmacy and the Dr. Adolph and Klara Brettler Center for Research in Molecular Pharmacology and Therapeutics, of the Hebrew University, Jerusalem. E.S. is the incumbent of the Dr. Leon and Dr. Mina Deutsch Chair in Psychopharmacology, and R.K. is the incumbent of the Richard and Jean Zarbin Chair in Medical Studies, at the Hebrew University, Jerusalem.

References

1. Evans PH. Free radicals in brain metabolism and pathology. Br Med Bull. 1993;49:577–87.
2. Halliwell B, Gutteridge JMC. Free radicals in biology and medicine. Oxford: Oxford University Press; 1989. p. 160–5.
3. Marshall LF. Head injury: recent past, present, and future. Neurosurgery. 2000;47:546–61.
4. Bayir H, Clark RS, Kochanek PM. Promising strategies to minimize secondary brain injury after head trauma. Crit Care Med. 2003;31:S112–7.
5. Nolan S. Traumatic brain injury: a review. Crit Care Nurs Q. 2005;28:188–94.
6. Nortje J, Menon DK. Traumatic brain injury: physiology, mechanisms, and outcome. Curr Opin Neurol. 2004;17:711–8.
7. Gentleman SM, Roberts GW, Gennarelli TA, Maxwell WL, Adams JH, Kerr S, Graham DI. Axonal injury: a universal consequence of fatal closed head injury? Acta Neuropathol. 1995;89:537–43.

8. Povlishock JT, Christman CW. The pathobiology of traumatically induced axonal injury in animals and humans: a review of current thoughts. J Neurotrauma. 1995;12:555–64.
9. Bramlett HM, Dietrich WD. Progressive damage after brain and spinal cord injury: pathomechanisms and treatment strategies. Prog Brain Res. 2007;161:125–41.
10. Yoles E, Schwartz M. Degeneration of spared axons following partial white matter lesion: implications for optic nerve neuropathies. Exp Neurol. 1988;153:1–7.
11. Harrop JS, Sharan AD, Vaccaro AR, Przybylski GJ. The cause of neurologic deterioration after acute cervical spinal cord injury. Spine. 2001;26:340–6.
12. Tymianski M, Tator CH. Normal and abnormal calcium homeostasis in neurons: a basis for the pathophysiology of traumatic and ischemic central nervous system injury. Neurosurgery. 1996;38:1176–95.
13. Shohami E, Beit-Yannai E, Horowitz M, Kohen R, Beit-Yannai E, Horowitz M, Kohen R. Oxidative stress in closed-head injury: brain antioxidant capacity as an indicator of functional outcome. J Cereb Blood Flow Metab. 1997;17:1007–19.
14. Shohami E, Ginis I, Hallenbeck JM. Dual role of tumor necrosis factor alpha in brain injury. Cytokine Growth Factor Rev. 1999;10:119–30.
15. Beattie MS, Farooqui AA, Bresnahan JC. Review of current evidence for apoptosis after spinal cord injury. J Neurotrauma. 2000;17:915–25.
16. Schwartz M, Yoles E. Self-destructive and self-protective processes in the damaged optic nerve: implications for glaucoma [comment]. Invest Ophthalmol Vis Sci. 2000;41:349–51.
17. Halliwell B. Oxidative stress and neurodegeneration: where are we now? J Neurochem. 2006;97:1634–58.
18. Kohen R, Yamamoto Y, Cundy KC, Ames BN. Antioxidant activity of carnosine, homocarnosine and anserine present in muscle and brain. Proc Natl Acad Sci USA. 1988;85:3175−9.
19. Kohen R, Nyska A. Oxidation of biological systems: oxidative stress phenomena, antioxidants, redox reactions, and methods for their quantification. Toxicol Pathol. 2002;30:620–50.
20. Dugan LL, Choi DW. Excitotoxicity, free radicals and cell membrane changes. Ann Neurol. 1994;35:517−21.
21. Gracy RW, Talent JM, Kong Y, Conrad CC. Reactive oxygen species: the unavoidable environmental insult? Mutat Res. 1999;428:17–22.
22. Levine RL, Stadtman ER. Oxidative modification of proteins during aging. Exp Gerontol. 2001;36:1495–502.
23. Davis KJ. Protein damage and degradation by oxygen radicals. I. General aspects. J Biol Chem. 1987;262:9895–901.
24. Rice-Evans C, Burdon R. Free radical-lipid interactions and their pathological consequences. Prog Lipid Res. 1993;32:71−110.
25. Beckman KB, Ames BN. Oxidative decay of DNA. J Biol Chem. 1997;272:19633–6.
26. Squadrito GL, Pryor WA. Oxidative chemistry of nitric oxide: the roles of superoxide, peroxynitrite, and carbon dioxide. Free Radic Biol Med. 1998;25:392–403.
27. Virag L, Szabo E, Bakondi E, Bai P, Gergely J, Hunyadi J, Szabo C. Nitric oxide-peroxynitrite-poly (ADP-ribose) polymerase pathway in the skin. Exp Dermatol. 2002;11:189–202.
28. Sinet PM, Heikkila RE, Cohen G. Hydrogen peroxide production in rat brain in vivo. J Neurochem. 1980;34:1421−8.
29. Reiter RJ. Oxidative processes and antioxidative defense mechanisms in the aging brain. FASEB J. 1995;9:526−33.
30. Chan PH. Role of oxidants in ischemic brain damage. Stroke. 1996;27:1124−9.
31. Raps SP, Lai JCK, Hertz I, Cooper AJL. Glutathione is present in high concentration in cultured astrocytes but not in cultured neurons. Brain Res. 1989;493:398−401.
32. Frei B, England L, Ames BN. Ascorbate is an outstanding antioxidant in human blood plasma. Proc Natl Acad Sci U S A. 1989;86:6377−81.

33. Brodie BB, Axelrode J, Shore PA, Udenfriend S. Ascorbic acid in aromatic hydroxylation. II. Products formed by reaction of substrates with ascorbic acid, ferrous ion, and oxygen. J Biol Chem. 1954;208:741–50.
34. Halliwell B. Vitamin C: antioxidant or prooxidant in vivo. Free Radic Res. 1996;25:439–54.
35. Boldyrev AA, Dupin A, Bunin YA, Babizhaev MA, Severin E. The antioxidative properties of carnosine, a natural histidine containing dipeptide. Biochem Int. 1987;15:1105–13.
36. Tabakman R, LazaroviciP, Kohen R. Neuroprotective effects of carnosine and homocarnosine on pheochromocytoma PC12 cells exposed to ischemia. J Neurosci Res. 2002;68:463–9.
37. Kalyankar G, Meister A. Enzymatic synthesis of carnosine and related β-alanyl and γ-aminobutyryl peptides. J Biol Chem. 1959;234:3210–8.
38. Horinishi H, Grillo M, Margolis FL. Purification and characterization of carnosine synthetase from mouse olfactory bulbs. J Neurochem. 1978;31:909–19.
39. Boldyrev AA. Does carnosine possess direct antioxidant activity? Int J Biochem. 1993;25:1101-7.
40. Boldyrev AA, Stvolinsky SL, Tyulina OV, Koshelev VB, Hori N, Carpenter DO. Biochemical and physiological evidence that carnosine is an endogenous neuroprotector against free radicals. Cell Mol Neurobiol. 1997;17:259–70.
41. Boldyrev AA, Johnson P, Wei Y, Tan Y, Carpenter DO. Carnosine and taurine protect rat cerebellar granular cells from free radical damage. Neurosci Lett. 1999;263:169–72.
42. Kohen R, Misgav R, Ginsburg I. The SOD like activity of copper:carnosine, copper:anserine and copper:homocarnosine complexes. Free Radic Res Commun. 1991;12–13(pt 1):179–85.
43. Pozo D, Reiter RJ, Calvo JR, Guerrero JM. Physiological concentration of melatonin inhibits nitric oxide synthase in rat cerebellum. Life Sci. 1994;55:455–60.
44. Wayner DDM, Burton GW, Ingold KU, Barclay LC, Locke SJ. The relative contributions of vitamin E, urate, ascorbate and proteins to the total peroxyl radical-trapping antioxidant activity of human blood plasma. Biochem Biophys Acta. 1987;924:408–19.
45. Benzie IFF, Strain JJ. Uric acid: friend or foe? Redox Rep. 1996;2:231–4.
46. Ames BN, Cathcart R, Schwiers E, Hochstein P. Uric acid provides an antioxidant defense in humans against oxidant- and radical-caused aging and cancer: a hypothesis. Proc Natl Acad Sci U S A. 1981;73:6858–62.
47. Davies KA, Sevanian A, Muallassah-Kelly F, Hochstein P. Uric acid-iron complexes. Biochem J. 1986;235:747–4.
48. Sevanian A, Davies KJA, Hochstein P. Serum urate as an anti-oxidant for ascorbic acid. Am J Clin Nutr. 1991;54:1129S–34S.
49. Glantz L, Avramovich A, Trembovler V, Gurvitz V, Kohen R, Eidelman LA, Shohami E. Ischemic preconditioning increases antioxidants in the brain and peripheral organs after cerebral ischemia. Exp Neurol. 2005;192:117–24.
50. Packer L, Witt EH, Tritschler HJ. Alpha-lipoic acid as a biological antioxidant. Free Radic Biol Med. 1995;19:227–50
51. Packer L, Tritschler H, Wessel K. Neuroprotection by the metabolic antioxidant α-lipoic acid. Free Radic Biol Med. 1997;22:359–78.
52. Nicholls DG, Ward MW. Mitochondrial membrane potential and neuronal glutamate excitotoxicity: mortality and millivolts. Trends Neurosci. 2000;23:166–74.
53. Sullivan PG, Thompson MB, Scheff SW. Cyclosporin A attenuates acute mitochondrial dysfunction following traumatic brain injury. Exp Neurol. 1999;160:226–34.
54. Singh IN, Sullivan PG, Deng Y, Mbye LH, Hall ED. Time course of post-traumatic mitochondrial oxidative damage and dysfunction in a mouse model of focal traumatic brain injury: implications for neuroprotective therapy. J Cereb Blood Flow Metab. 2006;26:1407–18.
55. Sullivan PG, Rabchevsky AG, Waldmeier PC, Springer JE. Mitochondrial permeability transition in CNS trauma: cause or effect of neuronal cell death? J Neurosci Res. 2005;79:231–9.

56. Beit-Yannai E, Kohen R, Horowitz M, Trembovler V, Shohami E. Changes in biological reducing activity in rat brain following closed head injury: a cyclic voltammetry study in normal and acclimated rats. J Cereb Blood Flow Metabol. 1997;17:273–9.
57. Beckman JS. The double-edged role of nitric oxide in brain function and superoxide-mediated injury. J Dev Physiol. 1991;15:53–9.
58. Halliwell B. Reactive oxygen species in living systems: source, biochemistry, and role in human disease. Am J Med. 1991;91:14S–22S.
59. Mattson MP. Modification of ion homeostasis by lipid peroxidation: roles in neuronal degeneration and adaptive plasticity. Trends Neurosci. 1998;21:53–7.
60. Khaldi A, Chiueh CC, Bullock MR, Woodward JJ. The significance of nitric oxide production in the brain after injury. Ann N Y Acad Sci. 2002;962:53–9.
61. Cherian L, Hlatky R, Robertson CS. Nitric oxide in traumatic brain injury. Brain Pathol. 2004;14:195–201.
62. Darwish RS, Amiridze N, Aarabi B. Nitrotyrosine as an oxidative stress marker: evidence for involvement in neurologic outcome in human traumatic brain injury. J Trauma. 2007;63:439–42.
63. Bard AJ, Faulkner LR. Electrochemical methods: fundamentals and application, vol. 9. New York: Wiley; 1980, p. 316–69.
64. Beit-Yannai E, Zhang R, Trembovler V, Samuni A, Shohami E. Cerebroprotective effect of stable nitroxide radicals in closed head injury in the rat. Brain Res. 1996;717:22–8.
65. Elangovan V, Kohen R, Shohami E. Neurological recovery from closed head injury is impaired in diabetic rats. J. Neurotrauma. 2000;17:1013–27.
66. Chen Y, Lomnitski L, Michaelson DM, Shohami E. Motor and cognitive deficits in apolipoprotein E-deficient mice after closed head injury. Neuroscience. 1997;80:1255–62.
67. Lomnitski L, Kohen R, Chen Y, Shohami E, Vogel T, Michaelson DM. Reduced levels of antioxidants in brain of apolipoprotein E deficient mice, following closed head injury. Pharmacol Biochem Behav. 1997;57:669–73.
68. Reiter RJ. The role of the neurohormone melatonin as a buffer against macromolecular oxidative damage. Neurochem Int. 1995;27:453–60.
69. Beni SM, Kohen R, Reiter RJ, Tan D-X, Shohami E. Melatonin-induced neuroprotection after closed head Injury is associated with increased brain antioxidants and attenuated late-phase activation of NF-κB and AP-1. FASEB J. 2004;18:149–51.
70. Shohami E, Gati I, Beit-Yannai E, Trembovler V, Kohen R. Closed head injury in the rat induces whole body oxidative stress: overall reducing antioxidant profile. J Neurotrauma. 1999;16:365–76.
71. Nickander KK, Schmelzer JD, Rohwer DA, Low PA. Effect of tocopherol deficiency on indices of oxidative stress in normal and diabetic peripheral nerve. J Neurol Sci. 1994;126:6–14.
72. Cotter MA, Love A, Watt MJ, Cameron NE, Dines KC. Effect of natural free radical scavengers on peripheral nerve and neurovascular function in diabetic rats. Diabetologia. 1995;38:1285–94.
73. Davis M, Mendelow AD, Perry RH, Chambers IR, James OF. Experimental stroke and neuroprotection in the aging rat brain. Stroke. 1995;26:1072–8.
74. Armstead WM. Age dependent NMDA contribution to impaired hypotensive cerebral hemodynamics following brain injury. Brain Res Dev Brain Res. 2002;139:19–28.
75. Bashore TR, Ridderinkhof KR. Older age, traumatic brain injury, and cognitive slowing: some convergent and divergent findings. Psychol Bull. 2002;128:151–98.
76. Goldstein FC, Levin HS. Cognitive outcome after mild and moderate traumatic brain injury in older adults. J Clin Exp Neuropsychol. 2001;23:739–53.
77. Butterfield D, Drake J, Pocernich C, Castenga A. Evidence of oxidative damage in Alzheimer's disease brain: central role for amyloid beta-peptide. Trends Mol Med. 2001;7:548–54.

78. Esposito E, Rotilio E, Di Matteo V, Di Giulio C, Cacchio M, Algeri S. A review of specific dietary antioxidants and the effects on biochemical mechanisms related to neurodegenerative processes. Neurobiol Aging. 2002;23:719–35.
79. Maier CM, Chan PH. Role of superoxide dismutases in oxidative damage and neurodegenerative disorders. Neuroscientist. 2002;8:323–34.
80. Harman D. Aging: theory based on free radical and radiation chemistry. J Gerontol. 1956;11:298–300.
81. Joseph JA, Denisova NA, Bielinski D, Fisher DR, Shukkit-Hale B. Oxidative stress protection and vulnerability in aging: putative nutritional implications for intervention. Mech Ageing Dev. 2000;116:141–53.
82. Moor E, Shohami E, Kanevsky E, Grigoriadis N, Symeonidou C, Kohen R. Impairment of the ability of the injured aged brain in elevating urate and ascorbate. Exp Gerontol. 2006;41:303–11.
83. Fridovich I. Superoxide dismutases. Annu Rev Biochem. 1975;44:147–59.
84. Huang CY, Fujimura M, Noshita N, Chang YY, Chan PH. SOD1 down-regulates NF-kappaB and c-Myc expression in mice after transient focal cerebral ischemia. J Cereb Blood Flow Metab. 2001;21:163–73.
85. Chan PH, Yang GY, Chen SF, Carlson E, Epstein CJ. Cold-induced brain edema and infarction are reduced in transgenic mice overexpressing CuZn-superoxide dismutase. Ann Neurol. 1991;29:482–6.
86. Lewen A, Sugawara T, Gasche Y, Fujimura M, Chan PH. Oxidative cellular damage and the reduction of APE/Ref-1 expression after experimental traumatic brain injury. Neurobiol Dis. 2001;8:380–90.
87. Mikawa S, Kinouchi H, Kamii H, Gobbel GT, Chen SF, Carlson E, Epstein CJ, Chan PH. Attenuation of acute and chronic damage following traumatic brain injury in copper, zinc-superoxide dismutase transgenic mice. J Neurosurg. 1996;85:885–91
88. Chan PH, Kawase M, Murakami K, Chen SF, Li Y, Calagui B, Reole L, Carlson E, Epstein CJ. Overexpression of SOD1 in transgenic rats protects vulnerable neurons against ischemic damage after global cerebral ischemia and reperfusion. J Neurosci. 1998;18:8292–9.
89. Murakami K, Kondo T, Epstein CJ, Chan PH. Overexpression of CuZn-superoxide dismutase reduces hippocampal injury after global ischemia in transgenic mice. Stroke. 1997;28:1797–804.
90. Yang GY, Chan P, Chen J, Carlson E, Chen SF, Weinstein P, Epstein CJ, Kamii H. Human copper-zinc superoxide dismutase transgenic mice are highly resistant to reperfusion injury after focal cerebral ischemia. Stroke. 1994;25:165–70.
91. Fujimura M, Morita-Fujimura Y, Copin J, Yoshimoto T, Chan PH. Reduction of copper, zinc-superoxide dismutase in knockout mice does not affect edema or infarction volumes and the early release of mitochondrial cytochrome c after permanent focal cerebral ischemia. Brain Res. 2001;889:208–13.
92. Kelner MJ, Bagnell R, Montoya M, Estes L, Uglik SF, Cerutti P. Transfection with human copper-zinc superoxide dismutase induces bidirectional alterations in other antioxidant enzymes, proteins, growth factor response, and paraquat resistance. Free Radic Biol Med. 1995;18:497–506.
93. Peled-Kamar M, Lotem J, Wirguin I, Weiner L, Hermalin A, Groner Y. Oxidative stress mediates impairment of muscle function in transgenic mice with elevated level of wild-type Cu/Zn superoxide dismutase. Proc Natl Acad Sci U S A. 1997;94:3883–7.
94. Gahtan E, Auerbach JM, Groner Y, Segal M. Reversible impairment of long-term potentiation in transgenic Cu/Zn-SOD mice. Eur J Neurosci. 1998;10:538–44.
95. Beni SM, Tsenter J, Alexandrovich AG, Galron-Krool N, Barzilai A, Kohen R, Grigoriadis N, Simeonidou C, Shohami E. CuZn-SOD deficiency, rather than overexpression, is associated with enhanced recovery and attenuated activation of NF-kappaB after brain trauma in mice. J Cereb Blood Flow Metab. 2006; 26:478–90.

96. de Haan JB, Crack PJ, Flentjar N, Iannello RC, Hertzog PJ, Kola I. An imbalance in antioxidant defense affects cellular function: the pathophysiological consequences of a reduction in antioxidant defense in the glutathione peroxidase-1 (Gpx1) knockout mouse. Redox Rep. 2003;8:69–79.
97. Rodriguez AM, Carrico PM, Mazurkiewicz JE, Melendez JA. Mitochondrial or cytosolic catalase reverses the MnSOD-dependent inhibition of proliferation by enhancing respiratory chain activity, net ATP production, and decreasing the steady state levels of H_2O_2. Free Radic Biol Med. 2000;29:801–13
98. Xiong Y, Shie F-S, Zhang J, Lee C-P, Ho Y-S. The protective role of cellular glutathione peroxidase against trauma-induced mitochondrial dysfunction in the mouse brain. J Stroke Cereb Dis. 2004;13:129–37.
99. Maines MD. Heme oxygenase: function, multiplicity, regulatory mechanisms, and clinical applications. FASEB J. 1988;2:2557–68.
100. Maines MD. The heme oxygenase system: a regulator of second messenger gases. Annu Rev Pharmacol Toxicol. 1997;37:517–54.
101. McCoubrey WK Jr, Maines MD. The structure, organization and differential expression of the gene encoding rat heme oxygenase-2. Gene. 1994;139:155–61.
102. Maines MD, Eke BC, Zhao X. Corticosterone promotes increased heme oxygenase-2 protein and transcript expression in the newborn rat brain. Brain Res. 1996;722:83–94.
103. Chang EF, Wong RJ, Vreman HJ, Igarashi T, Galo E, Sharp FR, Stevenson DK, Noble-Haeussleinl LJ. Heme oxygenase-2 protects against lipid peroxidation-mediated cell loss and impaired motor recovery after traumatic brain injury. J Neurosci. 2003;23:3689–96.
104. Hillered L, Kotwica Z, Ungerstedt U. Interstitial and cerebrospinal fluid levels of energy-related metabolites after middle cerebral artery occlusion in rats. Res Exp Med (Berl). 1991;191:219–25.
105. Solaroglu I, Okutan O, Kaptanoglu E, Beskonakli E, Kilinc K. Increased xanthine oxidase activity after traumatic brain injury in rats. J Clin Neurosci. 2005;12:273–75.
106. Ankel EG, Lai CS, Hopwood LE, Zivkovic Z. Cytotoxicity of commonly used nitroxide radical spin probes. Life Sci. 1987;40:495–8.
107. Berliner JL, Fujii H. Magnetic resonance imaging of biological specimens by electron paramagnetic resonance of nitroxide spin labels. Science. 1985;227:517–9.
108. Krishna MC, Russo A, Mitchell JB, Goldstein S, Dafni H, Samuni A. Do nitroxide antioxidants act as scavengers of O_2^{-}. or as SOD mimics? J Biol Chem. 1996;271:26026–31.
109. Samuni A, Krishna CM, Mitchell JB, Collins CR, Russo A. Superoxide reaction with nitroxides. Free Radic Res Commun. 1990;9:241–9.
110. Samuni A, Mitchell JB, DeGraff W, Krishna CM, Samuni U, Russo A. Nitroxide SOD-mimics: modes of action. Free Radic Res Commun. 1991;12–13:187–94.
111. Zhang R, Shohami E, Beit-Yannai E, Bass R, Trembovler V, Samuni A. Mechanism of brain protection by nitroxide radicals in experimental model of closed head injury. Free Radic Biol Med. 1998;24:332–40.
112. McDonald MC, Zacharowski K, Bowes J, Cuzzocrea S, Thiemermann C. Tempol reduces infarct size in rodent models of regional myocardial ischemia and reperfusion. Free Radic Biol Med. 1999;27:493–503.
113. Hahn SM, Tochner Z, Krishna CM, Glass J, Wilson L, Samuni A, Sprague M, Venzon D, Glatstein E, Mitchell JB, et al. Tempol, a stable free radical, is a novel murine radiation protector. Cancer Res. 1992;52:1750–3.
114. DeGraff WG, Krishna MC, Russo A, Mitchell JB. Antimutagenicity of a low molecular weight superoxide dismutase mimic against oxidative mutagens. Environ Mol Mutagen. 1992;19:21–6.
115. Mota-Filipe H, McDonald MC, Cuzzocrea S, Thiemermann C. A membrane-permeable radical scavenger reduces the organ injury in hemorrhagic shock. Shock. 1999;12:255–61.

116. Leker RR, Teichner A, Lavie G, Shohami E, Lamensdorf I, Ovadia H. The nitroxide antioxidant tempol is cerebroprotective against focal cerebral ischemia in spontaneously hypertensive rats. Exp Neurol. 2002;176:355–63.
117. Hall ED, Pazara KE, Braughler JM. Effect of tirilazad mesylate on postischaemic brain lipid peroxidation and recovery of extracellular calcium in gerbils. Stroke. 1991;22:361–6.
118. Hall ED, Yonkers PA, Andrus PK, Cox JW, Anderson DK. Biochemistry and pharmacology of lipid antioxidants in acute brain and spinal cord injury. J Neurotrauma. 1992;9:425–42.
119. Hall ED, Yonkers PA, Horan KL, Braughler JM. Correlation between attenuation of post-traumatic spinal cord ischaemia and preservation of vitamin E by the 21-aminosteroid U-74006: evidence for an in vivo antioxidant effect. J Neurotrauma. 1989;6:169–76.
120. Braughler JM, Pregenzer JF. The novel 21-aminosteroid inhibitors of lipid peroxidation: reaction with lipid peroxyl and phenoxyl radicals. Free Radic Biol Med. 1989;7:125–30.
121. Hall ED, Yonkers PA, McCall JM, Braughler JM. Effects of the 21-aminosteroid U-74006 on experimental head injury in mice. J Neurosurg. 1988;68:456–61.
122. Raub TJ, Barsuhn CL, Williams LR, Decker DE, Sawada GA, Ho NF. Use of a biophysical-kinetic model to understand the roles of protein binding and membrane partitioning on passive diffusion of highly lipophilic molecules across cellular barriers. J Drug Target. 1993;1:269–86.
123. Narayan RK, Michel ME, Ansell B, Baethmann A, Biegon A, Bracken MB, et al. Clinical trials in head injury. J Neurotrauma. 2002;19:503–57.
124. Empey PE, McNamara PJ, Young B, Rosbolt MB, Hatton J. Cyclosporin A disposition following acute traumatic brain injury. J Neurotrauma. 2006;23:109–16.

Chapter 8
Distinct Roles of Cyclooxygenase-1 and Cyclooxygenase-2 in Inflammatory and Excitotoxic Brain Injury

Saba Aïd, Sang-Ho Choi, Christopher D. Toscano, and Francesca Bosetti

Abstract The cyclooxygenases COX-1 and COX-2 metabolize arachidonic acid to prostaglandins (PGs) and thromboxanes and are thought to play a role in neuroinflammation and excitotoxicity, which are important components in the progression of neurodegenerative diseases. However, the exact role of each isoform in these processes remains unclear. This chapter reviews preclinical and clinical data on COX-1 and COX-2 inhibition in the neuroinflammatory and excitotoxic processes. Potential implications for clinical use in patients suffering from neurodegenerative disorders with a marked inflammatory component, such as Alzheimer's disease, are discussed.

Keywords Cyclooxygenase · Arachidonic acid · Inflammation · Excitotoxicity · Nonsteroidal antiinflammatory drugs · Alzheimer's disease · Prostaglandins

1 Cyclooxygenase-1 and -2 in the Central Nervous System

Cyclooxygenases (COX) catalyze the rate-limiting step in the conversion of arachidonic acid (AA) to prostaglandins and thromboxanes, lipid mediators involved in several physiological and pathological processes [1, 2] (Fig. 1). Inflammation associated with an increased expression of COX and elevated levels of prostaglandins have been implicated in a variety of acute and chronic neurological and neurodegenerative disorders [3–5], including Alzheimer's disease (AD) [6], Parkinson's disease [7], and amyotrophic lateral sclerosis [8].

F. Bosetti (✉)
Brain Physiology and Metabolism Section, National Institute on Aging, National Institutes of Health, Bethesda, MD 20892, USA
e-mail: frances@mail.nih.gov

N. Gadoth and H.H. Göbel (eds.), *Oxidative Stress and Free Radical Damage in Neurology*, Oxidative Stress in Applied Basic Research and Clinical Practice, DOI 10.1007/978-1-60327-514-9_8,

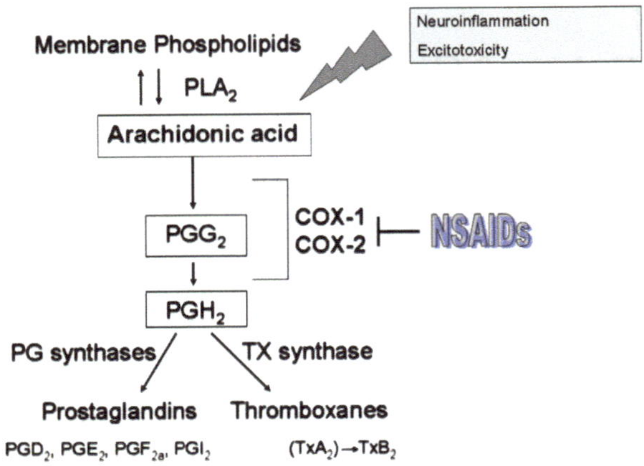

Fig. 1 Brain arachidonic acid cascade. Arachidonic acid is released from membrane phospholipids by a phospholipase A_2 (PLA_2) and then converted to bioactive prostanoids (prostaglandins and thromboxanes) by cyclooxygenases, COX-1 and -2, and terminal prostaglandin and thromboxane synthases. COX-1 and COX-2 are the pharmacological target of nonsteroidal antiinflammatory drugs (NSAIDs). Brain arachidonic acid release and metabolism can be activated by neuroinflammation and excitotoxicity, which are important components in several neurological and neurodegenerative diseases

Two distinct COX isoforms have been characterized, COX-1 and COX-2, which share 60% homology in their amino acid sequences and have comparable kinetics [1, 9]. However, the two isoforms differ in regulatory mechanisms, tissue distribution, and preferential coupling to upstream and downstream enzymes [10–12]. Although relatively little information has been reported about the transcriptional regulation of COX-1, few *cis*-regulatory elements, and no NF-κB binding site and TATA box, have been identified in the 5′-promoter region of the COX-1 gene [13,14]. In contrast to COX-1, the COX-2 gene contains at least six different *cis*-regulatory elements, including NF-κB, Sp1, NF-κB, interleukin (IL)-6, and CRE binding sites, interacting with *trans*-acting factors generated by multiple signaling pathways [13, 14]. The reason why two COX isoforms are necessary in mammals remains unclear; however, each isoform is specifically localized in discrete cell populations distributed in distinct areas of the central nervous system (CNS). Immunohistochemical studies demonstrated that COX-1 is predominantly expressed in glial cells, most of which were devoid [15] of glial fibrillary acidic protein (GFAP) immunoreactivity and thus appeared to be microglia, regardless of their activation state, and enriched in midbrain, pons, and medulla. In contrast, COX-2 is expressed in specific neuronal cells and glial cells, particularly in the limbic system [15–20]. Interestingly, COX-1 immunoreactivity is found in glial cells and some neurons in the brains of wild-type and COX-2-deficient mice but not in those of COX-1-deficient mice [17]. In the CNS, COX-1 has traditionally been considered as the constitutively expressed isoform, primarily responsible for homeostatic prostaglandin synthesis [21, 22]. In contrast, COX-2 expression can be induced by inflammatory or excitotoxic stimuli, and this isoform is thought to be the major source of inflammatory prostaglandins. Thus, COX-2 has been considered as the most appropriate therapeutic target for antiinflammatory drugs. However, recent studies have questioned this earlier perspective and indicated a pro-inflammatory role of COX-1 in the pathophysiology of acute and chronic neurological disorders [23, 24]. On the other hand, as well as being induced by inflammatory stimuli, in the CNS COX-2 is involved in major physiological functions, such as synaptic activity and memory acquisition, neurotransmitter release, blood flow regulation, and the sleep/wake cycle [25–30], which can explain its constitutive expression in several neuronal cell populations, especially in hippocampal and cortical glutamatergic neurons [10, 31, 32]. COX-2 involvement in neuronal activity has been linked to evidence that its upregulation is dependent on *N*-methyl D-aspartate (NMDA) receptor activity [33]and that it contributes to postsynaptic activity and long-term synaptic plasticity [27, 33, 34]. In particular, a recent study demonstrated that PGE_2 resulting from the activity of COX-2, but not from that of COX-1, is necessary for the induction of long-term potentiation and spatial learning in vivo, whereas COX-1 inhibition facilitated baseline synaptic transmission [30]. In this regard, several behavioral studies demonstrated that COX-2 inhibitors cause impairment in memory acquisition and consolidation [35, 36], in passive avoidance memory retention [37, 38], and in spatial memory retrieval [39]. On the other hand, in addition to their homeostatic role, COX-1-derived prostaglandins may also contribute to inflammation in certain pathological conditions. For instance, microglial activation, the hallmark of neuroinflammation, is presumed to be responsible for the increase in expression of COX-1 [18, 23, 32, 40, 41]. In spite of the intense research of the last decade, the exact role of COX-1 and -2 in neuroinflammation is still controversial.

2 Microglia as a Target for Regulation of the Neuroinflammatory Response

Microglia are the primary resident immune cells within the CNS. In normal healthy brain, microglia show a typical "resting" phenotype, but they rapidly react in response to a number of acute and chronic insults [42, 43]. Microglia are particularly sensitive to changes in the microenvironment and readily become activated in response to infection or injury [44, 45]. Microglial activation is generally determined by the gradual changes of their morphology from ramified resting state to activated amoeboid microglia and upregulation of surface molecules such as the major histocompatibility complex [46]. Besides this morphological transformation, activated microglia produce several pro-inflammatory and neurotoxic mediators, including complement, cytokines, chemokines, prostaglandins, and free radicals such as nitric oxide and superoxide [45, 47], which, in turn, may participate in a positive feedback loop to aggravate inflammation and contribute to delayed neuronal cell death [48]. Intervention studies in vivo and in vitro demonstrated that several of these activated microglia-derived factors contribute directly to neuronal injury [24, 49–51] and neurodegeneration [52]. In certain brain insults and neurodegenerative diseases, activated microglia are key cellular mediators of innate immune response and contribute to the neuroinflammatory processes for the progression of diseases [44, 53, 54]. However, whether microglia-mediated neuroinflammation is a consequence or the cause of neurodegeneration is still unclear. Because reactive microglia play a role in the pathogenesis of neurodegenerative diseases, understanding of the molecular mechanisms involved in microglial activation and microglial function is essential for designing rational therapeutic approaches for prevention or treatment of neurodegenerative disorders.

3 The Neuroinflammatory Hypothesis for Alzheimer's Disease

Alzheimer's disease (AD) is the most common neurodegenerative disease and the major cause of aging-related dementia. AD histopathology is characterized by extracellular deposits of β-amyloid peptides (neuritic or senile plaques) and intracellular accumulation of hyperphosphorylated *tau* protein (neurofibrillary tangles). Major neuroinflammatory changes in the AD brain consist of activated microglia and astrocytes surrounding the amyloid deposits and increased expression of several inflammatory mediators, including complement components, the chemokines macrophage inflammatory protein (MIP)-1α and MCP-1, the chemokines receptors CCR3 and CCR5, and the cytokines tumor necrosis factor (TNF)-α and transforming growth factor (TGF)-β [55, 56]. Few studies reported the presence of infiltrating leukocytes and macrophages in the AD brain, suggesting a possible disruption of the blood–brain-barrier [57, 58]. Supporting the neuroinflammatory hypothesis of AD, more than 20 independent epidemiological studies have demonstrated that early long-term administration of nonsteroidal antiinflammatory drugs (NSAIDs), usually to treat osteoarthritis, leads to a significant reduction in the risk of developing AD later in life [59–62].

4 Clinical Trials with COX Inhibitors in AD

Because of the beneficial effects of NSAIDs use on reducing AD risk in epidemiological studies, the idea of treating AD with NSAIDs has been tested in several clinical trials. Two early pilot studies showed promising results on the potential of NSAIDs to ameliorate AD symptoms. A small trial using indomethacin, a preferential COX-1 inhibitor, reported significant improvement in cognitive rating scores in the indomethacin-treated group, despite a 36% rate of withdrawal mainly because of gastric side effects [63]. Another small trial with diclofenac, a COX-2 selective inhibitor, in combination with misoprostol, a gastroprotectant, reported no differences between placebo and treatment groups in an intent-to-treat analysis.

Nevertheless, the authors underscored a nonsignificant trend toward reduced cognitive decline in the diclofenac/misoprostol treatment, despite the 50% withdrawal rate [64]. Unfortunately, subsequent large clinical trials using the nonselective COX inhibitor naproxen [65] and COX-2 selective inhibitors such as rofecoxib, celecoxib, and nimesulide have failed to show any improvement in slowing the progression of AD over 1 year [65–69] or 4 years [70]. Because of rising serious concerns on cardiovascular toxicity of coxibs, a large clinical trial (ADAPT) for the primary prevention of AD to evaluate a nonselective COX inhibitor, naproxen, and a selective COX-2 inhibitor, celecoxib, was halted in 2004 [68]. In the same year, the selective COX-2 inhibitor (Vioxx, rofecoxib) was voluntarily withdrawn from the U.S. market by the manufacturer because of retrospective reports of increased mortality rate resulting from cardiovascular thrombotic events that industry-sponsored trials failed to report on time [71]. The growing concerns about the cardiovascular toxicity of coxibs hindered appropriate follow-up studies to examine the potential therapeutic effect of these drugs in AD. Indeed, the failure of clinical trials to confirm epidemiological data may also be attributed to ineffective dose, inadequate timing (preventive vs. therapeutic approach), or inappropriate choice of NSAIDs (most of the trials focused on COX-2 selective inhibitors). Another explanation could be that targeting COX-2 during established dementia when the inflammatory cascade has already started is not helpful and may even worsen the outcome, as suggested by a recent trial of short-term use of naproxen and celecoxib in patients with mild dementia, in which the treated groups appeared to have a greater rate of AD compared to those given the placebo [68]. Moreover, another trial in patients with mild cognitive impairment found an increase rate of possible or probable AD diagnoses in patients assigned to rofecoxib compared to placebo [70]. In line with those clinical results, preclinical studies testing celecoxib and nimesulide, two selective COX-2 inhibitors, on transgenic mouse models of AD showed either no effect on the pathological burden or a substantial increase in $A\beta_{42}$, which is generally viewed as a key player in the progression of AD [62]. Specifically, celecoxib increased the production of $A\beta_{42}$ at a comparable level to the increases in $A\beta_{42}$ induced by Alzheimer disease-causing mutations [72].

Several recent articles questioned whether the potential efficacy of NSAIDs could be attributed to non-COX-dependent actions. For instance, a subset of NSAIDs known as selective $A\beta_{42}$-lowering agents is responsible for this apparent reduction in AD. In this subset, ibuprofen has been shown to lower β-amyloid in the brain independently of COX activity [73]. However, a recent study that pooled individual-level data from six prospective studies revealed that there was no apparent advantage in AD risk reduction for the subset of NSAIDs shown to selectively lower $A\beta_{42}$ [74]. Another possibility is that protection afforded by NSAIDs is mediated by their ability to activate the peroxisome proliferator-activated receptors (PPARs), which are known to modulate neuroinflammation and β-secretase activity in the brain [75]. Clinical trials to evaluate PPAR-γ agonist on AD risk have yielded promising results [76]. Finally, some NSAIDs have antiplatelet-aggregating effects, which may reduce cognitive decline through vascular mechanisms.

In summary, to date, use of NSAIDs (mostly COX-2 selective inhibitors) has not been shown to be beneficial in patients with mild to severe cognitive decline. However, follow-up trials with COX-1 preferential inhibitors, such as indomethacin, have not been carried out.

5 Pro-Inflammatory Role of COX-1 in Lipopolysaccharide-Induced Neurodegeneration: Pharmacologic and Genetic Studies

Because of its constitutive expression, COX-1 has been traditionally considered as the isoform responsible for the production of prostaglandins relevant to physiological processes, whereas the "inducible" COX-2 has been linked to inflammatory responses. Thus, it was thought that selective inhibition of COX-2 is antiinflammatory/neuroprotective without altering the physiological and homeostatic functions of COX-1, such as those linked to gastroprotection. This hypothesis led to the development and marketing of antiinflammatory drugs that selectively inhibit COX-2 but not COX-1 [77]. However, pharmacologic inhibition or genetic deletion of COX-2 determined only a partial, if any, reduction of prostaglandin levels at the sites of inflammation [31, 78, 79]. Therefore, COX-1 activity can contribute significantly to the total pool of prostaglandins produced in response to inflammation.

To date, the precise role and contribution of COX-1 to the inflammatory cascade in neurodegenerative diseases have not been clearly established. However, COX-1 mRNA expression was selectively upregulated in rat hippocampus during normal aging, possibly increasing the susceptibility of the aging brain to neuroinflammation [80]. To further understand the role of COX-1 in neuroinflammation, intracerebroventricular (i.c.v.) injection of lipopolysaccharide (LPS), a cell-wall component of gram-negative bacteria, was used to activate innate immune response in COX-1-deficient and wild-type mice [24]. LPS administration caused massive resident microglial activation and peripheral leukocyte infiltration into the CNS, which was accompanied by a robust and transient transcriptional activation of genes encoding pro-inflammatory cytokines, chemokines, prostaglandins, and free radical-generating enzymes. However, levels of brain prostaglandins (PGE_2, PGD_2, $PGF_{2\alpha}$, and TXB_2), the expression of enzymes involved in the AA metabolism, oxidative damage, and neuronal death in response to LPS were significantly attenuated in COX-1-deficient mice [24]. Mice pretreated with the COX-1-specific inhibitor SC-560 also showed reduced levels of prostaglandins and inflammatory cytokines and chemokines after LPS compared to wild-type mice [24]. Although one could speculate that the reduced neuroinflammatory response to LPS may be caused by a decreased COX-2 expression and PGE_2 production, a recent study indicated that COX-2-deficient mice, in contrast, showed increased glial activation and inflammatory markers after LPS injection compared to wild-type mice [51]. Supporting these observations, selective inhibition of COX-1 in LPS-stimulated human and rat microglia was very effective in reducing PGE_2 secretion [81, 82]. Although one could speculate that the reduced neuroinflammatory response to LPS may be caused by a decreased COX-2 expression and PGE_2 production, another study indicated that COX-2-deficient mice, in contrast, showed increased glial activation and inflammatory markers after LPS injection compared to wild-type mice [51]. Importantly, i.c.v. LPS did not induce mPGES-1 or COX-2 in neurons [83], supporting the idea that PGE_2 production occurred mainly in nonneuronal cells such as microglia. Moreover, it is possible that augmented PGE_2 generation might act in an autocrine and paracrine manner to further potentiate microglial function. COX-1-mediated inflammatory responses and neuronal injury are likely the consequence of downstream effects of one or more prostaglandins that affect cellular changes through activation of specific prostaglandin receptors and second messenger systems. Based on current data, it is difficult to determine whether the altered prostaglandin production is initiator for all subsequent phenomena. Further studies are needed to elucidate how COX-1-dependent prostaglandins could alter inflammatory responses.

Microglial activation coupled with peripheral leukocyte infiltration is a prominent feature of the complex cellular immune responses that lead to inflammation and oxidative stress and subsequently promote tissue damage. The disruption of the blood–brain barrier (BBB) is a key player in this event. Supporting this concept, administration of LPS increased BBB permeability in different animal models [84–86], and LPS-induced NO and PGE_2 production had a temporal association with BBB permeability alteration and leukocyte infiltration in the cerebrospinal fluid (CSF) [87]. Selective pharmacologic inhibition of COX-1, but not COX-2, significantly reduced TNF-α-induced BBB disruption and free radical production [88]. Overall, these findings indicate that COX-1 seems to be more important than COX-2 in mediating alteration of BBB permeability and migration of peripheral leukocytes. This notion is also supported by evidence that pretreatment with the COX-2 selective inhibitor rofecoxib was ineffective in preventing the disruption of tight junction proteins and BBB integrity in mice [89], and that indomethacin, a COX-1 preferential inhibitor, in contrast completely blocked LPS-induced permeability changes in cultured monolayers of brain microvessel endothelial cells [90]. Overall, these data suggest that COX-1 is an important player in the neuroinflammatory response and subsequent neuronal damage.

6 COX-2 May Mediate Neuroprotection During Neuroinflammation

In experimental models of brain inflammation, COX-2 genetic deletion or pharmacologic inhibition has been proven detrimental to neuronal survival. Specifically, pretreatment with COX-2 inhibitors or genetic deletion of COX-2 has been shown to increase seizure activity in response to kainate [91, 92] and to exacerbate endotoxin-induced ocular inflammation [93]. The expression of markers of inflammation and oxidative stress, such as cytokines (IL-1β, TNF-α), chemokines (MIP-1α), and NADPH oxidase subunit $p67^{phox}$ were upregulated after centrally injected LPS in mice genetically deficient in COX-2 or pretreated for 6 weeks with celecoxib, a selective COX-2 inhibitor [51]. Transcription of inflammatory genes (mPGES-1, TLR2, CD14, MCP-1) was reported to be increased also after the administration of a single dose of another selective COX-2 inhibitor, NS-398, in vascular-associated brain cells and parenchymal microglia in response to a systemic injection of LPS [94]. The inhibition of COX-2 activity has also been shown to exacerbate neuronal cell death in an experimental model associated with astrocyte and microglia activation such as LPS-induced neuroinflammation or thiamine deficiency-induced Wernicke's encephalopathy [51, 52, 95]. Thus, the protective or deleterious role of COX-2 in the brain is dependent on whether the stimulus is targeting primarily microglia (neuroinflammation) or neurons (MPTP- or ischemia-induced neurodegeneration) and also on the cellular localization of COX-2. In this respect, overexpression of COX-2 in neurons in transgenic mice did not modify the neuroinflammatory response to LPS (Aid et al., 2010).

COX-2 selective inhibition seems to modulate expression of tight junction proteins and vascular markers in vitro [96]. Quantitative magnetic resonance imaging showed that LPS-induced BBB disruption was increased in COX-2-deficient mice compared to wild-type mice, and that disruption was associated with an increased activity of matrix metalloproteinase-9 and -3, known to mediate BBB breakdown (Aid et al., 2010). Overall, these results suggest that glial or endovascular COX-2, but not neuronal COX-2, may account for a protective effect during acute neuroinflammation. It should be noted that acute neuroinflammation triggers COX-2 upregulation without any change in COX-1 expression [51, 52, 95], suggesting that the effects observed are not mediated by a compensatory induction of COX-1.

The mechanisms underlying the neuroprotective role of COX-2 during acute neuroinflammation remain elusive. Specific products derived from COX-2, but not from COX-1, activity may account for the neuroprotective role of COX-2 during activation of innate immunity in the brain. PGE_2 is the most studied COX-derived prostaglandin because of its known upregulation after COX-2 induction. Because prostaglandin PGE_2 acts via four distinct E-prostanoid (EP) receptors that are expressed on multiple cell types in the nervous system, opposite actions have been reported in the literature depending on the type of insult, the type of EP receptors involved, and the cell environment [97]. PGE_2, as well as other prostanoids, can modulate both pro- and antiinflammatory responses [98]. For instance, EP2 and EP4 receptors have been shown to differentially modulate TNF-α and IL-6 production induced by LPS in mouse peritoneal neutrophils [99], and indirect neuronal damage by CD14-dependent innate immune activation is completely suppressed in mice lacking the EP2 receptor [97].

Novel endogenous eicosanoids with antiinflammatory and pro-resolving actions were recently characterized [100], which unraveled the traditional view of COX-2 products as promoting only pro-inflammatory reaction. First, Gilroy and colleagues in 1999 showed that COX-2 inhibition, by reducing the production of PGE_2 and PGD_2, delays the onset of resolution in the carrageenan-induced pleurisy model [101], highlighting the importance of COX-2 in both the initiation and the resolution of inflammation. COX-2-derived PGE_2 and PGD_2 actively promoted the transcription of enzymes required for the production of arachidonic acid-derived antiinflammatory and pro-resolution lipid mediators, such as lipoxins, resolvins, and protectins. These new mediators have been shown to act at multiple levels by reducing neutrophil infiltration into the exudate, increasing monocyte recruitment to the exudate, stimulating macrophage uptake of apoptotic neutrophils, and promoting phagocyte exit from the exudate via the lymphatics [102]. Another specific role of COX-2, but not COX-1, is linked to the metabolism of endocannabinoids, anandamide, and 2-arachidonoylglycerol [103]. Although the number of studies on this topic is still limited, evidence suggests that the endocannabinoid system controls immune responses in the brain via multiple cellular and molecular targets and influences neurogenesis [104]. Clearly, further studies on COX-2-specific products during pro-inflammatory and pro-resolving phases are warranted to better understand the unique role of each COX isoform during acute, as well as chronic, neuroinflammation.

7 Distinct Roles of COX-1 and COX-2 in Excitotoxic Brain Injury

There are conflicting views about the role of COX-2 in neurodegeneration and neurotoxicity [105]. Despite the established role of COX-2 in the inflammatory process, COX-2 abrogation does not change the expression of inflammatory and oxidative markers after 1-methyl 4-phenyl 1,2,3,6-tetrahydropyridine (MPTP) [79], a neurotoxin that damages dopaminergic neurons in the substantia nigra and is used to model Parkinson's disease in rodents. Therefore, in injury models in which the insults directly damage the neurons, such as MPTP [79], quisquallic acid-induced damage [79, 106], and centrally injected NMDA-induced neurotoxicity and ischemic brain injury [107, 108], COX-2 inhibition appears to afford protection without mitigating inflammation.

On the other hand, studies attempting to investigate the role of COX isoforms in excitotoxicity by using nonsteroidal antiinflammatory drugs (NSAIDs) showed conflicting data, with COX-2 selective inhibition linked with either attenuation or potentiation of kainic acid (KA)-induced excitotoxic damage [109–113]. Systemic exposure to either KA or NMDA results in the generation of limbic seizures that progress to neuronal damage in the cerebral cortex, hippocampus, and other limbic structures in the brain because of activation of glutamatergic KA or NMDA receptors [114]. The exact mechanism of this excitotoxic neuronal damage has not been completely characterized, but a prevailing theory suggests that the over-excitation of the neurons allows significant concentrations of intracellular calcium to build, resulting in activation of calcium-dependent proteases [115]. KA causes an upregulation of $cPLA_2$ and COX-2 expression in hippocampal neurons [116] that is accompanied by increased production of prostaglandins [117]. To eliminate the confounding factor of timing of administration, the response to a single peripheral injection of a low dose of KA (10 mg/kg) was examined in COX-1- and COX-2-deficient mice and their respective wild types [91]. Although there was no difference in seizure intensity or neuronal damage between COX-1-deficient mice and their wild types, COX-2-deficient mice showed a significant increase in seizure intensity and neuronal damage in the hippocampal fields CA1/CA2 and CA3 after KA administration [91]. Chronic treatment with the COX-2 selective inhibitor celecoxib (6,000 ppm, through the diet) before KA also significantly increased seizure intensity and neuronal damage [91]. COX-2-deficient mice demonstrated increased seizure intensity and neuronal damage also in response to peripheral administration of NMDA, but not of lindane, which causes excitotoxicity by enhancing GABA disinhibition (Toscano et al., 2008), suggesting that COX-2-deficient mice are selectively vulnerable to directly acting glutamatergic agonists.

Although the exact mechanism responsible for the increased susceptibility of COX-2-deficient mice to glutamatergic agonists remains to be fully elucidated, a microarray study demonstrated that the expression of genes involved in GABAergic neurotransmission is altered in the brain of COX-2-deficient mice, suggesting a decrease in inhibitory GABAergic function [118]. Supporting this concept, the frequency of GABA inhibitory postsynaptic currents in CA1 pyramidal neurons was decreased in COX-2 deficient mice, indicating a decreased release of GABA by inhibitory GABAergic interneurons [91]. Because of a decreased GABAergic inhibitory modulation in response of excitotoxic stimuli, COX-2-deficient mice exhibit increased seizure activity compared to their wild-type controls (Fig. 2). These data also indicate that COX-2, but not COX-1, plays a key role in modulating neuronal excitability.

8 Summary and Conclusions

As shown in Fig. 3, because of its preferential localization in glial cells, particularly in microglia, which are the major responders to inflammatory stimuli, COX-1 may be playing a more important role than COX-2 in modulating the neuroinflammatory response. On the other hand, being predominantly expressed in neurons, COX-2 plays a major role in modulating neuronal excitability and in gating seizure intensity in response to excitotoxic stimuli.

Several recent studies identified unique and sometimes opposite roles of each COX isoform in neuroinflammation and excitotoxicity. Although the actual mechanisms by which COX-1 is involved in neuroinflammation and neurodegeneration remain to be elucidated, that the inhibition and ablation of COX-1 significantly attenuate the inflammatory response and neuronal loss suggests that COX-1 plays a key role in the neuroinflammatory cascade. Because neuroinflammation is a critical component of neurodegenerative diseases, a full understanding of the physiological and pathological role of COX-1 activity may help to develop better

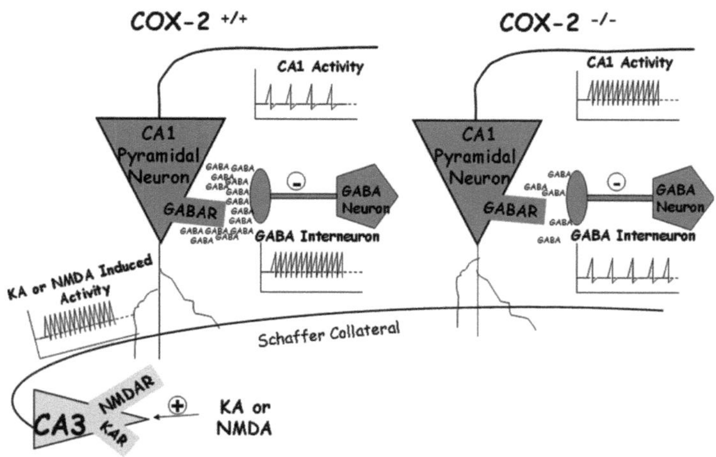

Fig. 2 Deficits in GABAergic neurotransmission in cyclooxygenase-2-deficient mice may contribute to increased seizure intensity. Action potentials from the CA3 pyramidal neurons travel down the Schaffer collaterals to impinge upon the apical dendrites of the CA1 pyramidal neurons. The frequency of these action potentials increase after kainate or NMDA stimulates the kainate receptors (KAR) or *N*-methyl D-aspartate (NMDA) receptors (NMDAR), respectively. This elevated frequency of stimulation of the CA1 apical dendrites results in depolarization of the CA1 pyramidal neurons and can increase the frequency of action potentials evoked in these cells. Under normal conditions, as seen in COX-$2^{+/+}$ mice, GABA interneurons can buffer this increase in activity by increasing the frequency of GABA release and therefore increase the stimulation of the GABA receptor (GABAR), a chloride channel on the CA1 pyramidal neurons that allows for hyperpolarization. However, in COX-$2^{-/-}$ mice the frequency of GABA interneuron activity is decreased by more than 50%. This reduction in GABA neurotransmission would be expected to increase the overall action potential firing rate of CA1 pyramidal neurons in response to increased apical dendrite stimulation. This diminished ability to buffer excitability by GABA interneurons may result in increased susceptibility of COX-2-deficient mice to excitotoxin-induced seizures

therapeutic strategies for the prevention or treatment of neurodegenerative disorders. Additionally, larger clinical trials with COX-1 preferential inhibitors, such as indomethacin [63], should be further explored.

On the other hand, preclinical data indicate a neuroprotective role of COX-2 in neuroinflammation and excitotoxicity. This new unrecognized role of COX-2 may in part explain the failure of clinical trials with selective COX-2 inhibitors in slowing the progression of AD disease. The potential beneficial effects of COX-2 in the phase of resolution and return to tissue homeostasis after an inflammatory event need to be considered when developing treatment strategies aimed at reducing neuroinflammation. Because of accumulating evidence of a COX-2 neuroprotective role, the usefulness of COX-2 inhibitors in treating AD, and other neurodegenerative diseases with an inflammatory component, such as Parkinson's disease and amyotrophic lateral sclerosis, may become questionable [26].

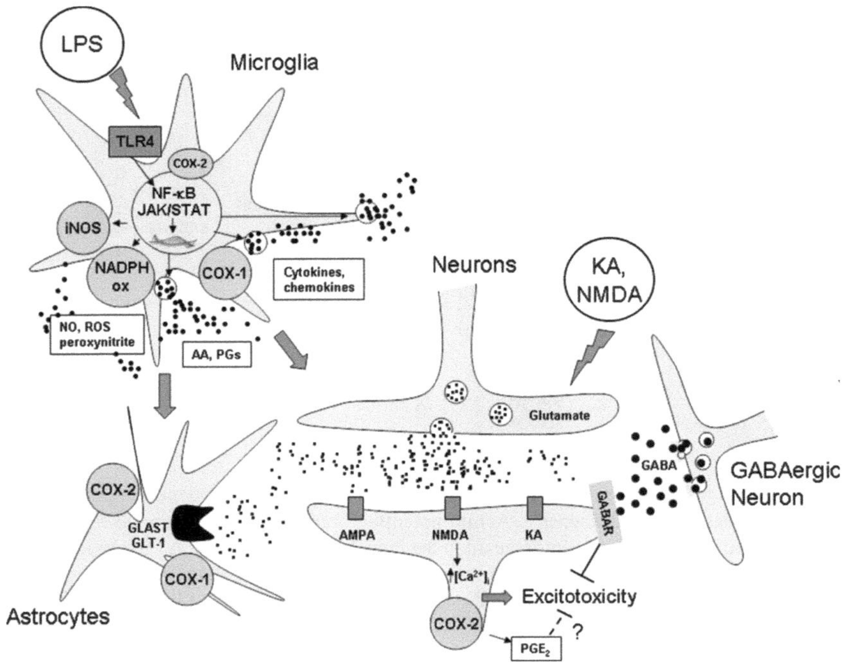

Fig. 3 Cellular distribution of COX-1 and COX-2 and their different roles in neuroinflammation and excitotoxicity. Microglia are key players in initiating brain inflammatory response. Through toll-like receptors 4 (TLR4), lipopolysaccharide (LPS) initiates the neuroinflammatory cascade by transcriptional activation of proinflammatory genes. Activated microglia produce oxygen and nitrogen reactive species (NO and ROS), arachidonic acid and prostaglandins (AA and PGs), and cytokines and chemokines, that propagate the inflammatory cascade to astrocytes and neurons. Because of its predominant localization on microglial cells, inhibition of COX-1 can significantly attenuate the inflammatory response. On the other hand, because of its predominant localization in postsynaptic neurons, COX-2 is the isoform that plays a major role in excitotoxic damage by kainic acid and *N*-methyl D-aspartate (NMDA). One possibility is that COX-2-derived products modulate the release of GABA by GABAergic interneurons

References

1. Smith WL, DeWitt DL, Garavito RM. Cyclooxygenases: structural, cellular, and molecular biology. Annu Rev Biochem. 2000;69:145–82.
2. Bazan NG, Colangelo V, Lukiw WJ. Prostaglandins and other lipid mediators in Alzheimer's disease. Prostaglandins Other Lipid Mediat. 2002;68-69:197–210.
3. Candelario-Jalil E, Fiebich BL. Cyclooxygenase inhibition in ischemic brain injury. Curr Pharm Des. 2008;14:1401–18.
4. Hoozemans JJ, Rozemuller JM, van Haastert ES, Veerhuis R, Eikelenboom P. Cyclooxygenase-1 and -2 in the different stages of Alzheimer's disease pathology. Curr Pharm Des. 2008;14:1419–27.
5. Asanuma M, Miyazaki I. Nonsteroidal anti-inflammatory drugs in experimental parkinsonian models and Parkinson's disease. Curr Pharm Des. 2008;14:1428–34.

6. Wenk GL. Neuropathologic changes in Alzheimer's disease. J Clin Psychiatry. 2003;64 Suppl. 9:7–10.
7. Hunot S, Hirsch E.C. Neuroinflammatory processes in Parkinson's disease. Ann Neurol. 2003;53 Suppl. 3:S49–58; discussion S-60.
8. Mhatre M, Floyd RA, Hensley K. Oxidative stress and neuroinflammation in Alzheimer's disease and amyotrophic lateral sclerosis: common links and potential therapeutic targets. J Alzheimers Dis. 2004;6:147–57.
9. Morita I, Schindler M, Regier MK, et al. Different intracellular locations for prostaglandin endoperoxide H synthase-1 and -2. J Biol Chem. 1995;270:10902–8.
10. Breder CD. Cyclooxygenase systems in the mammalian brain. Ann N Y Acad Sci. 1997;813:296–301.
11. Murakami M, Kudo I. Recent advances in molecular biology and physiology of the prostaglandin E_2-biosynthetic pathway. Prog Lipid Res. 2004;43:3–35.
12. Ueno N, Takegoshi Y, Kamei D, Kudo I, Murakami M. Coupling between cyclooxygenases and terminal prostanoid synthases. Biochem Biophys Res Commun. 2005;338:70–6.
13. Morita I. Distinct functions of COX-1 and COX-2. Prostaglandins Other Lipid Mediat. 2002;68–69:165–75.
14. Kang YJ, Mbonye UR, DeLong CJ, Wada M, Smith WL. Regulation of intracellular cyclooxygenase levels by gene transcription and protein degradation. Prog Lipid Res. 2007;46:108–25.
15. Tomimoto H, Shibata M, Ihara M, Akiguchi I, Ohtani R, Budka H. A comparative study on the expression of cyclooxygenase and 5-lipoxygenase during cerebral ischemia in humans. Acta Neuropathol. 2002;104:601–7.
16. Breder CD, Dewitt D, Kraig RP. Characterization of inducible cyclooxygenase in rat brain. J Comp Neurol. 1995;355:296–315.
17. Li S, Wang Y, Matsumura K, Ballou LR, Morham SG, Blatteis CM. The febrile response to lipopolysaccharide is blocked in cyclooxygenase-2(–/–), but not in cyclooxygenase-1(–/–) mice. Brain Res. 1999;825:86–94.
18. Yermakova AV, Rollins J, Callahan LM, Rogers J, O'Banion MK. Cyclooxygenase-1 in human Alzheimer and control brain: quantitative analysis of expression by microglia and CA3 hippocampal neurons. J Neuropathol Exp Neurol. 1999;58:1135–46.
19. Deininger MH, Meyermann R, Trautmann K, et al. Cyclooxygenase (COX)-1 expressing macrophages/microglial cells and COX-2 expressing astrocytes accumulate during oligodendroglioma progression. Brain Res. 2000;885:111–6.
20. Kroin JS, Takatori M, Li J, Chen EY, Buvanendran A, Tuman KJ. Upregulation of dorsal horn microglial cyclooxygenase-1 and neuronal cyclooxygenase-2 after thoracic deep muscle incisions in the rat. Anesth Analg. 2008;106:1288–95, table of contents.
21. Kaufmann WE, Andreasson KI, Isakson PC, Worley PF. Cyclooxygenases and the central nervous system. Prostaglandins. 1997;54:601–24.
22. Phillis JW, Horrocks LA, Farooqui AA. Cyclooxygenases, lipoxygenases, and epoxygenases in CNS: their role and involvement in neurological disorders. Brain Res Brain Res Rev. 2006;52:201–43.
23. Schwab JM, Beschorner R, Meyermann R, Gozalan F, Schluesener HJ. Persistent accumulation of cyclooxygenase-1-expressing microglial cells and macrophages and transient upregulation by endothelium in human brain injury. J Neurosurg. 2002;96:892–9.
24. Choi SH, Langenbach R, Bosetti F. Genetic deletion or pharmacological inhibition of cyclooxygenase-1 attenuate lipopolysaccharide-induced inflammatory response and brain injury. FASEB J. 2008;22:1491–501.
25. Hayaishi O, Matsumura H. Prostaglandins and sleep. Adv Neuroimmunol. 1995;5:211–6.
26. Minghetti L. Role of COX-2 in inflammatory and degenerative brain diseases. Subcell Biochem. 2007;42:127–41.
27. Kaufmann WE, Worley PF, Pegg J, Bremer M, Isakson P. COX-2, a synaptically induced enzyme, is expressed by excitatory neurons at postsynaptic sites in rat cerebral cortex. Proc Natl Acad Sci USA. 1996;93:2317–21.

28. Stefanovic B, Bosetti F, Silva AC. Modulatory role of cyclooxygenase-2 in cerebrovascular coupling. Neuroimage. 2006;32:23–32.
29. Ojeda SR, Urbanski HF, Junier MP, Capdevila J. The role of arachidonic acid and its metabolites in the release of neuropeptides. Ann N Y Acad Sci. 1989;559:192–207.
30. Cowley TR, Fahey B, O'Mara SM. COX-2, but not COX-1, activity is necessary for the induction of perforant path long-term potentiation and spatial learning in vivo. Eur J Neurosci. 2008.
31. Seibert K, Zhang Y, Leahy K, et al. Pharmacological and biochemical demonstration of the role of cyclooxygenase 2 in inflammation and pain. Proc Natl Acad Sci USA. 1994;91:12013–7.
32. Yasojima K, Schwab C, McGeer EG, McGeer PL. Distribution of cyclooxygenase-1 and cyclooxygenase-2 mRNAs and proteins in human brain and peripheral organs. Brain Res. 1999;830:226–36.
33. Yamagata K, Andreasson KI, Kaufmann WE, Barnes CA, Worley PF. Expression of a mitogen-inducible cyclooxygenase in brain neurons: regulation by synaptic activity and glucocorticoids. Neuron. 1993;11:371–86.
34. Chen C, Magee JC, Bazan NG. Cyclooxygenase-2 regulates prostaglandin E_2 signaling in hippocampal long-term synaptic plasticity. J Neurophysiol. 2002;87:2851–7.
35. Rall JM, Mach SA, Dash PK. Intrahippocampal infusion of a cyclooxygenase-2 inhibitor attenuates memory acquisition in rats. Brain Res. 2003;968:273–6.
36. Teather LA, Packard MG, Bazan NG. Post-training cyclooxygenase-2 (COX-2) inhibition impairs memory consolidation. Learn Mem. 2002;9:41–7.
37. Sato T, Ishida T, Irifune M, et al. Effect of NC-1900, an active fragment analog of arginine vasopressin, and inhibitors of arachidonic acid metabolism on performance of a passive avoidance task in mice. Eur J Pharmacol. 2007;560:36–41.
38. Holscher C. Inhibitors of cyclooxygenases produce amnesia for a passive avoidance task in the chick. Eur J Neurosci. 1995;7:1360–5.
39. Sharifzadeh M, Tavasoli M, Soodi M, Mohammadi-Eraghi S, Ghahremani MH, Roghani A. A time course analysis of cyclooxygenase-2 suggests a role in spatial memory retrieval in rats. Neurosci Res. 2006;54:171–9.
40. Candelario-Jalil E, Gonzalez-Falcon A, Garcia-Cabrera M, et al. Assessment of the relative contribution of COX-1 and COX-2 isoforms to ischemia-induced oxidative damage and neurodegeneration following transient global cerebral ischemia. J Neurochem. 2003;86:545–55.
41. Pepicelli O, Fedele E, Berardi M, et al. Cyclo-oxygenase-1 and -2 differently contribute to prostaglandin E2 synthesis and lipid peroxidation after in vivo activation of *N*-methyl-D-aspartate receptors in rat hippocampus. J Neurochem. 2005;93:1561–7.
42. Bruce-Keller AJ. Microglial-neuronal interactions in synaptic damage and recovery. J Neurosci Res. 1999;58:191–201.
43. Streit WJ. Microglia as neuroprotective, immunocompetent cells of the CNS. Glia. 2002;40:133–9.
44. Kreutzberg GW. Microglia: a sensor for pathological events in the CNS. Trends Neurosci. 1996;19:312–8.
45. Nelson PT, Soma LA, Lavi E. Microglia in diseases of the central nervous system. Ann Med. 2002;34:491–500.
46. Stence N, Waite M, Dailey ME. Dynamics of microglial activation: a confocal time-lapse analysis in hippocampal slices. Glia. 2001;33:256–66.
47. Block ML, Zecca L, Hong JS. Microglia-mediated neurotoxicity: uncovering the molecular mechanisms. Nat Rev Neurosci. 2007;8:57–69.
48. Hata AN, Breyer RM. Pharmacology and signaling of prostaglandin receptors: multiple roles in inflammation and immune modulation. Pharmacol Ther. 2004;103:147–66.

49. Liberatore GT, Jackson-Lewis V, Vukosavic S, et al. Inducible nitric oxide synthase stimulates dopaminergic neurodegeneration in the MPTP model of Parkinson disease. Nat Med. 1999;5:1403–9.
50. Dehmer T, Lindenau J, Haid S, Dichgans J, Schulz JB. Deficiency of inducible nitric oxide synthase protects against MPTP toxicity in vivo. J Neurochem. 2000;74:2213–6.
51. Aid S, Langenbach R, Bosetti F. Neuroinflammatory response to lipopolysaccharide is exucerbated in mice genetically deficient in cyclooxygenase-2. J Neuroinflammation. 2008;5:17.
52. Xie Z, Wei M, Morgan TE, et al. Peroxynitrite mediates neurotoxicity of amyloid beta-peptide1–42- and lipopolysaccharide-activated microglia. J Neurosci. 2002;22:3484–92.
53. McGeer PL, Itagaki S, Tago H, McGeer EG. Reactive microglia in patients with senile dementia of the Alzheimer type are positive for the histocompatibility glycoprotein HLA-DR. Neurosci Lett. 1987;79:195–200.
54. Streit WJ, Mrak RE, Griffin WS. Microglia and neuroinflammation: a pathological perspective. J Neuroinflammation. 2004;1:14.
55. McGeer PL, McGeer EG. Inflammation and the degenerative diseases of aging. Ann N Y Acad Sci. 2004;1035:104–16.
56. Heneka MT, O'Banion MK. Inflammatory processes in Alzheimer's disease. J Neuroimmunol. 2007;184:69–91.
57. Fiala M, Liu QN, Sayre J, et al. Cyclooxygenase-2-positive macrophages infiltrate the Alzheimer's disease brain and damage the blood-brain barrier. Eur J Clin Invest. 2002;32:360–71.
58. Matsumoto Y, Yanase D, Noguchi-Shinohara M, Ono K, Yoshita M, Yamada M. Blood-brain barrier permeability correlates with medial temporal lobe atrophy but not with amyloid-beta protein transport across the blood-brain barrier in Alzheimer's disease. Dement Geriatr Cogn Disord. 2007;23:241–5.
59. McGeer PL, Schulzer M, McGeer EG. Arthritis and anti-inflammatory agents as possible protective factors for Alzheimer's disease: a review of 17 epidemiologic studies. Neurology. 1996;47:425–32.
60. Breitner JC, Welsh KA, Helms MJ, et al. Delayed onset of Alzheimer's disease with nonsteroidal anti-inflammatory and histamine H2 blocking drugs. Neurobiol Aging. 1995;16:523–30.
61. Rich JB, Rasmusson DX, Folstein MF, Carson KA, Kawas C, Brandt J. Nonsteroidal anti-inflammatory drugs in Alzheimer's disease. Neurology. 1995;45:51–5.
62. McGeer PL, McGeer EG. NSAIDs and Alzheimer disease: epidemiological, animal model and clinical studies. Neurobiol Aging. 2007;28:639–47.
63. Rogers J, Kirby LC, Hempelman SR, et al. Clinical trial of indomethacin in Alzheimer's disease. Neurology. 1993;43:1609–11.
64. Scharf S, Mander A, Ugoni A, Vajda F, Christophidis N. A double-blind, placebo-controlled trial of diclofenac/misoprostol in Alzheimer's disease. Neurology. 1999;53:197–201.
65. Aisen PS, Schafer KA, Grundman M, et al. Effects of rofecoxib or naproxen vs placebo on Alzheimer disease progression: a randomized controlled trial. JAMA. 2003;289:2819–26.
66. Aisen PS, Schmeidler J, Pasinetti GM. Randomized pilot study of nimesulide treatment in Alzheimer's disease. Neurology. 2002;58:1050–4.
67. Reines SA, Block GA, Morris JC, et al. Rofecoxib: no effect on Alzheimer's disease in a 1-year, randomized, blinded, controlled study. Neurology. 2004;62:66–71.
68. Group AR, Lyketsos CG, Breitner JC, et al. Naproxen and celecoxib do not prevent AD in early results from a randomized controlled trial. Neurology. 2007;68:1800–8.
69. Soininen H, West C, Robbins J, Niculescu L. Long-term efficacy and safety of celecoxib in Alzheimer's disease. Dement Geriatr Cogn Disord. 2007;23:8–21.
70. Thal LJ, Ferris SH, Kirby L, et al. A randomized, double-blind, study of rofecoxib in patients with mild cognitive impairment. Neuropsychopharmacology. 2005;30:1204–15.

71. Psaty BM, Kronmal RA. Reporting mortality findings in trials of rofecoxib for Alzheimer disease or cognitive impairment: a case study based on documents from rofecoxib litigation. JAMA. 2008;299:1813–7.
72. Kukar T, Murphy MP, Eriksen JL, et al. Diverse compounds mimic Alzheimer disease-causing mutations by augmenting Abeta42 production. Nat Med. 2005;11:545–50.
73. Weggen S, Eriksen JL, Das P, et al. A subset of NSAIDs lower amyloidogenic Abeta42 independently of cyclooxygenase activity. Nature. 2001;414:212–6.
74. Szekely CA, Green RC, Breitner JC, et al. No advantage of A beta 42-lowering NSAIDs for prevention of Alzheimer dementia in six pooled cohort studies. Neurology. 2008;70:2291–8.
75. Sastre M, Dewachter I, Rossner S, et al. Nonsteroidal anti-inflammatory drugs repress beta-secretase gene promoter activity by the activation of PPARgamma. Proc Natl Acad Sci USA. 2006;103:443–8.
76. Heneka MT, Landreth GE. PPARs in the brain. Biochim Biophys Acta. 2007;1771:1031–45.
77. Teismann P, Ferger B. Inhibition of the cyclooxygenase isoenzymes COX-1 and COX-2 provide neuroprotection in the MPTP-mouse model of Parkinson's disease. Synapse. 2001;39:167–74.
78. Portanova JP, Zhang Y, Anderson GD, et al. Selective neutralization of prostaglandin E_2 blocks inflammation, hyperalgesia, and interleukin 6 production in vivo. J Exp Med. 1996;184:883–91.
79. Teismann P, Tieu K, Choi DK, et al. Cyclooxygenase-2 is instrumental in Parkinson's disease neurodegeneration. Proc Natl Acad Sci USA. 2003;100:5473–8.
80. Aid S, Bosetti F. Gene expression of cyclooxygenase-1 and Ca(2+)-independent phospholipase A(2) is altered in rat hippocampus during normal aging. Brain Res Bull. 2007;73:108–13.
81. Hoozemans JJ, Veerhuis R, Janssen I, van Elk EJ, Rozemuller AJ, Eikelenboom P. The role of cyclo-oxygenase 1 and 2 activity in prostaglandin E(2) secretion by cultured human adult microglia: implications for Alzheimer's disease. Brain Res. 2002;951:218–26.
82. Candelario-Jalil E, de Oliveira AC, Graf S, et al. Resveratrol potently reduces prostaglandin E_2 production and free radical formation in lipopolysaccharide-activated primary rat microglia. J Neuroinflammation. 2007;4:25.
83. Xia Y, Yamagata K, Krukoff TL. Differential expression of the CD14/TLR4 complex and inflammatory signaling molecules following i.c.v. administration of LPS. Brain Res. 2006;1095:85–95.
84. Veszelka S, Urbanyi Z, Pazmany T, et al. Human serum amyloid P component attenuates the bacterial lipopolysaccharide-induced increase in blood-brain barrier permeability in mice. Neurosci Lett. 2003;352:57–60.
85. Tomas-Camardiel M, Venero JL, Herrera AJ, et al. Blood–brain barrier disruption highly induces aquaporin-4 mRNA and protein in perivascular and parenchymal astrocytes: protective effect by estradiol treatment in ovariectomized animals. J Neurosci Res. 2005;80:235–46.
86. Rosenberg GA, Estrada EY, Mobashery S. Effect of synthetic matrix metalloproteinase inhibitors on lipopolysaccharide-induced blood-brain barrier opening in rodents: differences in response based on strains and solvents. Brain Res. 2007;1133:186–92.
87. Jaworowicz DJ Jr, Korytko PJ, Singh Lakhman S, Boje KM. Nitric oxide and prostaglandin E_2 formation parallels blood-brain barrier disruption in an experimental rat model of bacterial meningitis. Brain Res Bull. 1998;46:541–6.
88. Candelario-Jalil E, Taheri S, Yang Y, et al. Cyclooxygenase inhibition limits blood-brain barrier disruption following intracerebral injection of tumor necrosis factor-alpha in the rat. J Pharmacol Exp Ther. 2007;323:488–98.
89. Pu H, Hayashi K, Andras IE, Eum SY, Hennig B, Toborek M. Limited role of COX-2 in HIV Tat-induced alterations of tight junction protein expression and disruption of the blood-brain barrier. Brain Res. 2007;1184:333–44.

90. de Vries HE, Blom-Roosemalen MC, van Oosten M, et al. The influence of cytokines on the integrity of the blood-brain barrier in vitro. J Neuroimmunol. 1996;64:37–43.
91. Toscano CD, Ueda Y, Tomita YA, Vicini S, Bosetti F. Altered GABAergic neurotransmission is associated with increased kainate-induced seizure in prostaglandin-endoperoxide synthase-2 deficient mice. Brain Res Bull. 2008;75:598–609.
92. Baik EJ, Kim EJ, Lee SH, Moon C. Cyclooxygenase-2 selective inhibitors aggravate kainic acid induced seizure and neuronal cell death in the hippocampus. Brain Res. 1999;843:118–29.
93. Tuo J, Tuaillon N, Shen D, Chan CC. Endotoxin-induced uveitis in cyclooxygenase-2-deficient mice. Invest Ophthalmol Vis Sci. 2004;45:2306–13.
94. Blais V, Turrin NP, Rivest S. Cyclooxygenase 2 (COX-2) inhibition increases the inflammatory response in the brain during systemic immune stimuli. J Neurochem. 2005;95:1563–74.
95. Gu B, Desjardins P, Butterworth RF. Selective increase of neuronal cyclooxygenase-2 (COX-2) expression in vulnerable brain regions of rats with experimental Wernicke's encephalopathy: effect of nimesulide. Metab Brain Dis. 2008.
96. Germann B, Neuhaus W, Hofer-Warbinek R, Noe CR. Applying blood–brain barrier in vitro models to study the influence of drugs on endothelial cells: effects of selected COX-inhibitors. Pharmazie. 2008;63:303–7.
97. Shie FS, Montine KS, Breyer RM, Montine TJ. Microglial EP2 is critical to neurotoxicity from activated cerebral innate immunity. Glia. 2005;52:70–7.
98. Sugimoto Y, Narumiya S. Prostaglandin E receptors. J Biol Chem. 2007;282:11613–7.
99. Yamane H, Sugimoto Y, Tanaka S, Ichikawa A. Prostaglandin E(2) receptors, EP2 and EP4, differentially modulate TNF-alpha and IL-6 production induced by lipopolysaccharide in mouse peritoneal neutrophils. Biochem Biophys Res Commun. 2000;278:224–8.
100. Serhan CN, Chiang N. Endogenous pro-resolving and anti-inflammatory lipid mediators: a new pharmacologic genus. Br J Pharmacol. 2008;153(:uppl 1):S200–15.
101. Gilroy DW, Colville-Nash PR, Willis D, Chivers J, Paul-Clark MJ, Willoughby DA. Inducible cyclooxygenase may have anti-inflammatory properties. Nat Med. 1999;5:698–701.
102. Serhan CN, Chiang N, Van Dyke TE. Resolving inflammation: dual anti-inflammatory and pro-resolution lipid mediators. Nat Rev Immunol. 2008;8:349–61.
103. Kozak KR, Crews BC, Morrow JD, et al. Metabolism of the endocannabinoids, 2-arachidonylglycerol and anandamide, into prostaglandin, thromboxane, and prostacyclin glycerol esters and ethanolamides. J Biol Chem. 2002;277:44877–85.
104. Wolf SA, Ullrich O. Endocannabinoids and the brain immune system: new neurones at the horizon? J Neuroendocrinol. 2008;20 Suppl. 1:15–9.
105. Bosetti F. Arachidonic acid metabolism in brain physiology and pathology: lessons from genetically altered mouse models. J Neurochem. 2007.
106. Scali C, Prosperi C, Vannucchi MG, Pepeu G, Casamenti F. Brain inflammatory reaction in an animal model of neuronal degeneration and its modulation by an anti-inflammatory drug: implication in Alzheimer's disease. Eur J Neurosci. 2000;12:1900–12.
107. Iadecola C, Niwa K, Nogawa S, et al. Reduced susceptibility to ischemic brain injury and *N*-methyl-D-aspartate-mediated neurotoxicity in cyclooxygenase-2-deficient mice. Proc Natl Acad Sci USA. 2001;98:1294–9.
108. Hewett SJ, Silakova JM, Hewett JA. Oral treatment with rofecoxib reduces hippocampal excitotoxic neurodegeneration. J Pharmacol Exp Ther. 2006;319:1219–24.
109. Candelario-Jalil E, Ajamieh HH, Sam S, Martinez G, Leon Fernandez OS. Nimesulide limits kainate-induced oxidative damage in the rat hippocampus. Eur J Pharmacol. 2000; 390:295–8.
110. Kunz T, Oliw EH. Nimesulide aggravates kainic acid-induced seizures in the rat. Pharmacol Toxicol. 2001;88:271–6.
111. Kunz T, Oliw EH. The selective cyclooxygenase-2 inhibitor rofecoxib reduces kainate-induced cell death in the rat hippocampus. Eur J Neurosci. 2001;13:569–75.

112. Ciceri P, Zhang Y, Shaffer AF, et al. Pharmacology of celecoxib in rat brain after kainate administration. J Pharmacol Exp Ther. 2002;302:846–52.
113. Gobbo OL, O'Mara SM. Post-treatment, but not pre-treatment, with the selective cyclooxygenase-2 inhibitor celecoxib markedly enhances functional recovery from kainic acid-induced neurodegeneration. Neuroscience. 2004;125:317–27.
114. Wang Q, Yu S, Simonyi A, Sun GY, Sun AY. Kainic acid-mediated excitotoxicity as a model for neurodegeneration. Mol Neurobiol. 2005;31:3–16.
115. Sattler R, Tymianski M. Molecular mechanisms of glutamate receptor-mediated excitotoxic neuronal cell death. Mol Neurobiol. 2001;24:107–29.
116. Tu B, Bazan NG. Hippocampal kindling epileptogenesis upregulates neuronal cyclooxygenase-2 expression in neocortex. Exp Neurol. 2003;179:167–75.
117. Yoshikawa K, Kita Y, Kishimoto K, Shimizu T. Profiling of eicosanoid production in the rat hippocampus during kainate-induced seizure: dual-phase regulation and differential involvement of cox-1 and cox-2. J Biol Chem. 2006.
118. Toscano CD, Prabhu VV, Langenbach R, Becker KG, Bosetti F. Differential gene expression patterns in cyclooxygenase-1 and cyclooxygenase-2 deficient mouse brain. Genome Biol. 2007;8:R14.
119. Aid S, Parikh N, Palumbo S, Bosetti F. Neuronal overexpression of cyclooxygenase-2 does not alter the neuroinflammatory response during brain innate immune activation. Neurosci Lett. 2010;478(3): 113–8.
120. Aid S, Silva AC, Candelario-Jail E, Choi SH, Rosenberg GA, Bosetti F. Cyclooxygenase-1 and -2 differentially modulate lipoplysaccharide-induced blood-brain barrier disruption through matrix metalloproteinase activity. J Cereb Blood Flow Metab. 2010;30(2):370–80.
121. Toscano CD, Kingsley PJ, Marnett LJ, Bosetti F. NMDA-induced seizure indensity is enhanced in COX-2 deficient mice. Neurotoxicology. 2008;29(6):1114–20.

Chapter 9
Stroke and Oxidative Stress

Arianna Vignini

Abstract Stroke is a heterogeneous syndrome caused by multiple disease mechanisms, resulting in a disruption of cerebral blood flow with subsequent tissue damage. Stroke remains the third leading cause of death and the leading cause of disability in adults. In 25–40% of patients with ischemic stroke, neurological symptoms progress during the initial hours. Early clinical deterioration results in increased mortality and functional disability. The molecular mechanisms underlying early clinical worsening are still not well clarified. There is considerable evidence that reactive oxygen species (ROS), reactive nitrogen species (RNS), and oxidative state are important mediators to tissue injury in cerebral ischemia. In normal conditions, free radical levels are controlled by cellular endogenous antioxidants. In fact, oxidative stress occurs when there is an impairment or inability to balance antioxidant production with ROS and RNS levels. The brain is highly susceptible to oxidative stress because of its high consumption of body oxygen (~20%) to produce energy and free radicals, which in turn can cause damage to the main components of cells (DNA, lipids, proteins). In this chapter we focus our attention on the evidence of oxidative stress in ischemic stroke; particular attention is given to the current knowledge about the biomarkers of oxidative stress that can possibly be used to monitor the severity and outcome of stroke.

Keywords Stroke · Free radicals · Reactive nitrogen species · Reactive oxygen species

A. Vignini (✉)
Institute of Biochemistry, Marche Polytechnic University,
Via P. Ranieri 65, 60131 Ancona, Italy
e-mail: vignini2@libero.it

N. Gadoth and H.H. Göbel (eds.), *Oxidative Stress and Free Radical Damage in Neurology*, Oxidative Stress in Applied Basic Research and Clinical Practice,
DOI 10.1007/978-1-60327-514-9_9,

1 Introduction

Stroke is a heterogeneous syndrome caused by multiple disease mechanisms, resulting in a disruption of cerebral blood flow with subsequent tissue damage. The past decade has witnessed tremendous achievements in the ability to diagnose stroke, but its treatment is still unsatisfactory [1]. Stroke remains the third leading cause of death, particularly in the elderly. The mortality rate of stroke in the acute phase is as high as 20% [2], and it remains higher for several years after the acute event in stroke patients than in the general population [3]. In 25–40% of patients with ischemic stroke, neurological symptoms progress during the initial hours [4]. Early clinical deterioration results in increased mortality and functional disability [5]; up to 40% of survivors are not expected to recover independence with self-care and 25% become unable to walk independently. Because life expectancy is continuing to increase, the absolute number of individuals with stroke will further increase in the near future. However, the molecular mechanisms underlying early clinical worsening are still not well clarified. A complex cascade of molecular events is set in motion during cerebral ischemia and culminates in neuronal cell death. Improving our understanding of these events might help devise novel therapies to limit neuronal injury in stroke patients, a concept termed "neuroprotection."

Ischemic stroke accounts for about 75% of all cases whereas hemorrhagic stroke is responsible for almost 15% of all.strokes. It has also been estimated that up to 30% of all ischemic strokes will eventually undergo hemorrhagic transformation [6].

Oxidative stress is defined as the condition occurring when the physiological balance between oxidants and antioxidants is disrupted in favor of the former with potential damage for the organism [7]. Free radicals play an essential role in maintaining the physiological condition of the body. However, oxidative stress that is induced by an excess accumulation of the reactive oxygen and nitrogen species can damage basic components for cell function and survival.

In normal conditions, oxidant levels are controlled by cellular endogenous antioxidants, glutathione peroxidase, catalase and superoxide dismutase (SOD) [8]. A free radical is a very reactive atom with an unpaired electron, which can exist in a reduced or oxidized state. The majority of free radicals that damage biological systems are oxygen radicals and other reactive oxygen species (ROS), the main by-products formed in cells of aerobic organisms. The amount of free radical production is determined by the balance of many different factors, whereas the source of ROS formation is mainly constituted by mitochondria during electron transport in the oxidative phosphorylation chain.

An increased production of free radicals and other chemical species has been demonstrated in both ischemic and hemorrhagic stroke, and oxidative stress is proposed as a fundamental mechanism of brain damage in these conditions [9]. Similarly, ROS are important mediators to tissue injury in cerebral ischemia [10]. The brain is particularly sensitive to oxidative stress because its increased ROS levels promote free radical damage, mainly because of tissue lipid content, high oxygen consumption rate, and chemical reactions involving dopamine oxidation and glutamate [11]. Each of these effects can be found in almost all tissues, even if they can coexist in the same tissue.

2 Oxidative Stress in Ischemic and Hemorrhagic Stroke

Although ischemic and hemorrhagic strokes have different risk factors and pathophysiological mechanisms, there is evidence of an increased generation of free radicals and other reactive species in both conditions, leading to oxidative stress [12].

Ischemic stroke is the consequence of the interruption or severe reduction of blood flow in cerebral arteries. According to the degree of hypoperfusion, it is possible to identify an area with complete absence of flow, namely the *core*, where neuronal death occurs within a few minutes, and a surrounding area, called the *penumbra*, which suffers from a moderate reduction of blood flow and contains functionally impaired but still viable brain tissue. If blood flow is not restored within a relatively short time, the penumbra area will share the same fate as that of the core region. On the other hand, reperfusion can save the brain tissue but might potentially have negative consequences: upon reoxygenation, oxidative stress is rapidly built up and numerous nonenzymatic oxidation reactions take place in the cytosol and in cellular organelles [13].

From a theoretical point of view, the nervous system is particularly vulnerable to the deleterious effects of ROS; independent of the mechanism of ROS production, when their amount is significantly increased there is a high risk of neuronal damage as a result of the high susceptibility of the brain to oxidative stress. First, although the brain represents only 2% of body weight, it consumes about 20% of body oxygen to produce energy; second, the brain contains relatively high concentrations of readily peroxidizable lipids, such as polyunsaturated fatty acids (PUFAs), which represent one-third of brain fatty acids, and particularly those with many double bonds such as arachidonic acid (C20:4 n-6) and docosohexaenoic acid (C22:6 n-3), which are highly susceptible to lipid peroxidation; and third, the brain is relatively poorly endowed with protective antioxidant enzymes or substances with lower activity of glutathione peroxidase (GPx) and catalase (Cat) compared to other organs. Finally, many regions of the brain are highly enriched in iron [14] (Fig. 1).

Independent of the mechanisms responsible for ischemic stroke, ischemia causes a cascade of events that can increase free radical production via several different pathways.

Because free radicals are extremely reactive, they have a short half-life. Therefore, they are difficult to measure directly. Many studies have pursued an indirect approach to demonstrate free radical production during cerebral ischemia, measuring the products of free radical reaction with other molecules, such as DNA, lipids, or proteins, or measuring the levels of antioxidant, according to the hypothesis that changes in their level are the consequence of their reaction with radicals.

During ischemia and reperfusion, the altered function of the mitochondrial electron transport chain is a likely source of oxidative stress [15]. Under these

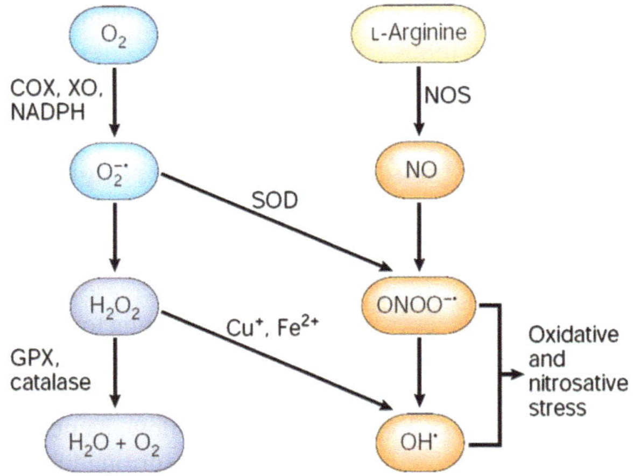

Fig. 1 Interactions between pathways that generate oxygen and nitrogen radicals. Combination of superoxide (O_2^-) and nitric oxide (NO) generates the potent anion, peroxynitrite ($ONOO^-$). Metal (Cu^+ and Fe^{2+})-catalyzed pathways can also produce the hydroxyl radical ($OH^\bullet$) from hydrogen peroxide (H_2O_2). COX cyclooxygenase; NADPH nicotinamide adenine dinucleotide phosphate; GPX glutathione peroxidase; NOS nitric oxide synthase; SOD superoxide dismutase; XO xanthine oxidase (from Lo et al. [79])

conditions there might be an increased formation of superoxide radical anions, as suggested by experimental studies. On one hand, knockout mice for mitochondrial superoxide dismutase (mSOD) genes suffer from larger brain lesions after focal ischemia [16], and, on the other hand, transgenic mice overexpressing this enzyme present with a smaller infarct size [17]. Moreover, the dismutation of the superoxide radical anion leads to the formation of hydrogen peroxide, which in turn can be the precursor of the reactive hydroxyl radical (OHI).

Morimoto et al. found an increased production of OHI beginning during the ischemic phase in the penumbra and further increasing during reperfusion in experimental transient focal ischemia [18]. Consistent with these results are those of Peters et al., who directly showed an increase in superoxide radical anions in the penumbral region not only during the initial ischemic phase but also after recirculation [19].

The production of free radicals is induced and sustained also by the modulation of general protein synthesis occurring after ischemia; in fact, the process is able to activate the transcription of proteins that are likely to play a damaging role, such as inducible nitric oxide synthase (iNOS) [20, 21]. Moreover, pro-inflammatory genes are expressed and several inflammatory mediators are released; this eventually can lead to the accumulation of blood-borne inflammatory cells, such as neutrophils and monocytes/macrophages, that might further promote oxidative stress.

Concerning hemorrhagic stroke, several mechanisms have been suggested to contribute to increased free radical generation, although they have been less extensively investigated than in ischemic stroke. The rupture of blood vessels following hemorrhagic stroke immediately results in accumulation of blood in the extravascular space, which in turn causes direct tissue destruction and compression of the surrounding tissue by mass effect from the hematoma and development of edema. These events increase intracranial pressure [22] and impair both cerebral perfusion and venous drainage, thereby potentially causing a condition of cerebral ischemia, although this is currently controversial [23]. Moreover, delayed damage can occur as a consequence of the release of vasoactive and toxic substances from extravasated blood breakdown products, of thrombin, and leukocyte infiltration [24].

The inflammatory response is likely to contribute to injury in cerebral hemorrhage [25]. Iron has a major role in free radical generation after brain hemorrhage. Brain iron is usually contained in heme enzymes or bound to ferritin. Large amounts of iron become free after intracerebral hemorrhage as breakdown products of hemoglobin, released as a consequence of red blood cells lysis [26]. Free iron is able to greatly enhance OHI production via the Fenton reaction [27].

Finally, subarachnoid hemorrhage also is associated with increased oxidative stress that might be directly involved in the pathogenesis of its more severe complication, that is, cerebral vasospasm [28].

2.1 *Stroke and Na^+/K^+-ATPase*

The main change affecting neurons during ischemia is the exhaustion of the high-energy phosphate compound adenosine triphosphate (ATP) by the lack of the substrates for its production, that is, oxygen and glucose; these depend on glycolysis and oxidative phosphorylation for energy production and maintaining ionic gradient [29]. In fact, following energy depletion, within 2 min from the onset of a stroke, neurons and glia lose membrane potential because of Na^+/K^+-ATPase pump failure [30]. The Na^+/K^+-ATPase is an integral membrane-bound protein that plays a key role in cellular osmotic regulation through the maintenance of the trans-membrane gradients of Na^+ and K^+. It is also responsible for about one-third of the total intracellular energy generation [31].

The energy failure causes membrane depolarization, and this impairment in turn compromises transmembranous ionic gradients, causing an influx of extracellular Ca^{2+} through voltage-sensitive Ca^{2+} channels and uncontrolled release of excitatory amino acids, such as glutamate, in the extracellular space [20]. Such excessive release of glutamate is also responsible for the further increase in the concentration of Ca^{2+} in the cytoplasm [32, 33]. The final result of ischemia is intracellular Ca^{2+} accumulation from multiple sources inducing the activation of a variety of Ca^{2+}-stimulated enzymes, such as proteases, lipases, nucleases, protein kinases, and nitric oxide synthase (NOS).

During the ischemic phase and early reperfusion, some Ca^{2+}-dependent enzymes produce oxygen free radicals while neuronal nitric oxide synthase (nNOS) generates nitric oxide (NO). Moreover, Na^+/K^+-ATPase is also a marker of membrane function. Being an integral membrane protein, it greatly depends for its activity on the chemicophysical properties of the microenvironment where it is embedded. Moreover, in association with membrane fluidity, it represents an index of membrane functionality, being involved in the modulation of phospholipid and protein interactions [34].

Recently, it was reported that a decrease in platelet Na^+/K^+-ATPase activity was found in patients with acute stroke as compared to controls. Moreover, data regarding platelet membrane fluidity suggest that it is significantly lower in stroke patients than in control subjects, with a significant negative correlation between platelet Na^+/K^+-ATPase activity and anisotropy [31].

Because platelet membrane fluidity, strictly related to microviscosity, is a critical determinant of platelet aggregation and secretion, its decrease in stroke patients led the authors to hypothesize its role in the pathogenesis of ischemic damage. Furthermore, the study shows a decrease in the platelet Na^+/K^+-ATPase activity in patients that might result from subconformational changes in the protein depending on its microenvironment. As the pump represents the key enzyme of cellular osmotic regulation, its inhibition leads to altered internal and external membrane ionic concentrations [31]. This unbalanced ionic presence causes derangement in cellular enzymatic systems, leading to membrane depolarization, which in turn could compromise platelet survival.

2.2 Nitric Oxide and Stroke

NO is an omnipresent intercellular messenger in all vertebrates, modulating blood flow, thrombosis, and neural activity. The biological production of NO is also important for nonspecific host defense, but NO itself is unlikely to directly kill intracellular pathogens and tumors. NO is a colorless gas with pleiotropic effects in different tissues, mediating a wide range of physiological and pathophysiological processes; it is synthesized by NO synthases through the enzymatic oxidation of the guanidino group of L-arginine [35].

The brain proved to be a rich source of NO synthesis and was the source of the first NO synthase (NOS) to be cloned and purified. Thus, it is called nNOS (neuronal NOS) or NOS1. The second NOS to be cloned was isolated from macrophages and known as NOS2 or iNOS (inducible NOS) because it is readily induced in many tissues by proinflammatory cytokines. In contrast to the neuronal or endothelial NOS isoenzymes, NOS2 is not regulated by changes in intracellular calcium. However, careful examination of purified preparations of iNOS revealed that calmodulin was so tightly bound that it copurified with iNOS. Consequently, the enzyme appeared to be unresponsive to physiologically relevant changes in calcium concentrations. Because NOS2 can be strongly induced by proinflammatory stimuli, it is often called a high-output source of NO.

The first source of NO identified, endothelial NO synthase, was the last to be cloned and is known as eNOS or NOS3. NOS3 binds to plasma membranes and is typically associated with caveolin. It is strongly activated by the entry of calcium through membrane-bound receptors and is also regulated by phosphorylation. NOS3 is also found in neurons and other tissues in addition to endothelium (Fig. 2) [36].

NO is often considered to be just another signaling molecule. However, it is important to consider how NO communicates information to understand why NO

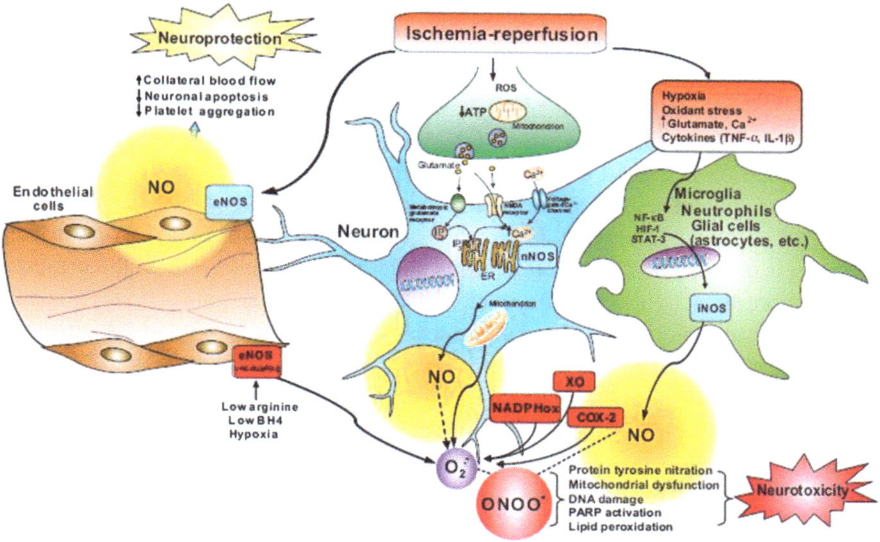

Fig. 2 Roles of NO and peroxynitrite in the pathophysiology of stroke. Brain ischemia and reperfusion lead to transient stimulation of the activity of endothelial NO synthase (eNOS), resulting in brief increases in endothelial NO generation, associated with neuroprotective actions in stroke. In parallel, ischemic energy depletion and oxidant (ROS) production trigger the release of glutamate, which results in neuronal calcium overload from extracellular (activation of calcium channels) and intracellular (phosphoinositol-3-kinase-endoplasmic reticulum signaling) sources. Calcium overload results in prolonged synthesis of NO, caused by stimulated activity of the neuronal isoform of NO synthase (nNOS). Enhanced NO generation also depends on the induced expression of inducible NOS (iNOS) in various types of reactive inflammatory cells, and upon the activation of several cell signaling pathways in response to hypoxia, cytokines, oxidants, and glutamate. During the same period of time, superoxide production is enhanced by uncoupling of eNOS, mitochondrial dysfunction, and the stimulated activity of NADPH oxidase, xanthine oxidase, and cyclooxygenase-2 (COX-2). Formation of peroxynitrite is then markedly favored, damaging lipids, proteins, and DNA and triggering the activation of poly(ADP-ribose) polymerase (PARP), which all contribute significantly to neurotoxicity in stroke (from Pacher et al. [36])

has so many physiological roles in vivo. The production of cyclic guanosine monophosphate (cGMP) by guanylate cyclase is the major signal transduction mechanism of NO. NO can diffuse from the site where it is synthesized into surrounding cells, where it will activate soluble guanylate cyclase in the target tissue to produce cGMP. In turn, cGMP activates cGMP-dependent kinases in the target tissue, which modulates intracellular calcium levels to influence many diverse activities in the target tissues. NO is a small hydrophobic molecule that crosses cell membranes without the need for channels or receptors as readily as molecular oxygen and carbon dioxide. Because NO is freely permeable through membranes, NO will repeatedly diffuse in as well as out of a cell over the time span of a second. Because the hydrophobicity of NO will allow slightly faster diffusion in a lipid membrane than in water, membranes do not provide an effective barrier to NO [36].

NO is a well-known vasorelaxant agent, but it also plays an important role as a neurotransmitter when produced by neurons; endothelium-derived NO is an important endogenous mediator of cerebral blood flow and cerebrovascular protection [37, 38] whose production is increased after ischemia [39, 40]. Moreover, it is also involved in defense functions when it is produced by immune and glial cells.

Inflammation-activated glia has been detected in several diseases of the central nervous system and has shown the ability to kill neuronal cells in the vicinity via the release of cytotoxic mediators. Glia, when activated, can express iNOS, producing high levels of NO, which can kill neurons under certain conditions. In the presence of inflammatory activation of glia, even low levels of NO can induce neuronal death under hypoxic conditions [41]. This process is mediated by glutamate after NO inhibition of respiration in competition with oxygen, involving also ROS and proinflammatory cytokines.

NO is thermodynamically unstable and tends to react with other molecules, resulting in the oxidation, nitrosylation, or nitration of proteins, thus affecting many cellular mechanisms. When NO is produced in excessive amounts, it changes from a physiological neuromodulator to a neurotoxic factor.

Several studies showed that NO can have both protective and damaging roles, depending on the cell type producing it and on the phase of ischemia. This deeply debated dual role of NO in cerebral ischemia (i.e., neurotoxic or neuroprotective) exists in experimental stroke models [39, 42]. In fact, in the early phase of ischemic damage, NO synthesized from eNOS may protect the ischemic brain by acutely increasing blood flow whereas during the postischemic period NO produced by both nNOS and iNOS may facilitate cell death [43].

During cerebral ischemia, high concentrations of NO are generated by nNOS and by the activation of iNOS in macrophages and other cell types. The current knowledge is that nNOS-generated NO may reach toxic levels in focal ischemia and particularly during reperfusion, thereby contributing to neuronal death. NO also participates in inflammatory and cytotoxic actions that contribute to neuronal death [39, 44]. It has also been observed that NO overproduction can be either the result of nNOS activation following persistent stimulation of excitatory amino acid receptors and/or iNOS induction by different stimuli, such as endotoxin or cytokines [45].

Finally, NO might exacerbate ischemic damage by inhibiting the mitochondrial respiratory chain, although only at very high concentrations [39], and by releasing iron from ferritin [27]. NO overproduction may combine with superoxide anion (O_2^-) to produce peroxynitrite ($ONOO^-$), which is involved in cellular dysfunction [46]. Peroxynitrite is formed when the two free radicals react in a near diffusion-limited reaction [47]. Under physiological conditions, the production of peroxynitrite will be low and oxidative damage minimized by endogenous antioxidant defenses.

Consequently, pathological conditions can greatly increase the production of peroxynitrite. Even the generation of a moderate flux of peroxynitrite over long periods of time will result in substantial oxidation and potential destruction of host cellular constituents, leading to the dysfunction of critical cellular processes, disruption of cell signaling pathways, and the induction of cell death through both apoptosis and necrosis [48]. Hence, the production of peroxynitrite can be instrumental in the development of many pathological processes in vivo. The half-life of peroxynitrite is short (~10–20 ms) but it is sufficient to cross biological membranes, diffuse one to two cell diameters [49], and allow significant interactions with most critical biomolecules [50].

$ONOO^-$ can further be protonated, forming the highly reactive peroxynitrous acid. Peroxynitrous acid can directly increase brain damage, reacting with several important biological molecules, or can be converted into OHI and other ROS, as well as reactive nitrogen species (RNS) [51].

Once formed in the ischemic brain, peroxynitrite may exert its toxic effects through multiple mechanisms, including lipid peroxidation, mitochondrial damage, protein nitration, and oxidation, depletion of antioxidant reserves (especially glutathione), and DNA damage followed by the activation of the nuclear enzyme poly (ADP-ribose) polymerase (PARP), as summarized in Fig. 1.

Evidence for the generation of peroxynitrite from NO and superoxide in stroke has been obtained by the concomitant demonstration that (1) nitrotyrosine rapidly accumulates in the brain following transient or permanent ischemia and (2) the nitrotyrosine accumulation is markedly prevented by strategies such as blocking NO or superoxide generation, as well as by treatments aimed at directly scavenging peroxynitrite [52].

Finally, in a recent paper, a significant decrease in plasma NO concentration in acute stroke patients was reported as compared to controls; this might most likely be caused by the increased expression of iNOS by the effect of a thrombotic attack. Moreover, the same authors found a significant increase of peroxynitrite concentration in patients with cerebral ischemia compared to controls as a consequence of the reaction of NO with O_2^- and a higher content of nitrotyrosine, which is a marker of nitrated tyrosines in proteins as compared to controls [53].

2.3 *Lipid Peroxidation and Stroke*

Lipid peroxidation is one of the major consequences of free radical-mediated injury to the brain. The peroxidation of polyunsaturated fatty acids, mostly in membrane phospholipids, is a chain reaction that can continue until substrate is completely consumed or termination occurs by antioxidants. Lipid peroxidation produces structural and functional damage to membranes as well as several secondary products (Fig. 3).

Membrane damage derives from the generation of fragmented fatty acid chains, lipid–lipid cross-links, endocyclization to produce novel fatty acid esters, and lipid–protein cross-links, which causes changes in fluidity and permeability of the cell membranes. Consequently, the activity of membrane-bound enzymes, the binding of molecules to receptors, cellular interactions, nutrient transport, and the function of second messenger systems will be impaired [54]. Secondary products of lipid peroxidation are either fragments of the original molecules or isoprostanes, when arachidonic acid is involved [55].

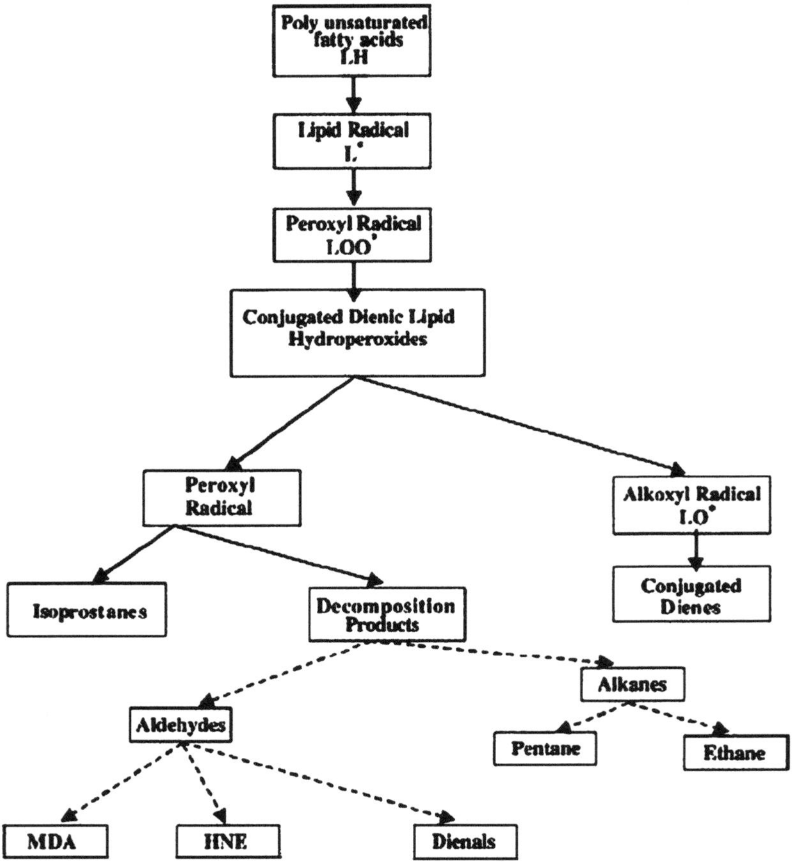

Fig. 3 The products and pathways relating to lipid peroxidation. *LH* polyunsaturated fatty acids; *L** lipid radical; *LOO** peroxyl radical; *LO** alkoxyl radical; *MDA* malondialdehyde; *HNA* hydroxynonenal (from Dotan et al. [80])

Lipid peroxidation products are a widely accepted group of oxidative stress indices. Initially, lipid peroxidation leads to the production of conjugated dienic hydroperoxides. These unstable substances decompose either into various aldehydes, such as malondialdehyde (MDA), 4-hydroxynonenal (HNE), and dienals, or into alkanes as pentane and ethane. Several studies have demonstrated that MDA levels are higher in acute stroke patients than in controls [56, 57].

The results concerning MDA level changes over time in stroke patients are controversial: some researchers observed higher erythrocyte MDA levels in the very early phase of stroke with subsequent decline to the levels of controls [57], while others observed an MDA increase only some days after acute stroke [58].

In a recent paper, it has been demonstrated that the modifications of lipid and lipoprotein levels in plasma of stroke patients are associated with an increase in lipid hydroperoxides and with a decrease in the activity of paraoxonase (PON1) [59], an esterase/lactonase associated to HDL, that exerts an antiinflammatory role and a protective effect against atherosclerosis and oxidative damage of lipoproteins and cells [59, 60]. The authors suggest that the lower PON1 activity in patients could result from the presence of circulating inhibitors, such as lipid peroxidation products. This hypothesis is supported by the negative correlation established between PON1 activity and levels of lipid hydroperoxides in the plasma of controls. The correlation confirms that subjects with lower PON1 activity are more exposed to oxidative damage compared to subjects with higher PON1 activity, in agreement with previous studies [61, 62]. The same trend was reported by Gariballa et al., who found that thiobarbituric acid-reactive substances (TBARs) levels were constantly higher in acute ischemic stroke patients compared with controls [63].

Moreover, oxidized low density lipoproteins (oxLDL) have also been used as biomarkers of lipid peroxidation in humans, because their source might be oxidized phospholipids released from brain tissue into circulation. Uno et al. demonstrated that patients with cerebral infarction, but not those with cerebral hemorrhage, had higher oxLDL concentrations compared to controls, reaching the highest peak on the third day after stroke and then rapidly declining to control levels. The levels were particularly high in those patients who suffered from cortical ischemic infarcts. It has been suggested that the source of oxLDL might be oxidized phospholipids released from brain tissue into circulation [64]. Higher oxLDL levels can be found in the brain of patients who died a few days after ischemic stroke but not in those who had a stroke more than 1 month before death [65].

Oxidative stress and impairment of the antioxidant system in the plasma of patients may play a role in ischemic stroke. The decrease in antioxidant molecules and/or enzymes could permit further brain tissue damage.

2.4 DNA Oxidation and Stroke

The damage of nuclear DNA during cerebral ischemia may be caused by two different mechanisms: oxidative modification and endonuclease-mediated DNA fragmentation [66]. Although DNA degradation would take place in the late stage of cell death, oxidative DNA damage is likely to represent an early event, as suggested by experimental studies in animal models of ischemic stroke [66–68].

Several different types of oxidative DNA damage have been shown, but DNA base lesions, particularly 8-hydroxy-2-deoxy-guanosine (8-OHDG), are the most commonly studied biomarker for oxidative DNA damage because of high specificity and relative abundance in DNA. In particular, plasma levels of 8-OHDG were found to be increased in an animal model of ischemic stroke, with a significant association with the brain content of 8-OHDG [68]. Regarding oxidative damage to DNA in humans, plasma levels of 8-OHDG were found to be increased after ischemic stroke, with a significant association with brain content of 8-OHDG [67].

2.5 *Protein Oxidation and Stroke*

With respect to protein oxidation, there is a lack of studies assessing the presence and levels of biomarkers of ROS damage against proteins and amino acids in human ischemic or hemorrhagic stroke. There is only one study on a small number of stroke patients, which showed that they did not differ from controls in the level in protein carbonyls [69].

Increased brain protein oxidation, however, has been found in Alzheimer's disease as well as in other neurodegenerative disorders [70]. In an animal study, 3-nitrotyrosine levels increased both in the core and in the surrounding penumbra during ischemia, with further increment during reperfusion only in the latter [71].

Moreover, protein carbonyls were increased in a porcine model of intracerebral hemorrhage [72]. Polidori et al. have recently showed that the protein carbonyl and the dityrosine contents of immunoglobulin G are similarly increased in patients with Alzheimer's disease and vascular dementia [73]. The choice of evaluating carbonyl and dityrosine content in immunoglobulin G rather than total oxidized proteins is justified by several considerations: circulating immunoglobulins are the second most prevalent serum protein; albumin carries lipids, thereby rendering difficult the discrimination between primary and secondary oxidation; furthermore, the carbonyl content of immunoglobulin G has been previously shown to be sensitive to dietary antioxidant supplementation [74]; and finally, immunoglobulins have a half-life of 15 days, making them a good short-term marker of oxidative stress.

3 Conclusion

Stroke is the main cause of disability and mortality in Western countries. Ischemic stroke accounts for about 75% of all cases whereas hemorrhagic stroke is responsible for almost 15% of all strokes. It has also been estimated that up to 30% of all ischemic strokes will eventually undergo hemorrhagic transformation. Brain ischemia, and especially the condition of ischemia and reperfusion occurring after stroke, has been shown to be associated with free radical-mediated reactions potentially leading to neuronal death [12].

Extensive evidence from experimental studies supports a role for free radical generation and oxidative injury in the pathogenesis of stroke. The large increase in glutamate following ischemia may contribute to free radical generation by excitotoxic mechanisms [18, 75].

Plasma levels of 8-OHDG were found to be increased in an animal model of ischemic stroke, with a significant association with brain content of 8-OHDG [68]. Regarding oxidative damage to DNA in humans, plasma levels of 8-OHDG were found to be increased after ischemic stroke, with a significant association with brain content of 8-OHDG [67]. Evidence for lipid peroxidation in cerebral ischemia comes from a large number of studies showing increased TBARS and fluorescent lipid peroxidation products in brain and peripheral tissues after ischemia [56, 76, 77]. A correlation has been observed between MDA levels and infarct size, clinical stroke severity, and patient outcome [57, 78]. oxLDL have also been used as biomarkers of lipid peroxidation in humans as their source might be oxidized phospholipids released from brain tissue into circulation. With respect to protein oxidation, there is a lack of studies assessing the presence and levels of biomarkers of oxidative damage against proteins and amino acids in ischemic or hemorrhagic stroke in humans. One study did not find any difference in protein carbonyls between a small cohort of stroke patients and controls [69].

In conclusion, the interrelationship between stroke and oxidative stress, even if thoroughly studied, needs further investigations to clarify the molecular mechanisms at the basis of this pathological status.

Acknowledgments I thank Prof. Laura Mazzanti for being my teacher in life and science; Profs. Leandro Provinciali and Mauro Silvestrini for their clinical support; Drs. Laura Nanetti, Francesca Raffaelli, Cinzia Moroni, and Monica D'Angelo for being my lab mates; and "Fondazione Cavaliere del Lavoro Mario Magnetto" for supporting our research in stroke.

References

1. Caplan LR. Treatment of acute stroke: still struggling. JAMA. 2004;292(15):1883–5.
2. Sarti C, Rastenyte D, Cepaitis Z, Tuomilehto J. International trends in mortality from stroke, 1968 to 1994. Stroke. 2000;31(7):1588–601.
3. Ayala C, Greenlund KJ, Croft JB, Keenan NL, Donehoo RS, Giles WH, Kittner SJ, Marks JS. Racial/ethnic disparities in mortality by stroke subtype in the United States, 1995–1998. Am J Epidemiol. 2001;154(11):1057–63.
4. Toni D, Fiorelli M, Gentile M, Bastianello S, Sacchetti ML, Argentino C, Pozzilli C, Fieschi C. Progressing neurological deficit secondary to acute ischemic stroke: a study on predictability, pathogenesis, and prognosis. Arch Neurol. 1995;52(7):670–5.
5. Davalos A, Cendra E, Teruel J, Martinez M, Genis D. Deteriorating ischemic stroke: risk factor and prognosis. Neurology. 1990;40(12):1865–9.
6. Lyden PD, Zivin JA. Hemorrhagic transformation after cerebral ischemia: mechanisms and incidence. Cerebrovasc Brain Metab Rev. 1993;5(1):1–16.
7. Sies H. Oxidative stress: introductory remarks. In: Oxidative stress. London: Academic, 1985. p. 1–8.
8. Borg J, London J. Copper/zinc superoxide dismutase overexpression promotes survival of cortical neurons exposed to neurotoxins in vitro. J Neurosci Res. 2000;70(2):180–9.
9. Cuzzocrea S, Riley DP, Caputi AP, Salvemini D. Antioxidant therapy: a new pharmacological approach in shock, inflammation, and ischemia/reperfusion injury. Pharmacol Rev. 2001;53(1):135–59.
10. Chan PH. Role of oxidants in ischemic brain damage. Stroke. 1996;27(6):1124–9.
11. Mattson MP, Duan W, Pedersen WA, Culmsee C. Neurodegenerative disorders and ischemic brain diseases. Apoptosis. 2001;6(1–2):69–81.
12. Alexandrova M, Bochev P, Markova V, Bechev B, Popova M, Danovska M, Simeonova V. Dynamics of free radical processes in acute ischemic stroke: influence on neurological status and outcome. J Clin Neurosci. 2004;11(5):501–6.
13. Schaller B, Graf R. Cerebral ischemia and reperfusion: the pathophysiologic concept as a basis for clinical therapy. J Cereb Blood Flow Metab. 2004;24(4):351–71.

14. Kedar NP. Can we prevent Parkinson's and Alzheimer's disease? J Postgrad Med. 2003;49(3):236–45.
15. Sims NR, Anderson MF. Mitochondrial contributions to tissue damage in stroke. Neurochem Int. 2002;40(6):511–26.
16. Murakami K, Kondo T, Kawase M, Li Y, Sato S, Chen SF, Chan PH. Mitochondrial susceptibility to oxidative stress exacerbates cerebral infarction that follows permanent focal cerebral ischemia in mutant mice with manganese superoxide dismutase deficiency. J Neurosci. 1998;18(1):205–13.
17. Keller JN, Kindy MS, Holtsberg FW, St Clair DK, Yen HC, Germeyer A, Steiner SM, Bruce-Keller AJ, Hutchins JB, Mattson MP. Mitochondrial manganese superoxide dismutase prevents neural apoptosis and reduces ischemic brain injury: suppression of peroxynitrite production, lipid peroxidation, and mitochondrial dysfunction. J Neurosci. 1998;18(2): 687–97.
18. Morimoto T, Globus MY, Busto R, Martinez E, Ginsberg MD. Simultaneous measurement of salicylate hydroxylation and glutamate release in the penumbral cortex following transient middle cerebral artery occlusion in rats. J Cereb Blood Flow Metab. 1996;16(1):92–9.
19. Peters O, Back T, Lindauer U, Busch C, Megow D, Dreier J, Dirnagl U. Increased formation of reactive oxygen species after permanent and reversible middle cerebral artery occlusion in the rat. J Cereb Blood Flow Metab. 1998;18(2):196–205.
20. Lipton P. Ischemic cell death in brain neurons. Physiol Rev. 1999;79(4):1431–568.
21. Lee JM, Zipfel GJ, Choi DW. The changing landscape of ischaemic brain injury mechanisms. Nature (Lond). 1999;399(6738 Suppl):A7–14.
22. Nath FP, Jenkins A, Mendelow AD, Graham DI, Teasdale GM. Early hemodynamic changes in experimental intracerebral hemorrhage. J Neurosurg. 1986;65(5):697–703.
23. Qureshi AI, Wilson DA, Hanley DF, Traystman RJ. No evidence for an ischemic penumbra in massive experimental intracerebral hemorrhage. Neurology. 1999;52(2):266–72.
24. Xue M, Del Bigio MR. Intracortical hemorrhage injury in rats: relationship between blood fractions and brain cell death. Stroke. 2000;31(7):1721–7.
25. Hickenbottom SL, Grotta JC, Strong R, Denner LA, Aronowski J. Nuclear factor-kappaB and cell death after experimental intracerebral hemorrhage in rats. Stroke. 1999;30(11):2472–7.
26. Wu J, Hua Y, Keep RF, Schallert T, Hoff JT, Xi G. Oxidative brain injury from extravasated erythrocytes after intracerebral hemorrhage. Brain Res. 2002;953(1–2):45–52.
27. Selim MH, Ratan RR. The role of iron neurotoxicity in ischemic stroke. Ageing Res Rev. 2004;3(3):345–53.
28. Pluta RM. Delayed cerebral vasospasm and nitric oxide: review, new hypothesis, and proposed treatment. Pharmacol Ther. 2005;105(1):23–56.
29. Martin RL, Lloyd HG, Cowan AI. The early events of oxygen and glucose deprivation: setting the scene for neuronal death? Trends Neurosci. 1994;17(6):251–7.
30. Katsura K, Kristian T, Siesjo BK. Energy metabolism, ion homeostasis, and cell damage in the brain. Biochem Soc Trans. 1994;22(4):991–6.
31. Nanetti L, Vignini A, Raffaelli F, Moroni C, Silvestrini M, Provinciali L, Mazzanti L. Platelet membrane fluidity and Na^+/K^+ ATPase activity in acute stroke. Brain Res. 2008;1205:21–6.
32. Siesjo BK, Agardh CD, Bengtsson F. Free radicals and brain damage. Cerebrovasc Brain Metab Rev. 1989;1(3):165–211.
33. Choi DW, Rothman SM. The role of glutamate neurotoxicity in hypoxic-ischemic neuronal death. Annu Rev Neurosci. 1990;13:171–82.
34. Zakim D. Interface between membrane biology and clinical medicine. Am J Med. 1986;80 (4):645–7.
35. Marletta MA, Hurshman AR, Rusche KM. Catalysis by nitric oxide synthase. Curr Opin Chem Biol. 1998;2(5):656–63.
36. Pacher P, Beckman JS, Liaudet L. Nitric oxide and peroxynitrite in health and disease. Physiol Rev. 2007;87(1):315–424.
37. Loscalzo J. Nitric oxide and vascular disease. N Engl J Med. 1995;333(4):251–3.

38. Rudic RD, Sessa WC. Nitric oxide in endothelial dysfunction and vascular remodelling: clinical correlates and experimental links. Am J Hum Genet. 1999;64(3):673–7.
39. Keynes RG, Garthwaite J. Nitric oxide and its role in ischaemic brain injury. Curr Mol Med. 2004;4(2):179–91.
40. Warner DS, Sheng H, Batinic-Haberle I. Oxidants, antioxidants and the ischemic brain. J Exp Biol. 2004;207(pt 18):3221–31.
41. Mander P, Borutaite V, Moncada S, Brown GC. Nitric oxide from inflammatory-activated glia synergizes with hypoxia to induce neuronal death. J Neurosci Res. 2005;79(1–2):208–15.
42. Kader A, Frazzini VI, Solomon RA, Trifiletti RR. Nitric oxide production during focal cerebral ischemia in rats. Stroke. 1993;24(11):1709–16.
43. Dawson VL, Dawson TM. Nitric oxide neurotoxicity. J Chem Neuroanat. 1996;10 (3–4):179–90.
44. Warner DS, Sheng H, Batinic-Haberle I. Oxidants, antioxidants and the ischemic brain. J Exp Biol. 2004;207(pt 18):3221–31.
45. Guix FX, Uribesalgo I, Coma M, Munoz FJ. The physiology and pathophysiology of nitric oxide in the brain. Prog Neurobiol. 2005;76(2):126–52.
46. Szabo C. Multiple pathway of peroxynitrite cytotoxicity. Toxicol Lett. 2003;140–141:105–12.
47. Beckman JS, Koppenol WH. Nitric oxide, superoxide, and peroxynitrite: the good, the bad, and the ugly. Am J Physiol. 1996;271(5 pt 1):C1424–37.
48. Virag L, Szabo E, Gergely P, Szabo C. Peroxynitrite-induced cytotoxicity: mechanism and opportunities for intervention. Toxicol Lett. 2003;140–141:113–24.
49. Denicola A, Souza JM, Radi R. Diffusion of peroxynitrite across erythrocyte membranes. Proc Natl Acad Sci USA. 1998;95(7):3566–71.
50. Pryor WA, Squadrito GL. The chemistry of peroxynitrite: a product from the reaction of nitric oxide with superoxide. Am J Physiol. 1995;268(5 pt 1):L699–722.
51. Beckman JS. Peroxynitrite versus hydroxyl radical: the role of nitric oxide in superoxide-dependent cerebral injury. Ann N Y Acad Sci. 1994;738:69–75.
52. Takizawa S, Hirabayashi H, Fukuyama N, Nakazawa H, Shinohara Y. Peroxynitrite production in cerebral ischemia. Rinsho Shinkeigaku. 1999;39(12):1295–7.
53. Nanetti L, Taffi R, Vignini A, Moroni C, Raffaelli F, Bacchetti T, Silvestrini M, Provinciali L, Mazzanti L. Reactive oxygen species plasmatic levels in ischemic stroke. Mol Cell Biochem. 2007;303(1–2):19–25.
54. Braughler JM, Hall ED. Central nervous system trauma and stroke. Biochemical considerations for oxygen radical formation and lipid peroxidation. Free Radic Biol Med. 1989;6 (3):2289–301.
55. Roberts LJ, Morrow JD. Measurement of F(2)-isoprostanes as an index of oxidative stress in vivo. Free Radic Biol Med. 2000;28(4):505–13.
56. Bolokadze N, Lobjanidze I, Momtselidze N, Solomonia R, Shakarishvili R, McHedlishvili G. Blood rheological properties and lipid peroxidation in cerebral and systemic circulation of neurocritical patients. Clin Hemorheol Microcirc. 2004;30(2):99–105.
57. Demirkaya S, Topcuoglu MA, Aydin A, Ulas UH, Isimer AI, Vural O. Malondialdehyde, glutathione peroxidase and superoxide dismutase in peripheral blood erythrocytes of patients with acute cerebral ischemia. Eur J Neurol. 2001;8(1):43–51.
58. Sharpe PC, Mulholland C, Trinick T. Ascorbate and malondialdehyde in stroke patients. Ir J Med Sci. 1994;163(11):488–91.
59. Ferretti G, Bacchetti T, Masciangelo S, Nanetti L, Mazzanti L, Silvestrini M, Bartolini M, Provinciali L. Lipid peroxidation in stroke patients. Clin Chem Lab Med. 2008;46(1):113–7.
60. Navab M, Yu R, Gharavi N, Huang W, Ezra N, Lotfizadeh A, Anantharamaiah GM, Alipour N, Van Lenten BJ, Reddy ST, Marelli D. High-density lipoprotein: antioxidant and anti-inflammatory properties. Curr Atheroscler Rep. 2007;9(3):244–8.
61. Ferretti G, Bacchetti T, Busni D, Rabini RA, Curatola G. Protective effect of paraoxonase activity in high-density lipoproteins against erythrocyte membranes peroxidation:

a comparison between healthy subjects and type 1 diabetic patients. J Clin Endocrinol Metab. 2004;89(6):2957–62.
62. Ferretti G, Bacchetti T, Principi F, Di Ludovico F, Viti B, Angeleri VA, Danni M, Provinciali L. Increased levels of lipid hydroperoxides in plasma of patients with multiple sclerosis: a relationship with paraoxonase activity. Mult Scler. 2005;11(6):677–82.
63. Gariballa SE, Hutchin TP, Sinclair AJ. Antioxidant capacity after acute ischaemic stroke. Q J Med. 2002;95(10):685–90.
64. Uno M, Kitazato KT, Nishi K, Itabe H, Nagahiro S. Raised plasma oxidised LDL in acute cerebral infarction. J Neurol Neurosurg Psychiatr. 2003;74(3):312–6.
65. Shie FS, Neely MD, Maezawa I, Wu H, Olson SJ, Jürgens G, Montine KS, Montine TJ. Oxidized low-density lipoprotein is present in astrocytes surrounding cerebral infarcts and stimulates astrocyte interleukin-6 secretion. Am J Pathol. 2004;164(4):1173–81.
66. Cui J, Holmes EH, Greene TG, Liu PK. Oxidative DNA damage precedes DNA fragmentation after experimental stroke in rat brain. FASEB J. 2000;14(7):955–67.
67. Liu H, Uno M, Kitazato KT, Suzue A, Manabe S, Yamasaki H, Shono M, Nagahiro S. Peripheral oxidative biomarkers constitute a valuable indicator of the severity of oxidative brain damage in acute cerebral infarction. Brain Res. 2004;1025(1–2):43–50.
68. Nagayama T, Lan J, Henshall DC, Chen D, O'Horo C, Simon RP, Chen J. Induction of oxidative DNA damage in the peri-infarct region after permanent focal cerebral ischemia. J Neurochem. 2000;75(4):1716–28.
69. Chang CY, Lai YC, Cheng TJ, Lau MT, Hu ML. Plasma levels of antioxidant vitamins, selenium, total sulfhydryl groups and oxidative products in ischemic-stroke patients as compared to matched controls in Taiwan. Free Radic Res. 1998;28(1):15–24.
70. Butterfield DA, Kanski J. Brain protein oxidation in age-related neurodegenerative disorders that are associated with aggregated proteins. Mech Ageing Dev. 2001;122(9):945–62.
71. Fukuyama N, Takizawa S, Ishida H, Hoshiai K, Shinohara Y, Nakazawa H. Peroxynitrite formation in focal cerebral ischemia-reperfusion in rats occurs predominantly in the peri-infarct region. J Cereb Blood Flow Metab. 1998;18(2):123–9.
72. Hall AC, Lucas FR, Salinas PC. Axonal remodeling and synaptic differentiation in the cerebellum is regulated by WNT-7a signaling. Cell. 2000;100(5):525–35.
73. Polidori MC, Mattioli P, Aldred S, Cecchetti R, Stahl W, Griffiths H, Senin U, Sies H, Mecocci P. Plasma antioxidant status, immunoglobulin G oxidation and lipid peroxidation in demented patients: relevance to Alzheimer disease and vascular dementia. Dement Geriatr Cogn Disord. 2004;18(3–4):265–70.
74. Carty JL, Bevan R, Waller H, Mistry N, Cooke M, Lunec J, Griffiths HR. The effects of vitamin C supplementation on protein oxidation in healthy volunteers. Biochem Biophys Res Commun. 2000;273(2):729–35.
75. Yang CS, Lin NN, Tsai PJ, Liu L, Kuo JS. In vivo evidence of hydroxyl radical formation induced by elevation of extracellular glutamate after cerebral ischemia in the cortex of anesthetized rats. Free Radic Biol Med. 1996;20(2):245–50.
76. Sakamoto A, Ohnishi ST, Ohnishi T, Ogawa R. Relationship between free radical production and lipid peroxidation during ischemia-reperfusion injury in the rat brain. Brain Res. 1991;554 (1–2):186–92.
77. White BC, Daya A, DeGracia DJ, O'Neil BJ, Skjaerlund JM, Trumble S, Krause GS, Rafols JA. Fluorescent histochemical localization of lipid peroxidation during brain reperfusion following cardiac arrest. Acta Neuropathol. 1993;86(1):1–9.
78. Polidori MC, Cherubini A, Stahl W, Senin U, Sies H, Mecocci P. Plasma carotenoid and malondialdehyde levels in ischemic stroke patients: relationship to early outcome. Free Radic Res. 2002;36(3):265–8.
79. Lo EH, Dalkara T, Moskowitz MA. Mechanisms, challenges and opportunities in stroke. Nat Rev Neurosci. 2003;4(5):399–415.
80. Dotan Y, Lichtenberg D, Pinchuk I. Lipid peroxidation cannot be used as a universal criterion of oxidative stress. Prog Lipid Res. 2004;43(3):200–27.

Chapter 10
Free Radicals in Epilepsy

Tali Siman-Tov and Natan Gadoth

Abstract The brain is a susceptible target for free radical damage because its oxidative metabolism is increased compared to other body parts. Furthermore, its decreased antioxidant ability in the presence of high concentrations of unsaturated fatty acids adds to its high vulnerability to oxidative stress. Epilepsy is a relatively common neurological disorder, especially during infancy and early childhood and during aging. Although anticonvulsant drug treatment can suppress seizures in the majority of patients, 15–20% of epileptic patients are drug resistant and thus suffer from uncontrolled seizures during their lifetime, causing not only reduced life quality but also secondary brain damage. Mitochondrial dysfunction leading to excessive free radical production has been implicated in brain damage and neuronal death. In this chapter, we review the oxidant/antioxidant system in human and experimental models of epilepsy and summarize briefly recent knowledge about the mechanisms involved in oxidative stress and neuronal death associated with seizure activity.

Keywords Epileptogenesis · Excitotoxicity · Hyperexcitability · Neuronal death

1 Introduction

Epilepsy, from the Greek *epilepsia* meaning "to seize upon" or "taking hold of," is a chronic brain disorder characterized by spontaneous recurrent seizures. Seizures and epilepsy have been documented since the earliest civilizations. The earliest

N. Gadoth (✉)
Sackler Faculty of Medicine, Tel Aviv University, Tel Aviv, 69978, Israel
and
Department of Neurology, Mayanei Hayeshua Medical Center, Bnei-Brak, 51544, Israel
e-mail: gadoth@post.tau.ac.il

N. Gadoth and H.H. Göbel (eds.), *Oxidative Stress and Free Radical Damage in Neurology*, Oxidative Stress in Applied Basic Research and Clinical Practice,
DOI 10.1007/978-1-60327-514-9_10,

descriptions of epilepsy appear in Mesopotamian writings, which date back to the fifth millennium BC. For years, persons with epileptic seizures were believed to be possessed or seized by demons or evil spirits. Hippocrates in 400 BC. was the first to suggest that epilepsy is a brain disorder, and today we assume that a seizure represents a period of abnormal synchronous excitation of a neuronal population. Epilepsy is defined by a state of recurrent, spontaneous (unprovoked) seizures, at least two episodes 24 h apart [1–3].

Epilepsy is a complex disease with diverse clinical manifestations. Signs and symptoms of seizures depend on the location of the primary epileptic discharge and the extent and pattern of its propagation within the brain. Generalized seizures denote the occurrence of seizures throughout the brain resulting from a generalized lowering of seizure threshold. Examples of generalized seizures are tonic-clonic (grand mal), absence (petit mal), or myoclonic seizures. Partial seizures begin in a localized brain region, and their signs and symptoms depend on the region involved (e.g., visual, motor or sensory symptoms, in occipital, pre-central, or post-central gyrus involvement, respectively). Partial seizures may or may not affect the level of consciousness (complex and simple partial seizures, respectively). Complex partial seizures frequently originate from the temporal lobe (temporal lobe epilepsy) [4]. Status epilepticus (SE) is a state of continuous seizure activity lasting for more than 5 min or rapidly recurrent seizures without full recovery of consciousness between them. SE can be convulsive or nonconvulsive. The type of seizure leading to SE can vary [5].

Much of our knowledge about the pathophysiology of epileptic seizures comes from seizure induction in animal models (Table 1). Basically, it is believed that the initiation and propagation of a seizure occur as a result of an imbalance between depolarizing and hyperpolarizing influences in a large interconnected network of neurons. The transition from normal to epileptiform behavior is probably caused by a combination of enhanced connectivity, enhanced excitatory transmission, a failure of inhibitory mechanisms, and changes in intrinsic neuronal properties [4].

The lifetime likelihood of having at least one febrile or nonfebrile epileptic seizure is about 9%, and the prevalence of active epilepsy is 0.8% [6]. The incidence of epilepsy is double peaked, higher in children younger than 5 years of age and in

Table 1 Animal models of epileptogenesis mentioned in the current chapter

Model	Putative mechanism	Species
Kindling	Repeated subthreshold induced seizures	Rat
Pentylenetetrazol	GABA-A antagonist	Rat
Bicuculline	GABA-A antagonist	Cat
Picorotoxin	GABA-A antagonist	Rat
N-methyl-d-asparate (NMDA)	Glutamate agonist	Rat
Kainate (kainic acid)	Glutamate agonist	Rat
Pilocarpine	Cholinergic agonist	Rat
Pilocarpine + Lithium	Cholinergic agonist + glutamate and aspartate release	Rat
$FeCl_3$	Iron-induced cellular damage	Rat

individuals older than 65 years. In general, three main etiologies may contribute to the onset and progression of epilepsy: (1) genetic causes; (2) developmental disorders and/or central nervous system (CNS) malformation; and (3) brain insult or injury. Interactions between these three factors may influence the progression of the disease [2]. Acquired epilepsy accounts for approximately 60% of all cases and is usually preceded by an insult, such as head trauma, stroke, infection, childhood prolonged febrile seizures, or an episode of nonfebrile prolonged seizures or SE. Epilepsy may develop years and decades after the insult, which might suggest that a progression of changes in response to the initial insult is critical to the development of the epileptic condition. The process in which a cascade of cellular and molecular changes, induced by a brain insult, leads to the occurrence of recurrent seizures is called epileptogenesis [2, 6, 7].

Among the events that occur in response to the initial insult, neuronal death has received significant attention as the propagating factor that may lead to the epileptic condition. Accumulating evidence suggests that neuronal cell death may be both a cause and a consequence of epileptic seizures. The evidence that seizures cause brain injury comes from the demonstration that intense seizure activity associated with SE can cause hippocampal damage [6]. Ammon's horn sclerosis was first described in 1880, when Sommer reported hippocampal CA1 pyramidal cell loss with or without CA3 and end-folium involvement [8]. Later, neurodegeneration has also been reported in the amygdala, entorhinal, perirhinal, and parahippocampal cortices as well as in extratemporal regions, such as the thalamus and cerebellum [6]. Cell death in epilepsy is believed to involve excitotoxicity, whereby excessive glutamate causes overstimulation of the postsynaptic *N*-methyl-D-asparate (NMDA) receptors; this leads to accumulation of intracellular calcium, followed by cell injury and/or death [9]. The idea that neuronal death can cause epilepsy is supported by the fact that surgical removal of a damaged hippocampus improves the condition of epilepsy patients [10]. Other cellular processes such as neurogenesis, gliosis, axonal sprouting, axonal injury, dendritic remodeling, angiogenesis, inflammation, alterations in extracellular matrix, and alterations in ion channels have been implicated in epileptogenesis [6]. However, the association of epileptogenesis with neuronal cell death is the more studied and established so far.

Free radicals (Table 2) are known to play a significant role in processes that result in tissue damage and cell death. Free radicals are capable of damaging all components of the cell, including nucleic acids, lipids, proteins, and carbohydrates, leading to progressive decline in physiological function and ultimately cell death. The brain is particularly vulnerable to free radical damage because of its high metabolic rate, high level of oxygen consumption, lower antioxidant defenses compared to other organs, and high concentration of polyunsaturated fatty acids (PUFA), which are particularly susceptible to lipid peroxidation under oxidative stress [11]. Among brain cells, neurons are particularly vulnerable to oxidative insults due to low levels of reduced glutathione, a major brain endogenous antioxidant (Table 3) [12].

Both reactive oxygen species (ROS) and reactive nitrogen species (RNS) have been implicated in epileptogenesis [7]. A major source for ROS is the process

Table 2 Free radicals

Reactive oxygen species (ROS)	
Superoxide anion radical	$O_2{\bullet}^-$
Hydrogen peroxide	H_2O_2
Hydroxyl radical	$HO{\bullet}$
Hypochlorous acid	HOCl
Reactive nitrogen species (RNS)	
Nitric oxide	NO
Peroxynitrite	$OONO^-$
Nitroyl	$NO{\bullet}$
Nitrogen dioxide	NO_2

Table 3 Enzymes and metabolites involved in antioxidant mechanisms

Cytosolic copper-zinc superoxide dismutase (SOD1)
Mitochondrial manganese superoxide dismutase (SOD2)
Extracellular superoxide dismutase (SOD3)
Superoxide reductases
Catalase
Glutathione peroxidases (GPX)
Selenium
Glutathione
Vitamin C
Vitamin E
Carotenes
Lipoic acid
Ubiquinol
Melatonin
Adenosine
Bilirubin
Albumin
Uric acid

of oxidative phosphorylation within the mitochondria. Association between mitochondrial dysfunction and epilepsy is well supported by the fact that epilepsy is a central manifestation of many syndromes attributed to mitochondrial dysfunction, such as myoclonic epilepsy with ragged red fibers (MERRF) and Leigh disease [13]. RNS refer to nitric oxide (NO) and its derivatives (see Table 2). The role of NO in epilepsy is controversial, as both pro- and anticonvulsant effects were shown using different models. However, other RNS are implicated in processes associated with seizures, particularly peroxynitrite, which is formed by the reaction of NO with superoxide [14].

In addition, it is suggested that free radical formation is induced by seizure activity [11]. Thus, free radicals might be associated with epilepsy as both cause and consequence.

2 Free Radical-Induced Cell Damage

Cell injury induced by free radicals may involve damage at different levels.

2.1 Nucleic Acids

1. *DNA*: Free radicals may trigger breaks in DNA strands or directly modify purine and pyridine bases, leading to deletions and other mutations, which may result in aberrant gene expression and cell death [11]. In addition, DNA damage activates the DNA repair enzyme *poly-ADP-ribose polymerase-1* (PARP-1).PARP-1 overactivation depletes its substrate, nicotinamide adenine dinucleotide, slowing the rate of glycolysis, electron transport, and ATP formation, eventually leading to functional impairment or cell death [15]. Moreover, activation of PARP-1 induces release of the apoptosis-inducing factor (AIF) from mitochondrial membranes and its translocation to the nuclei, where it participates in DNA fragmentation. Endonucleases are also activated and are responsible for larger-scale DNA fragmentation [8]. Common laboratory markers of oxidative damage to DNA are *8-hydroxy-2-deoxyguanosine* (8-OHdG) and *8-hydroxyguanosine* (8-OHG) [16].
2. *Mitochondrial DNA*: mtDNA is a particularly vulnerable target for free radical damage because of diminished repair mechanisms and lack of histones, showing a tenfold greater mutation rate than nuclear DNA. mtDNA disruption may lead to mitochondrial dysfunction that might result in disturbed cell function at different level; this includes electron transport chain failure and additional ROS production, which can result in a vicious cycle of oxidative stress [7, 17].
3. *RNA*: RNA is even more susceptible to oxidative damage than DNA, because it is mostly single stranded, not protected by hydrogen bonding, and less protected by proteins. Damage to RNA may result in errors in proteins or dysregulation of gene expression [18].

2.2 Lipid

Polyunsaturated lipids in lipoproteins and membranes are highly susceptible to oxidative damage, which leads to lipid peroxidation. The unsaturated hydroperoxides generated by PUFA peroxidation can break down to form reactive aldehydes; the best known is malondialdehyde (MDA). Reactive aldehydes can bind covalently to proteins, thereby altering their function and inducing cellular damage. Peroxidation products (peroxyl radicals) are more stable than ROS and are able to diffuse even across cells and react with other cellular constituents to cause diffuse cellular damage [14]. Because they are stable, peroxidation end products are valuable laboratory markers: MDA, *hexanoyllysine adduct* (HEL), *acrolein-lysine adduct* (ACR), and *4-hydroxy nonenal* (4-HNE). Another useful marker is F_2-*isoprostanes* (F_2-IsoPs), F_2-like prostaglandins that are produced by arachidonic acid peroxidation [14, 17].

2.3 Protein

Free radicals can oxidize both the backbone and side chains of proteins.

Backbone modifications can lead to the formation of peroxyl radicals with a chain reaction that result in protein cross-linking and/or peptide bond cleavage. The oxidation of side-chain residues gives rise to a multitude of products, which can in turn react with amino acid side chains to produce carbonyl functions. These oxidative modifications may disturb the function of enzymes, receptors, neurotransmitters, and structural proteins [14]. Inactivation of the enzyme aconitase is a useful marker for oxidative damage to proteins. The presence of a labile iron motif in the iron–sulfur center of aconitase renders it sensitive to free radicals, particularly superoxide and peroxynitrite. The oxidation of the enzyme results in its inactivation [7]. Peroxynitrite modifications of proteins occur mainly in tyrosine and tryptophan residues. 3-Nitrotyrosine, the main product of protein nitration, is considered an important marker for peroxynitrite formation [19].

2.4 Carbohydrates

Oxidation of monosaccharide sugars results in the formation of oxaldehydes, which can contribute to protein aggregation. Oxidation of carbohydrate polymers may cause depolymerization and disturbed function of the involved polymers [20]. Sugar oxidation products can react with proteins to generate advanced glycation end products, which are known as useful markers of oxidative damage [17].

3 Free Radicals as a Consequence of Epileptic Seizures

Free radical level was found to be increased following seizures in multiple models and by different methods, including indirect measurement of free radical production, markers of nucleic acid, lipid, and protein oxidative damage and evaluation of antioxidant defense mechanisms. Below are a few examples of supportive evidence:

1. Increased free radical level was found in the cerebrospinal fluid (CSF) overlying the seizure focus in a bicuculline-induced seizure cat model [21].

2. Frantseva et al. [22] applied topical bicuculline to hippocampal slices taken from 7-day-old rats and recorded rhythmic synchronous activity, which is believed to resemble seizure-like activity in situ [23]. There was also an increase in free radical generation within 10–15 min. This effect was more prominent in CA3, where neurons subsequently degenerated [22].
3. In a lithium-pilocarpine model of SE in rats, Peterson et al. [24] showed increased production of superoxide by hydroethidium staining. Interestingly, the increase in superoxide formation was significant in the parietal, piriform, and perirhinal cortices, lateral amygdala, and specific regions in the thalamus, but not within the hippocampus.
4. Hydroxyl radicals were reported to accumulate after pentylenetetrazol (PTZ)-induced seizures and kindling in a rat model [25].
5. In a picrotoxin-induced seizure model in rats, Rajasekaran [26] showed significant increase in NO levels in the frontal cortex, hippocampus, and midbrain at the onset of generalized seizures. Treatment with the neuronal NO synthase (nNOS) inhibitor *7-nitroindazole* suppressed the seizure-induced NO production.
6. Kainate administration to adult rats produced high levels of 8-OHdG, a marker of oxidative DNA damage. mtDNA was reported to be the major source for 8-OHdG in these experiments. The documented increase of 8-OHdG preceded overt cell death [7].
7. Kainate-induced SE in adult rats was found to be associated with an increase in F_2-*IsoP*, a marker of lipid peroxidation. This increase was reported in the CA3 region and preceded cell damage [7, 9].
8. In a kindling model of rats, Frantseva et al. [27] showed increased production of the lipid peroxidation markers MDA and 4-HNE, as well as increased cell death throughout the hippocampus, which were attenuated by antioxidative therapy with vitamin E and glutathione.
9. Kainate-induced seizures in rats inactivated mitochondrial aconitase but not fumarase, which is assumed to be resistant to oxidative stress. This inactivation was specific for mitochondrial aconitase, implicating the major role of mitochondria in seizure-induced oxidative stress [7]. Maximal inactivation of the mitochondrial aconitase occurred several hours after the cessation of SE and before the death of hippocampal neurons. Kainate-induced mitochondrial aconitase inactivation and hippocampal neuronal loss were attenuated in transgenic mice overexpressing the mitochondrial superoxide dismutase (Sod2) and exacerbated in mice partially deficient in Sod2 [7].
10. Inactivation of mitochondrial aconitase and alpha-ketoglutarate dehydrogenase, another oxidant-sensitive enzyme, was demonstrated in a rat model of pilocarpine-induced seizures [7].
11. In the bicuculline-induced seizure model described above (2), free radical scavengers (vitamin E and glutathione) significantly reduced seizure-induced neuronal death without affecting the epileptiform electrical activity [22].

There is also some evidence to support the emergence of oxidative stress following brief seizures. Erakovic et al. [28] reported an acute decrease in regional brain antioxidant levels following electroconvulsive shock in rats. They showed reduced SOD and glutathione peroxidase (GPX) activities in the hippocampus and the frontal cortex 2 h after a single electroconvulsive shock. Arnaiz et al. [29] reported increase in lipid peroxidation following 3-mercaptopropionic acid-induced seizures in rats. An increase up to 40% in lipid peroxidation was observed as early as 3–6 min after seizure onset.

Interestingly, the neonatal brain seems to be relatively resistant to seizure-induced free radical production and neuronal injury. This advantage over the adult brain is attributed to the uncoupling protein-2 (UCP-2), which is highly expressed in the neonatal brain, due to the fat-rich maternal milk. The uncoupling reaction causing the dissociation of ATP production from oxygen consumption is thought to reduce ROS formation by reducing mitochondrial membrane potential [30].

In the aforementioned studies, evidence for oxidative stress was frequently reported to precede overt cell death. These studies and others strongly suggest that free radicals play a crucial role in seizure-induced neuronal death. Both necrosis and apoptosis were implied in this neuronal death process [31]. Thus, antioxidant and antiapoptotic agents might have a role in preventing seizure-induced neuronal cell death, including mesial temporal sclerosis and its putative consequences, cognitive impairment and epilepsy.

4 Free Radicals as a Cause of Epileptic Seizures

Brain insults such as trauma, hypoxia, and aging are considered epileptogenic. These conditions are known to be associated with oxidative stress. The role of free radicals in epileptogenesis was thoroughly investigated in cases with brain trauma. Seizures following brain trauma are classified according to their timing: "immediate seizures" occur within 24 h following brain injury, "early seizure" within the first week, and "late seizures" after a week. Immediate and early seizures are most likely the result of the trauma itself and are not considered posttraumatic epilepsy (PTE) [32]. The incidence of PTE following severe head trauma with cortical injury is 7–39%. When the dura is penetrated, the incidence reaches 20–57%. Other factors that may affect the incidence of PTE are the severity of cortical contusions, presence of intracerebral hematoma, loss of consciousness for more than 24 h, midline shift greater than 5 mm, and surgical repair of depressed fractures. The occurrence of an early seizure also increases the risk for PTE [32, 33].

Brain trauma causes a cascade of structural, physiological, and biochemical changes, such as altered cerebral blood flow and vasoregulation, disruption of the blood–brain barrier, increased intracranial pressure, focal or diffuse ischemic hemorrhage, inflammation, necrosis and disruption of fiber tracts and blood vessels. However, hemorrhage and iron accumulation were suggested to have a central role in the cascade culminating in PTE [32]. In 1978, Willmore et al. reported an animal model of posttraumatic epilepsy. They showed that a single injection of $FeCl_3$ solution into the rat cerebral cortex can induce epileptic discharges seen in the electrocorticogram within 15 min [32]. Rosen et al. [34] showed that convulsive seizures are induced by intracortical injection of purified bovine hemoglobin (Hb) within a few days. It was then suggested that after intracranial hemorrhage red blood cells break down and release hemoglobin and iron ions, which then generate ROS by iron- or Hb-mediated reactions. Increased production of ROS induces a cascade of cellular and molecular changes, as already described, that might lead to hyperexcitability of neurons or to neuronal death. For instance, lipid peroxidation can result in disruption of membranes and subcellular organelles. One possible consequence is impairment of the Na^+/K^+-ATPase activity, which normally maintains ionic gradients of neuronal membranes. A decrease in Na^+/K^+-ATPase activity might decrease the convulsive threshold. Injury to membranes may also lead to an increase in the release of excitatory neurotransmitters, such as glutamate and aspartate, or to a decrease of inhibitory neurotransmitters, such as GABA. This kind of changes may contribute to hyperexcitability of neurons [35]. In addition, excessive release of excitatory amino acids may trigger excitotoxicity, which involves secondary formation of NO and ROS that can contribute to further cellular damage and seizures (Fig. 1).

Thus, the association between free radicals and epileptogenesis is well supported by the example of PTE. Further evidence to support this association comes from both clinical and laboratory studies:

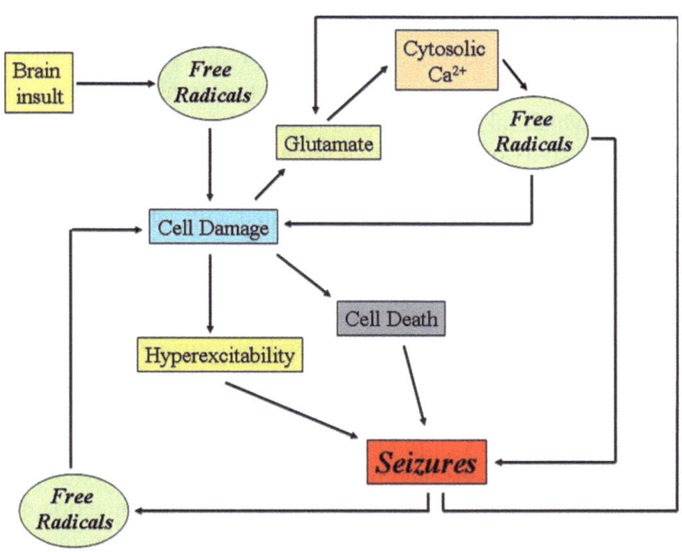

Fig. 1 A schematic outline of possible mechanisms of the role of free radicals in epilepsy: seizures might result in accumulation of free radicals. Increased free radical level, secondary to seizures or other brain insults, is associated with cell damage and cell death. Cell death in itself is considered a cause of seizures. In addition, cell damage might cause hyperexcitability, leading to lowered seizure threshold, and/or excitotoxicity, leading to further free radical release, culminating in a vicious cycle

1. As mentioned above, mitochondrial disorders that are expected to involve oxidative stress are frequently associated with epilepsy. Moreover, impairment of mitochondrial function was found in the seizure focus of human and experimental epilepsy [13]. A specific deficiency in complex I was reported in temporal foci of patients with temporal lobe epilepsy [36]. Recent proteomic studies have clearly indicated that certain mitochondrial components are altered following seizure activity, implying that brain injury secondary to prolonged seizures may be explained, at least partially, by such alterations. The Rieske protein (RP) may be such a component. This protein is a high-potential iron–sulfur cluster that is a part of the mitochondrial cytochrome *bc1* complex (ubiquinol:cytochrome *c* reductase, or complex III). It catalyzes electron transfer from ubiquinol to cytochrome *c* but is also involved in proton translocation, peptide processing, and superoxide generation. Junker et al. [37] have shown that seizure activity in kindled rats results in pronounced alterations in RP, in the form of a shift toward a more alkaline pH range in its isoelectric point. Furthermore, although RP was shown by immunohistochemistry to be present in the CA2 region of the hippocampus of control and kindled animals, it could not be detected in the hippocampal CA1 region, which is known as extremely susceptible to damage after seizure activity in humans and in rats, especially when seizures are accompanied by hypoxia. Another important observation made by the same authors was the asymmetrical expression of RP protein in the subgranular zone of the dentate gyrus, a region known to be involved in neurogenesis [37].
2. In patients with progressive myoclonic epilepsy, the activity of the cytosolic superoxide dismutase (SOD1) was reported to be low [38]. Mitochondrial manganese superoxide dismutase (SOD2) was found to be downregulated in the cerebral cortex of patients with epilepsy in contrast to nonepileptic subjects [39].
3. GPX activity, which takes part in antioxidant defense, and selenium, which is critical for the activity of GPX, were reported to be low in children with epilepsy [40]. Administration of selenium was reported to reduce seizure frequency in two patients with intractable seizures associated with selenium deficiency and in patients treated by total parenteral nutrition [40].
4. Vitamin E, vitamin C, adenosine, and melatonin are additionally well-known endogenous antioxidants. It was reported that pretreatment with vitamin E can prevent the development of iron-induced epileptiform activity in rats. Vitamin E was also reported to significantly delay the onset of electroencephalographic seizures induced by intracerebral injection of ferrous chloride [41]. A phosphodiester of vitamin E and C, *alpha-tocopheryl-L-ascorbate-2-O-phosphate*, was reported to reduce ferric-induced production of markers of oxidative stress and to delay or suppress epileptic discharges [41]. Systemic injection of adenosine was reported to prevent audiogenic-, kainate-, and picrotoxine-induced seizures. In iron-induced seizures in rats, intraperitoneal injection of adenosine, 30 min before iron injection, delayed or suppressed the occurrence of epileptiform discharges [41]. Melatonin, the pineal hormone that regulates circadian rhythm, was found to protect cells against oxidative damage. It was reported to scavenge free radicals, stimulate GPX activity, and inhibit NOS activity. In a rat model of

iron-induced seizures, administration of melatonin caused inhibition of epileptic activity by suppressing peroxidation [42].

5. Mice with a partial deficiency of Sod2 ($Sod2^{-/+}$) provide a good model for sublethal chronic elevation of mitochondrial superoxide. These mice appear normal at birth but develop age-dependent oxidative stress and seizures. Before the appearance of spontaneous seizures, these mice showed increased susceptibility to kainate-induced seizures and hippocampal cell loss [7]. $Sod2^{-/+}$ mice showed an age-dependent decrease in the hippocampal expression of glial glutamate transporters that are crucial for maintaining a low level of synaptic glutamate and are known to be sensitive to oxidative damage. It is suggested that H_2O_2, generated in excess in the presence of superoxide, could oxidize glial glutamate transporters and lead to their decreased expression. The disturbance of glutamate balance leading to hyperexcitability might be responsible for epileptogenesis in this animal model [7].
6. Theophylline is a potent drug for the treatment of bronchial asthma and chronic obstructive pulmonary disease. Its use is restricted for reasons of cardiotoxicity and neurotoxicity, including recurrent seizures. One of the major metabolites of theophylline is a substrate for xanthine oxidase, which is an important player in pathways of ROS generation. In a mice model, aminophylline, an ethylenediamine salt of theophylline, was shown to induce seizures and to increase MDA levels and NO metabolites levels. Pretreatment with the NOS inhibitors *Nω-nitro-L-arginine methyl ester* (L-NAME) and *7-nitroindazole* markedly reduced seizures in this model, implicating a role for free radicals, including NO, in aminophylline/theophylline-induced seizures [43]. The pro-epileptogenic role of NO was also demonstrated with the hyperbaric O_2-induced seizures rat model. Nitrotyrosine, a marker of peroxynitrite formation, was markedly elevated in the hyperbaric O_2 animals. Administration of the nNOS inhibitor, 7-nitroindazole, significantly delayed the onset of hyperbaric O_2-induced seizures [44]. Proconvulsant effects of NO has also been reported in other models of induced seizures, e.g., by NMDA, arginine, and PTZ [41]. However, other studies reported anticonvulsant effects of NO, in many different models, using kainate, PTZ, bicuculline, lithium-pilocarpine, NMDA, and other known agents for seizure induction [41]. Furthermore, increase of the duration and severity of generalized seizures was reported after inhibition of NOS [45]. Thus, the role of NO in epileptogensis is still controversial as both anticonvulsant and pro-convulsant roles were suggested by many different studies [45].

5 Antiepileptic Therapy and Free Radicals

Long-term use of antiepileptic drugs (AEDs) has been proposed to increase free radical formation and cause oxidative damage in neurons. As already discussed, this might lead to enhancement of seizure activity through increased hyperexcitability and/or the induction of neuronal damage.

A few reports support the role of valproic acid (VPA) in exacerbation of oxidative stress. Reduced total antioxidant capacity was reported during treatment with VPA. Serum uric acid and albumin, which are considered endogenous antioxidant molecules, were found to be reduced in children treated with VPA [46]. Other studies showed enhanced lipid peroxidation and increased levels of 8-OHdG in epileptic patients treated with VPA [46–48]. Ayciceck and Iscan [46] also found marked increase in serum total peroxide levels in children treated by carbamazepine and higher lipid hydroperoxide in children treated with phenobarbital (PB). Significantly elevated levels of lipid hydroperoxide were observed in epileptic patients treated by phenytoin [49]. Musavi and Kakkar [50] reported free radical-mediated changes following treatment by diazepam in rats. Gabapentine and topiramate were demonstrated to initiate an oxidative process in primary cultures of rat cortical astrocytes. Both ROS and NO were shown to be involved [51].

However, other studies claim a neuroprotective role for AEDs through antioxidative pathways. VPA was shown to be protective against oxidative stress in both in vitro and in vivo models of epilepsy [52, 53]. It was suggested that VPA may increase levels of glutathione and by that means induce its neuroprotective effect [53]. Diazepam and PB were reported to prevent seizure-induced increase in NO and lipid peroxidation [26]. Topiramate was shown to attenuate lipid peroxidation levels and act as an antioxidant [54]. It was suggested that topiramate may affect voltage-gated calcium channels and lead to reduced cytosolic calcium levels that protect cells from ROS production by mitochondria [54]. Zonisamide, which is believed to exert its antiepileptic effect through blockage of voltage-sensitive sodium and T-type calcium channels, was reported to reduce activity of hydroxyl and NO radicals. It was assumed that its anticonvulsant effect also involves protection from free radical damage [55].Thus, AEDs have been shown to contribute to both pro- and antioxidant activity. Their role in exacerbation of oxidative stress is still to be investigated.

A ketogenic diet has been known for many years as an important alternative for the treatment of medically refractory epilepsy, especially in children. The exact mechanism by which the ketogenic diet exerts its antiepileptic effect is still unknown. However, it is suggested that the ketogenic diet has a neuroprotective role through enhancement of ATP synthesis and suppression of ROS production. Ketogenic diet is initiated by a fasting period, followed by a gradual transition to a fat-based diet with a 3:1 to 4:1 ratio of fat to carbohydrate plus protein. Ketone bodies, which are the hallmark of a high-fat diet, provide an alternative substrate to glucose for energy utilization. They are also known to enhance mitochondrial respiration and ATP production. In addition, ketone bodies might decrease ROS production through NADH oxidation and inhibition of mitochondrial permeability transition. ROS levels were significantly reduced in the hippocampus of mice treated with a ketogenic diet. It was also reported that a ketogenic diet induces UCP expression, which is associated with increased energy production and reduced ROS generation. The protective effect of UCP on seizures was already mentioned in relation to the relative resistance of the immature brain to ROS production following seizures. Interestingly, a few studies suggested that the ketogenic diet is more effective in immature animals and in infants [56, 57].

References

1. Victor M, Ropper AH. Adams and Victor's principles of neurology, 7th ed. New-York: McGraw-Hill; 2001. p. 331–2.
2. Scharfman HE. The neurobiology of epilepsy. Curr Neurol Neurosci Rep. 2007;7:348–54.
3. Riggs AJ, Riggs JE. Epilepsy's role in the historical differentiation of religion, magic, and science. Epilepsia. 2005;46:452–3.
4. McCormick DA, Contreras D. On the cellular and network bases of epileptic seizures. Annu Rev Physiol. 2001;63:815–46.
5. Feen ES, Bershad EM, Suarez JI. Status epilepticus. South Med J. 2008;101:400–6.
6. Pitkanen A, Lukasiuk K. Molecular and cellular basis of epileptogenesis in symptomatic epilepsy. Epilepsy Behav. 2009;14(Suppl 1):S16–25.
7. Patel M. Mitochondrial dysfunction and oxidative stress: cause and consequence of epileptic seizures. Free Radic Biol Med. 2004;37:1951–62.
8. Fujikawa DG. Prolonged seizures and cellular injury: understanding the connection. Epilepsy Behav. 2005;7(Suppl 3):S3–11.
9. Cock HR. The role of mitochondria and oxidative stress in neuronal damage after brief and prolonged seizures. Prog Brain Res. 2002;135:187–96.
10. Shin C, McNamara JO. Mechanism of epilepsy. Annu Rev Med. 1994;45:379–89.
11. Choi BH. Oxygen, antioxidants and brain dysfunction. Yonsei Med J. 1993;34:1–10.
12. Dringen R, Gutterer JM, Hirrlinger J. Glutathione metabolism in brain metabolic interaction between astrocytes and neurons in the defense against reactive oxygen species. Eur J Biochem. 2000;267:4912–6.
13. Kunz WS. The role of mitochondria in epileptogenesis. Curr Opin Neurol. 2002;15:179–84.
14. Poon HF, Calabrese V, Scapagnini G, Butterfield DA. Free radicals: key to brain aging and heme oxygenase as a cellular response to oxidative stress. J Gerontol A Biol Sci Med Sci. 2004;59:478–93.
15. Facchinetti F, Dawson VL, Dawson TM. Free radicals as mediators of neuronal injury. Cell Mol Neurobiol. 1998;18:667–82.
16. Hayashi M. Oxidative stress in developmental brain disorders. Neuropathology. 2009;29:1–8.
17. Sayre LM, Moreira PI, Smith MA, Perry G. Metal ions and oxidative protein modification in neurological disease. Ann Ist Super Sanita. 2005;41:143–64.
18. Nunomura A, Moreira PI, Takeda A, Smith MA, Perry G. Oxidative RNA damage and neurodegeneration. Curr Med Chem. 2007;14:2968–75.
19. Murray J, Oquendo CE, Willis JH, Marusich MF, Capaldi RA. Monitoring oxidative and nitrative modification of cellular proteins; a paradigm for identifying key disease related markers of oxidative stress. Adv Drug Deliv Rev. 2008;60:1497–503.
20. Gupta VK, Mallika V, Gupta Y, Srivastava DK. Oxygen derived free radicals in clinical context. Ind J Clin Biochem. 1992;7:3–10.
21. Bauknight GC Jr, Wei EP, Kontos HA. Superoxide production in experimental seizures in cats. Stroke. 1992;23:1512–4.
22. Frantseva MV, Perez Velazquez JL, Hwang PA, Carlen PL. Free radical production correlates with cell death in an in vitro model of epilepsy. Eur J Neurosci. 2000;12:1431–9.
23. Traub RD, Miles R. Multiple modes of neuronal population activity emerge after modifying specific synapses in a model of the CA3 region of the hippocampus. Ann N Y Acad Sci. 1991;627:277–90.
24. Peterson SL, Morrow D, Liu S, Liu KJ. Hydroethidine detection of superoxide production during the lithium-pilocarpine model of status epilepticus. Epilepsy Res. 2002;49:226–38.
25. Rauca C, Zerbe R, Jantze H. Formation of free hydroxyl radicals after pentylenetetrazol-induced seizure and kindling. Brain Res. 1999;847:347–51.
26. Rajasekaran K. Seizure-induced oxidative stress in rat brain regions: blockade by nNOS inhibition. Pharmacol Biochem Behav. 2005;80:263–72.

27. Frantseva MV, Perez Velazquez JL, Tsoraklidis G, et al. Oxidative stress is involved in seizure-induced neurodegeneration in the kindling model of epilepsy. Neuroscience. 2000;97:431–5.
28. Erakovic V, Zupan G, Varljen J, Radosevic S, Simonic A. Electroconvulsive shock in rats: changes in superoxide dismutase and glutathione peroxidase activity. Mol Brain Res. 2000;76:266–74.
29. Arnaiz SL, Travacio M, Llesuy S, Arnaiz GRD. Regional vulnerability to oxidative stress in a model of experimental epilepsy. Neurochem Res. 1998;1477–87.
30. Sullivan PG, Dube C, Dorenbos K, Steward O, Baram TZ. Mitochondrial uncoupling protein-2 protects the immature brain from excitotoxic neuronal death. Ann Neurol. 2003;53:711–7.
31. Arzimanoglou A, Hirsch E, Nehlig A, et al. Epilepsy and neuroprotection: an illustrated review. Epileptic Disord. 2002;4:173–82.
32. Willmore LJ, Ueda Y. Posttraumatic epilepsy: hemorrhage, free radicals and the molecular regulation of glutamate. Neurochem Res. 2009;34:688–97.
33. Pitkanen A, McIntosh TK. Animal models of post-traumatic epilepsy. J Neurotrauma. 2006;23:241–61.
34. Rosen AD, Frumin NV. Focal epileptogenesis after intracortical hemoglobin injection. Eur Neurol. 1979;66:277–84.
35. Gupta YK, Gupta M. Post traumatic epilepsy: a review of scientific evidence. Indian J Physiol Pharmacol. 2006;50:7–16.
36. Kunz WS, Kudin AP, Vielhaber S, et al. Mitochondrial complex I deficiency in the epileptic focus of patients with temporal lobe epilepsy. Ann Neurol. 2000;48:766–73.
37. Junker H, Spate K, Suofu Y, et al. Proteomic identification of the involvement of the mitochondrial Rieske protein in epilepsy. Epilepsia. 2005;46:339–43.
38. Ben-Menachem E, Kyllerman M, Marklund S. Superoxide dismutase and glutathione peroxidase function in progressive myoclonus epilepsies. Epilepsy Res. 2000;40:33–9.
39. Eun JP, Choi HY, Kwak YG. Proteomic analysis of human cerebral cortex in epileptic patients. Exp Mol Med. 2004;36:185–91.
40. Ashrafi MR, Shams S, Nouri M, et al. A probable causative factor for an old problem: selenium and glutathione peroxidase appear to play important roles in epilepsy pathogenesis. Epilepsia. 2007;48:1750–5.
41. Mori A, Yokoi I, Noda Y, Willmore LJ. Natural antioxidants may prevent posttraumatic epilepsy: a proposal based on experimental animal studies. Acta Med Okayama. 2004;58:111–8.
42. Kabuto H, Yokoi I, Ogawa N. Melatonin inhibits iron-induced epileptic discharges in rats by suppressing peroxidation. Epilepsia. 1998;39:237–43.
43. Gulati K, Ray A, Pal G, Vijayan VK. Possible role of free radicals in theophylline-induced seizures in mice. Pharmacol Biochem Behav. 2005;82:241–5.
44. Chavko M, Auker CR, McCarron RM. Relationship between protein nitration and oxidation and development of hyperoxic seizures. Nitric Oxide. 2003;9:18–23.
45. Ferraro G, Sardo P. Nitric oxide and brain hyperexcitability. In Vivo. 2004;18:357–66.
46. Aycicek A, Iscan A. The effects of carbamazepine, valproic acid and phenobarbital on the oxidative and antioxidative balance in epileptic children. Eur Neurol. 2007;57:65–9.
47. Chang TK, Abbott FS. Oxidative stress as a mechanism of valproic acid-associated hepatotoxicity. Drug Metab Rev. 2006;38:627–39.
48. Schulpis KH, Lazaropoulou C, Regoutas S, et al. Valproic acid monotherapy induces DNA oxidative damage. Toxicology. 2006;217:228–32.
49. Mahle C, Dasgupta A. Decreased total antioxidant capacity and elevated lipid hydroperoxide concentrations in sera of epileptic patients receiving phenytoin. Life Sci. 1997;61:437–43.
50. Musavi S, Kakkar P. Diazepam induced early oxidative changes at the subcellular level in rat brain. Mol Cell Biochem. 1998;178:41–6.
51. Cardile V, Pavone A, Renis M, Maci T, Perciavalle V. Effects of Gabapentin and topiramate in primary rat astrocyte cultures. Neuroreport. 2001;12:1705–8.

52. Wang JF, Azzam JE, et al. Valproate inhibits oxidative damage to lipid and protein in primary cultured rat cerebrocortical cells. Neuroscience. 2003;116:485–9.
53. Cui J, Shao L, Young LT, Wang JF. Role of glutathione in neuroprotective effects of mood stabilizing drugs lithium and valproate. Neuroscience. 2007;144:1447–53.
54. Naziroglu M, Kutluhan S, Yilmaz M. Selenium and topiramate modulates brain microsomal oxidative stress values, Ca^{2+}-ATPase activity, and EEG records in pentylentetrazol-induced seizures in rats. J Membr Biol. 2008;225:39–49.
55. Leppik IE. Zonisamide: chemistry, mechanism of action, and pharmacokinetics. Seizure. 2004;13(Suppl 1:)S5–9; discussion S10.
56. Kim do Y, Rho JM. The ketogenic diet and epilepsy. Curr Opin Clin Nutr Metab Care. 2008;11:113–20.
57. Bough KJ, Rho JM. Anticonvulsant mechanisms of the ketogenic diet. Epilepsia. 2007;48:43–58.

Chapter 11
Neurological Disorders Associated with Iron Misdistribution: The Therapeutic Potential of Siderophores

Or Kakhlon, Bill Breuer, Arnold Munnich, and Z. Ioav Cabantchik

Abstract The association between iron accumulation and oxidative cell damage in specific regions of the brain has raised the possibility of using iron chelation as a therapeutic strategy in neurodegenerative disorders. In this review, Friedreich's ataxia (FRDA) is described as a paradigm of a disorder in which cell damage is associated with misdistribution of cell iron resulting from aberrant utilization of the metal by mitochondria. Frataxin-deficient cells, serving as an in vitro model of FRDA, were used as a model for testing the effects of a chelator with the ability to sequester cell iron as well as to donate it to biological acceptors inside or outside cells (termed here a siderophore). The clinically used siderophore deferiprone (DFP) was shown to reduce mitochondrial labile iron accumulation and restore cell functions affected by frataxin deficiency. In a translation of these findings to the clinical setting, moderate doses of DFP were administered to selected patients with FRDA or neurodegeneration with brain iron accumulation (NBIA) as part of open-label studies. Treatment with DFP resulted in a significant reduction in foci of brain iron accumulation and in some functional improvements, with only minor changes in body iron stores. The prospects of using drug-mediated iron relocation vs. chelation are critically discussed in relation to other neurodegenerative disorders in which accumulated (toxic) iron has been implicated as a causative factor in regional oxidative damage, such as Parkinson's disease or Alzheimer's disease.

Keywords Iron misdistribution · Iron relocation · Siderophores · Chelators · Neurological disorders · Friedreich's ataxia · Neurodegeneration with brain iron accumulation · Alzheimer's disease · Parkinson disease

---

Z.I. Cabantchik (✉)
Adelina and Massimo Della Pergola Chair Life Sciences Institute,
Hebrew University of Jerusalem,
Safra Campus-Givat Ram, Jerusalem 91904, Israel
e-mail: ioav@cc.huji.ac.il

N. Gadoth and H.H. Göbel (eds.), *Oxidative Stress and Free Radical Damage in Neurology*, Oxidative Stress in Applied Basic Research and Clinical Practice,
DOI 10.1007/978-1-60327-514-9_11, © Springer Science+Business Media, LLC 2011

Abbreviations

AD	Alzheimer's disease
APP	Amyloid precursor protein
DFP	Deferiprone
FRDA	Friedreich ataxia
ISC	Iron sulfur clusters
LIP	Labile iron pool
NBIA	Neurodegeneration with brain iron accumulation
PD	Parkinson's disease
ROI	Reactive oxygen intermediates
ROS	Reactive oxygen species

1 Introduction

Iron accumulation has been observed in affected areas of brains of patients with various neurodegenerative disorders and has been suggested to be the sine qua non of neuronal damage [1]. Hitherto the causes of iron accumulation in specific regions of the brain, the chemical nature of the accumulated metal, and, most importantly, the causal relationship between iron accumulation and neuronal damage, have all remained largely circumstantial [2–7]. Evidently, iron-mediated damage is attributed not to the bulk of accumulated iron but rather to a relatively small fraction appearing in redox-active forms, which can catalyze the formation of noxious radicals from partially reduced oxygen species. The major evidence for the involvement of iron in various forms of brain damage is the protection provided by treatment with iron chelators [8–11]. These observations have provided the conceptual basis and justification for developing therapeutic approaches for neurodegenerative disorders based on chelators that cross the blood–brain barrier (BBB) and presumably remove labile iron from regional iron deposits. However, considering that regionally accumulated iron represents only a minor fraction of the total body iron and that iron accumulation in a particular area of the brain might occur at the expense (if not as a result) of iron depletion elsewhere, we opted for referring to the disorders as associated with or resulting from misdistribution of iron rather than from accumulation per se. Therefore, the therapeutic approach to those disorders should be based on (a) selective sequestration of labile, reactive iron without perturbing the bulk of safely complexed iron; (b) safe redistribution of the metal within the target cell, tissue, or organism; and (c) sparing of other essential metals such as copper and zinc [12, 13].

In the present chapter we first discuss the cellular and molecular basis of iron toxicity, which results in conditions of misregulated iron metabolism at the level of the whole organism, specific tissues, or individual cells. We then discuss FRDA as a paradigm of a disorder that is etiologically associated with cell damage caused by misdistribution of iron in a distinct area of the brain [14]. We describe recent attempts to (a) define biochemically the nature of iron misdistribution and its functional consequences in cell models of FRDA; (b) identify agents capable of selectively relocating cell iron (siderophores) to restore cellular functions affected by iron misdistribution; and (c) translate the knowledge on siderophore action gained in cell models to the clinical setting, first in FRDA and subsequently in neurodegeneration with brain iron accumulation (NBIA) patients. Finally, we discuss the prospects of chelating brain iron with drugs for treating more widespread neurodegenerative disorders such as Parkinson's disease (PD) [15–17], in which iron has been implicated in regional oxidative damage. As part of this discussion, we compare the classical approach based on removal and excretion of excess iron from the body with the alternative concept of iron relocation, aimed at redistributing labile iron from sites where it is potentially toxic to sites where it is required or can be safely stored.

The toxic side of iron metabolism. The human organism is endowed with molecular machineries for acquiring metals from the environment and distributing them among tissues in a controlled and balanced manner [18, 19]. At the cellular level, acquisition of iron for metabolic use from the surrounding fluids is tightly regulated and poised to prevent excessive accumulation while ensuring the maintenance of cellular levels within a homeostatic range. As a rule, most of the cell iron is secured in cage-like structures associated with proteins and only a minor fraction is maintained readily available for metabolic utilization, mostly as a labile iron pool (LIP) [20–22]. However, the maintenance of intracellular pools of labile iron carries with it an inherent risk. If labile iron interacts with reactive oxygen intermediates (ROIs), mainly superoxide anions resulting from incomplete reduction of O_2 during respiration and with hydrogen peroxide resulting from the dismutation of superoxide, it can potentially generate noxious reactive oxygen species (ROS) [23]. To minimize excessive ROS formation and ensuing damage, cells actively maintain LIP at relatively low levels and biochemically eliminate ROIs in various cell compartments [24, 25]. Such measures become insufficient when excessive iron accumulation leads to a concomitant rise in LIP or when ROIs are overproduced or inadequately removed, as in inflammation [26]. In those cases, ROS leads to oxidative damage of critical cell components, often resulting in cell death.

Well-described paradigms of iron toxicity exist in conditions of systemic iron overload (e.g., hemochromatosis, thalassemia) where labile iron complexes present in plasma gain unrestricted access to cells via nonselective pathways, causing massive intracellular iron accumulation (especially in liver, heart, and endocrine glands) [27, 28]. However, another subtle modality of iron accumulation has recently been recognized in genetic disorders that affect cell iron utilization (e.g., trafficking and incorporation into proteins) and result in regional misdistribution of the metal. In such cases, a faulty component of the cell's iron-handling machinery results on the one hand in a deficiency of essential iron-dependent proteins and on the other hand in the accumulation of "unused" iron in a given cell compartment or tissue. Examples of this modality of iron accumulation are the inherited disorders FRDA, caused by frataxin deficiency [29, 30], X-linked sideroblastic anemia with ataxia (XLSA/A), caused by ABCB7 deficiency [31–34], and NBIA, caused by a deficiency in either pantothenate kinase 2 (PANK2) or in *PLA2G6* encoding a calcium-independent group VI phospholipase A2 that is important in phospholipid remodeling, arachidonic acid release, leukotriene and prostaglandin synthesis, and apoptosis [35]. An analogous situation with a different etiology, but a similar end result, is found in acquired anemia of chronic disease (ACD), where iron misdistribution is manifested systemically as iron accumulation in the reticuloendothelial system and deprivation in interstitial fluids and plasma [26, 36].

2 Friedreich's Ataxia and Iron

FRDA, a degenerative disease with autosomal recessive inheritance affecting 1 in 40,000 births, is characterized by progressive limb and gait ataxia, areflexia, pyramidal signs in the legs, and hypertrophic cardiomyopathy [6, 37–39]. The disease is caused by a GAA repeat expansion in the first intron of the nuclear-encoded gene for the protein frataxin, resulting in deficiency of this mitochondrial protein. The long GAA repeats apparently interfere with transcription by forming aberrant DNA structures or by promoting repressive heterochromatin forms [40]. The silencing of the frataxin gene can be reversed in experimental systems by histone deacetylase inhibitors [41], which is one of the pharmacological approaches currently under investigation. The severity of the disease appears to correlate with the degree of frataxin deficiency [6]. Frataxin has been claimed to have several functions, including iron storage and protecting mitochondria from oxidative damage [30], but its best defined role is the delivery of iron to the machinery responsible for the synthesis of iron–sulfur clusters (ISCs) [30, 42, 43], which serve as prosthetic groups for a variety of enzymes, collectively termed iron–sulfur cluster proteins (ISPs). Inefficient ISP formation in FRDA causes a combined aconitase and respiratory chain (complex I–III) deficiency [44] and leads to mitochondrial accumulation of labile iron [45, 46], which in turn promotes oxidative damage.

Much of the information on the biochemical consequences of frataxin deficiency has been obtained in studies with cellular models including genetically manipulated yeasts [39], fibroblasts isolated from patient skin [47–49], immortalized lymphoblasts from patients [49], or other cell types in which frataxin was suppressed ectopically [50–52]. Most studies in these models focused on a set of properties reflecting discrete changes in energy production and resistance to oxidative stress. It is generally agreed that frataxin-deficient cells show a marked reduction in mitochondrial membrane potential (MMP) [53], respiration rate, and ATP levels [54], impaired antioxidant capacity manifested as increased levels of ROS, and signs of oxidative damage detectable as increased protein carbonylation [51, 55] and apoptotic index [50]. A study in which the biochemical and cellular effects of frataxin deficiency and the associated misdistribution of cell iron were comprehensively analyzed in a single system was recently published [56].

Histopathological examination of biopsies and magnetic resonance imaging (MRI) studies of patients have shown that iron accumulates regionally in heart muscle [57, 58], spinocerebellar tracts (dentate nuclei), and the spinal cord [29, 44, 59] of FRDA patients, often colocalizing with structurally damaged areas [7, 14]. The iron detected histochemically is ferric, non-heme iron (usually ferritin or hemosiderin), based on Perl's acid ferrocyanide staining of biopsies, fixed cells, or tissues [58]. The appearance of Prussian blue-colored forms (or brown-black forms if these are subsequently enhanced by treatment with diaminobenzidine + hydrogen peroxide) identifies abnormally high levels of non-heme iron, which, depending on the microscopic resolution, can be revealed in whole cells or in their organelles, primarily mitochondria. In patients, the identification of iron accumulation in particular organs is done by MRI, based on the ability of clustered iron to cause a local nonhomogeneity in the magnetic field and thereby decrease the MRI signal of surrounding water [60, 61]. This effect leads to variations in the apparent transverse relaxation rate (R2*), or its inverse, T2*, the indices used to quantify MRI signal decay. Thus, assuming a homogeneous external magnetic field, changes in R2* reflect variations in local iron concentration. In FRDA patients, changes in R2* have been attributed to regional iron accumulation in the dentate nuclei of the cerebellum, with a possible, although less likely, signal contribution from fluctuating deoxyhemoglobin levels resulting from regional changes in blood perfusion [60].

2.1 FRDA as a Paradigm of Misdistribution of Cell Iron

The pathophysiological role of iron in FRDA has been a matter of debate [62]. This question pertains to the issue of whether iron accumulates in the mitochondria of frataxin-deficient cells at all [49, 52, 63] and to the possible role of iron accumulation in oxidative damage observed both in cells and in animal models [29, 30, 64, 65]. The conflicting conclusions could arise either from differences between the model systems employed, namely, how closely they reproduce the FRDA phenotype, or from the methods, namely, sensitivity of the assays for tracing labile iron in particular compartments, or from both.

The notion that iron is improperly distributed in frataxin-deficient FRDA cells was originally based on the observation that mitochondrial accumulation of redox-active iron is accompanied by depletion of the cytosolic LIP [29, 30] (reviewed in [34]). The latter has been inferred from the reduced levels of cytosolic iron regulatory proteins found in mammalian cells [66] and the activation of the iron regulon in frataxin-deficient yeasts [67–69]. Experimental in vivo support for this assumption was recently provided by a study of cardiomyopathy in conditional frataxin-knockout mice, where cytosolic iron depletion in cardiomyocytes was shown to be associated with an increased rate of total cellular iron uptake and accumulation in the mitochondria [70].

At the cellular level, labile iron accumulation in mitochondria and its concomitant depletion in the cytosol were observed in frataxin-deficient cells probed with fluorescent sensors of labile iron and ROS targeted to specific cell compartments [56]. The cause of iron accumulation in mitochondria and iron depletion from the cytosol in frataxin-deficient cells has been assumed to be the failure to export processed forms of iron from the mitochondria, either in labile forms or "packaged" in heme or ISCs, but this has not been established [71]. The fact that frataxin deficiency causes an increase in mitochondrial LIP and a reduction in cytosolic LIP is consistent with the presumed function of frataxin and the interdependence between the two pools [34, 72]. However, more importantly, it provides an opportunity for assessing the redistribution of iron as a tool for correcting the cell properties impaired by the misdistributed metal [56]. Regardless of the underlying causes of cellular iron misdistribution in FRDA, any attempt to remove iron needs to take into account that cytosolic iron depletion by chelators may be just as detrimental as excessive mitochondrial iron accumulation because it could promote a futile cycle of increased cellular iron uptake. One way to resolve this dilemma experimentally is to use frataxin-deficient cell models to compare the beneficial effects of iron-binding agents with different properties: iron chelators, which irretrievably sequester cellular LIPs, vs. siderophores (i.e., iron carriers), which chelate iron, but also donate it to selected acceptors, analogously to microbial or plant siderophores [73].

2.2 *Restoration of Functions by a Siderophore in FRDA Models*

The classical approach to coping with accumulation of iron has been the administration of iron chelators, which prevent the toxic effects of iron excess and, in most cases, bring about its excretion [74–76]. Most of the experience with chelation has been gained from treating patients with systemic iron overload, namely, patients with hemosiderosis who show iron accumulation in all body fluids and tissues [74–78]. The objective of such treatments is bulk removal of iron from the system. In contrast, diseases of regional iron accumulation are often accompanied by systemic or regional iron deficiency that affects specific iron-dependent processes. In these diseases, it may be essential to conserve iron and ideally even to render it available for metabolic reuse. In the case of FRDA, mere removal of excess iron from the mitochondria may not alleviate defects that stem from inadequate delivery of iron to the ISC synthesis machinery, which is also located primarily in the mitochondria [71]. In spite of this apparent paradox, the effect of iron-chelating drugs on FRDA models has been explored in numerous studies, with variable success [66, 70, 79, 80]. Oxidatively stressed frataxin-deficient cells have been shown to be rescued by treatment with the coenzyme Q10 analogue idebenone, an antioxidant and electron donor to the respiratory chain [81], as well as the classical iron-depleting chelator deferrioxamine [10]. Furthermore, a combination of the membrane-penetrating chelator pyridoxal isonicotinoyl hydrazone with deferrioxamine prevented cardiac iron loading and partially alleviated cardiac hypertrophy in mice with cardiac frataxin deficiency, although it did not prevent decreased cardiac function [70]. The rationale of this approach, based on an earlier study [82], was that the chelator pyridoxal isonicotinoyl hydrazone rapidly reaches all intracellular LIPs and apparently transfers the chelated iron to extracellular circulating deferrioxamine, producing optimal iron extraction from cells and ultimate excretion from the organism.

An alternative therapeutic approach to FRDA is based on iron *redistribution*, which entails chelation of a particular LIP, relieving cells or mitochondria from foci of accumulated labile iron and transfer of the metal, directly or indirectly, to endogenous acceptors and possibly to other compartments inside or outside the cells [83]. Ideally, the ultimate acceptors of siderophore-redistributed iron should be iron-requiring proteins, such as ISC-containing enzymes. The approach utilizes agents such as the oral chelator 1,2-dimethyl-3-hydroxypyrid-4-one (deferiprone, DFP) [77, 84], which fulfills the criteria for a siderophore based on its ability to perform the following:

(a) Gain access to cells, effectively scavenge intracellular LIPs [85], and exit cells as an iron chelate [83].
(b) Selectively bind iron in the various intracellular LIPs [83] and thereby reduce iron-dependent free radical formation [13, 86, 87].
(c) Cause minimal losses of other essential metals such as zinc [88, 89].
(d) Spare extracellular transferrin-bound iron and even transfer chelated iron to apotransferrin; this has been demonstrated both in vitro [83, 89] and in vivo [90][1].
(e) Donate iron for metabolic reutilization, as recently shown in a cellular in vitro system of hemoglobin synthesis [83]. To what extent DFP is able to donate iron for intracellular reutilization in the brain might also depend on the local concentration of transferrin, which is lower in the cerebrospinal fluid (CSF) than in the circulation by at least an order of magnitude [92, 93].

A further advantage of DFP in relationship to neurodegenerative diseases is its ability to cross the BBB, as experimentally shown in perfused rats [94–96]. Although no effects on central nervous system (CNS) function have been reported in thalassemia patients treated with DFP [97, 98], dosages of the drug must be selected so as avoid any adverse effects on enzymes involved in dopamine metabolism [96, 99].

[1]In iron overload conditions, which often lead to >75% transferrin saturation and poor iron binding capacity in the plasma, the DFP-iron complexes are excreted via the urine. On the other hand, under conditions of normal transferrin saturation (25–35%), essentially all of the DFP-bound iron in plasma is transferred to circulating transferrin and is not lost from the system. Therefore, DFP administration in moderate doses is not likely to lead to significant iron excretion in subjects with normal iron saturation [90, 91].

The capacity of DFP to act as a siderophore was assessed in frataxin-deficient cells as a model of cellular iron misdistribution [56]. At pharmacological concentrations the agent markedly reduced the mitochondrial LIP, diminished the increased ROS generation, oxidative damage, mitochondrial redox potential, and apoptotic index of the cells, and restored the reduced MMP, ATP production, and respiration. Presumably, the biochemical basis for the recovery of these multiple, diverse functions is the resumption of ISC synthesis leading to restoration of the levels of ISPs required for energy production. Indeed, DFP was shown to enhance the regeneration of the activity of the ISP enzyme aconitase, a process that is defective in frataxin-deficient cells. Two other clinically used chelators, deferrioxamine and deferasirox, known for their higher iron-binding affinities and their reduced capacity for donating iron to cellular acceptors, restored the impaired functions to a limited extent [56]. The study demonstrated for the first time that a chemical agent can act as a "frataxin surrogate" in the reconstitution of ISP activity and the correction of the defective energetic parameters observed in frataxin-deficient cells. Similar to frataxin, DFP could act in FRDA either by detoxification of LIP, thus slowing down oxidative destruction of ISCs, or by iron donation, facilitating ISC repair or synthesis, or both (Fig. 1).

2.3 Clinical Application of Siderophores in FRDA and NBIA

The recognition of FRDA as a disease of regional misdistribution of iron that affects a plethora of cell properties restorable in vitro by siderophore treatment has led to an initial efficacy-tolerability Phase I–II open study of DFP [90]. Nine FRDA adolescents were selected, who had been treated for years with idebenone, as a possible cardioprotective agent [99–101], but who showed no neuromuscular improvements. The cohort completed a 6-month treatment series with 20–30 mg/kg/day DFP and was followed periodically for neurological and hematological parameters. Iron accumulation in dentate nuclei was assessed using R2* MRI. DFP treatment significantly and selectively reduced iron accumulation in cerebellar dentate nuclei, as well as improving the ICARS score in the youngest patients, without causing systemic iron deficiency or affecting the hematological status [90]. Based on this preliminary work a multicenter randomized double-blind (DFP vs. placebo) study has been initiated on a larger cohort of patients that included walking adolescents and nonwalking adults, using ataxia scales as primary clinical end points [102].

Another candidate disease that may potentially respond to siderophores is NBIA, a clinically and genetically heterogeneous group of progressive neurological disorders characterized by high basal ganglia iron accumulation and axonal dystrophy [103–105]. To date, two genetic forms have been associated with mutations in pantothenate kinase 2 (PANK2) and phospholipase 2G6 (*PLA2G6)*, both of which encode proteins that are critical to organellar or membrane integrity [103]. Cases of

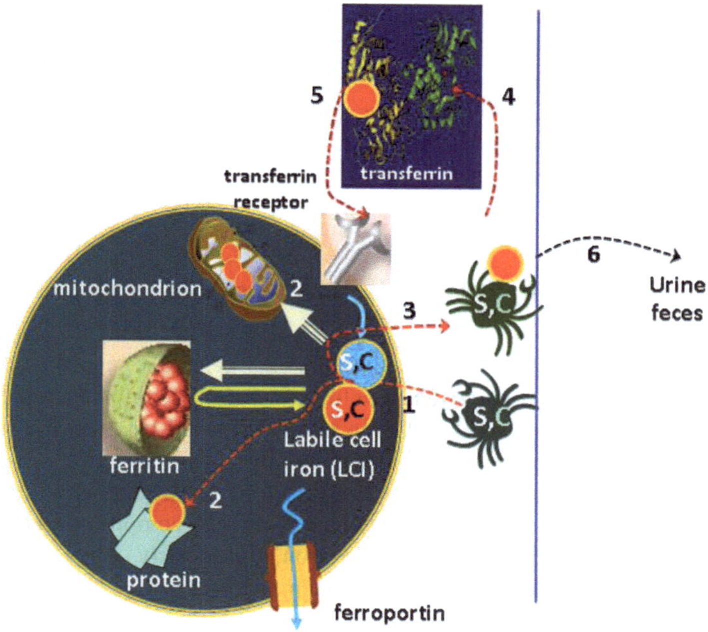

Fig. 1 Mode of action of a siderophore as iron-relocating agent vs. a chelator as iron scavenger agent. A siderophore (*S*) permeates into cells (*1*) and binds labile cell iron (*LCI*) [*blue circle*, Fe(II); *red circle*, iron(III)] and either relocates it intracellularly (*2*) or exits the cell (*3*) and transfers the iron to external transferrin (*4*), which in turn cycles the metal back into the cell (*5*). An iron chelator (*C*) permeates into cells (*1*) and binds labile cell iron (*LCI*) and exits the cell (*3*) and eventually the organism (*6*). Indicated are the transferrin receptor that is involved in iron uptake by receptor-mediated endocytosis of circulating holo-transferrin

NBIA caused by mutations in the gene that encodes for pantothenate kinase 2 (PANK2) are classified as pantothenate kinase-associated neurodegeneration (PKAN) and are characterized by accumulation of iron in the globus pallidus and substantia nigra, discernible by T_2*-weighted brain MRI [103]. Axonal cell death in PKAN is thought to be primarily caused by decreased levels of coenzyme A leading to impaired energy and lipid production. Accumulation of secondary metabolites such as cysteine in the mitochondria, with the ensuing deposition of reactive iron, is thought to be a secondary factor that further exacerbates the oxidative stress and cell damage [5]. The reason for the occurrence of an estimated threefold increase in iron content, particularly in the globus pallidus and substantia nigra, is not known. It has been suggested that because these areas are normally iron enriched relative to the rest of the brain, they are highly prone to excessive iron deposition caused by iron binding by the accumulated cysteine [35, 103].

The first reported case of successful treatment of NBIA with DFP was published recently [106]. DFP administration to a 61-year-old NBIA patient for 8 months led to a gradual, significant amelioration of various neurological symptoms related to the gait, involuntary movements, blepharospasm, and choreic dyskinesias. Brain

MRI revealed reduced iron deposition in the basal nuclei. Whether the effects of DFP in this case can be attributed to net removal of iron ("chelation") or its transfer ("redistribution") from areas of deposition to other areas capable of iron absorption, is unclear. Nevertheless, it is notable that adverse events or abnormal hematological values were not observed during treatment, indicating that systemic iron homeostasis was largely preserved.

2.4 Possible Side Effects of Chelators and Siderophores

Concern about side effects in chelator treatments generally revolves around the question of whether the expected benefits of chelation outbalance the risk of generating iron deficiency [107, 108]. In iron-overloaded patients, iron chelation regimens are adjusted according to the degree of iron overload and suspended when serum ferritin is decreased to an acceptable level [107]. Any chelation treatment of individuals with normal metabolism, let alone those with compromised iron metabolism, must be carried out with moderation, particularly when applied over extended periods of time with chelators of high iron-binding affinity and tissue accessibility. This concern is exemplified by chelator treatment of aceruloplasminemia, which is accompanied by anemia in the background of massive iron overload in the brain and liver. Deferrioxamine [109] and deferasirox [110] removed excess iron from the liver but aggravated the disease-related anemia. In non-iron-overloaded humans and in experimental animals, moderate chelation treatments appear to be tolerated [13, 111], presumably because of enhanced iron recycling and absorption from the diet. Thus, although chelators can potentially induce a state of intracellular iron deficiency in the organism, that deficiency can be compensated by increased iron acquisition from serum transferrin by upregulated iron-uptake mechanisms [18, 112, 113]. Theoretically, the potential complications of overchelation may be less pronounced in treatments based on siderophores.

An example of a less predictable side effect of chelation with DFP specifically is the risk of approximately 1–2% of iron-overloaded patients developing neutropenia/agranulocytosis [106]. Although a slightly higher incidence was found in the only trial of DFP in FRDA patients published so far [90], the sample number was too small to draw definitive conclusions as to the significance of the observations. Irrespectively, the neutropenia/agranulocytosis condition is readily reversible by suspending DFP administration [107].

Another, less evident side effect might ensue following the fast withdrawal from circulation of highly permeant chelators. First, a time-lag in the readjustment of iron acquisition mechanisms activated by the chelator may lead to a temporary rebound and transient increase in cellular LIP [114]. Second, it is theoretically possible that labile, redox-active iron forms may be generated under in vivo conditions by a decrease in the ratios of chelator to iron [114]. One prophylactic maneuver to reduce the impact of such sporadic events could be the coadministration of an antioxidant. Although the validity of such an approach remains to be demonstrated clinically, idebenone, which is thought to enhance the antioxidant capacity of cells by compensating for deficient electron transport [10, 47, 48] is already in common use in FRDA and has been claimed to be beneficial at high doses [99]. The antioxidant coadministration strategy may also have additional benefits because antioxidants could also synergize with DFP by rendering iron more accessible to chelation (via its reduction) and relocation [9, 86, 115, 116].

3 Iron Accumulation in Other Neurological Disorders

The anomalous levels of iron detected in specific areas of the brain in various pathologies have been presumed to be linked to iron dysregulation; however, it has remained a matter of debate whether that property should be regarded as proof that iron accumulation has a causative role or is an exacerbating factor in neurodegeneration [6]. In most cases regional iron accumulation might be a secondary outcome of the neurological disorder because of its late appearance/detection in a variety of disparate diseases such as Alzheimer's disease (AD), PD, progressive supranuclear palsy (PSP), and multiple system atrophy with strionigral degeneration (MSA) [117]. In a cardiac model of FRDA in the mouse, heart damage was identified before any blatant accumulation of iron [58]. In the NBIA group of syndromes, including neuroaxonal dystrophy [35, 118], mutations in different genes coding for distinctly different enzymes affect cellular properties that are not directly linked to iron accumulation in general and to labile forms in particular. In fact, the iron species identified in many postmortem brain sections are mostly found in forms sufficiently stable to withstand fixation procedures and associated with the iron storage protein ferritin. Nonetheless, as briefly described next, various studies dealing with PD, AD, and brain inflammation have provided a body of experimental evidence implicating iron as a major contributor to these pathologies [13, 78, 111, 119, 120].

3.1 Parkinson's Disease

The appearance of iron accumulation in the substantia nigra is increasingly accepted as a hallmark of PD. The evidence points toward an increase in labile chelatable iron in the brain of PD patients that is (1) unaccompanied by a parallel elevation in ferritin [15], (2) involved in promoting α-synuclein aggregation, the main mediator of neurotoxicity in the disease [120–123], and (3) chelatable by deferrioxamine, which prevents the process in vitro [8]. Iron is also known to form complexes with the naturally occurring, catechol group-containing neuromelanin, which accumulates over time, particularly in the substantia nigra. These complexes may be a source of redox-active iron, which could then mediate oxidative stress and inflammation [124].

3.2 Alzheimer's Disease

The principal contribution of iron to neuronal damage in AD is assumed to derive from the enhanced translation of the amyloid precursor protein (APP), whose mRNA carries an iron-responsive element (IRE) in the 5′-untranslated end [111]. The presumed sequence of events is that increased neuronal cytosolic LIP stimulates increased APP synthesis, which then gives rise to increased levels of Aβ peptide via proteolytic processing and ultimately to toxic Aβ aggregates [16, 125]. However, the causes of the predicted rise in LIP and the involvement of extracellular iron in the formation and deposition of extracellular Aβ plaques are open to conjecture. In vitro, metals have been shown to promote Aβ peptide precipitation, while chelators have been shown to prevent and even reverse Aβ peptide aggregation [126]. The known capacity of the Aβ peptide for binding metals such as zinc, copper, and iron [127] can lead to localized, redox-active metal accumulation on Aβ aggregates, with potential for generating ROS and ensuing oxidative stress. However, iron has been claimed to be predominantly located in neuritic processes within plaques and associated with intracellular ferritin rather than with extracellular Aβ [120]. It has also been suggested that sequestration of iron in ferritin, a recognized defense mechanism, may even cause localized iron deprivation, with consequent disruption of normal iron metabolism [128]. Circumstantial evidence for the involvement of iron in AD has been provided by studies of AD patients with the HFE mutation [128], a condition associated with hemochromatosis and systemic iron overload [129]. The HFE mutations C282Y and H63D have been associated with increased oxidative stress and severity of AD, as assessed neuropathologically [130, 131].

3.3 Brain Inflammation

Both AD and PD elicit local inflammatory responses, as evidenced by upregulation of inflammatory mediators and microglial activation [132]. Iron accumulation in itself can aggravate or even engender localized inflammatory responses, mediated by the release of various pro-inflammatory cytokines, including interleukin-1β, nitric oxide [133, 134], tumor necrosis factor-α, and interleukin-6 [135, 136]. The link between iron and inflammation in PD models has been reported to be the activation of NF-κB [135, 136]. The combination of inflammation and iron accumulation can be especially damaging in the brain for various reasons, which include (a) the low iron-binding capacity of the CSF [92, 93], which may be insufficient to prevent the tendency for extracellular iron to become redox active as well as undergo deposition at cell surfaces; (b) the high oxygen consumption of the CNS (20% of the total body consumption), necessarily accompanied by high ROI generation [13], and (c) variable levels of antioxidants other than ascorbate. Both intracellular and extracellular reduced ascorbate levels in the brain are high (reaching the mM range), sufficient to provide a measure of antioxidant protection, but at the same time maintain iron in a reduced and redox-active form [13]. Therefore, iron restriction by chelators has been envisaged as a promising strategy for curbing the inflammatory cycle in the brain [137].

3.4 *Treatment of PD and AD with Iron Chelators*

Considerable effort has been devoted to demonstrating the ability of various iron chelators to either provide protection against the onset of PD and AD or delay the course of the diseases. The effectiveness of classical iron-chelating agents has been assessed in a variety of cell and animal models [12, 13, 138, 139]. A promising variation on the chelator-based approach has employed multifunctional compounds, equipped with high-affinity iron-binding moieties in combination with radical scavenging groups, monoamine oxidase inhibitors [111], or neuroprotective peptides [140]. However, with the possible exception of epigallocatechin gallate, a component of green tea and apparently a brain-permeable iron chelator [141], all chelators in preclinical evaluation for efficacy in particular neurodegenerative disorders remain to be evaluated for hematological side effects resulting from systemic iron deprivation.

The three iron chelators in current clinical use for treating systemic iron overload – deferrioxamine, deferasirox, and DFP – are first-line candidates in attempts to chelate brain iron. An early clinical study (that was apparently not followed up) evaluated the long-term efficacy of deferrioxamine chelation in AD, showing a delay in neurodegeneration [142]. Similarly, there have been case reports of neurotherapeutic efficacy of deferrioxamine in aceruloplasminemia [143] (reviewed in [144]), although these were not corroborated [109]. The orally absorbable DFP and deferasirox provide alternatives to deferrioxamine that are not only more convenient and practical but also have the ability to rapidly cross membranes in the iron-bound and free forms. DFP has been reported to ameliorate autoimmune encephalomyelitis [145]. Deferasirox has recently been reported to alleviate liver iron overload and improve neurological symptoms in a case of nonclassical aceruloplasminemia without detectable cerebral iron accumulation [110]. In another patient with aceruloplasminemia and MRI-detectable brain iron accumulation, but without neurological symptoms, DFP failed to relieve the brain iron load [109].

As the foregoing examples indicate, chelation of excess brain iron is still at the trial-and-error stage. However, if chelator efficacy studies are to be undertaken in AD or PD, choosing appropriate dosages might be crucial, as a red line exists, even with DFP, between causing iron withdrawal from the organism rather than redistribution of the chelated metal in bioavailable form. With this consideration in mind, the DFP doses of 20–30 mg/kg chosen for treating FRDA patients [90] and NBIA [106] were only a fraction (1/3–1/4) of the doses usually given to thalassemia major patients.

Assuming the iron-related damage in PD and AD is predominantly intracellular, chelators should be permeant and able to access intracellular LIPs to reduce cell iron loads and exert notable functional effects. However, as already shown for FRDA, a strategy of iron relocation rather than mere chelation might also be applicable to AD and PD and other brain iron accumulation disorders. For that purpose, the iron-redistribution effect could conceivably be enhanced by increasing the levels of acceptors for the chelated iron, such as by coadministration of DFP and intracranial or intra-CSF injections of human apo-transferrin.

4 Summary and Future Directions

Relieving selected brain areas of accumulated iron has been considered a therapeutic target in various neurodegenerative disorders [146 ,147]. Removal by chelation might be a necessary step for eliminating potentially toxic iron but a step insufficient on its own for restoring affected functions or preventing their deterioration. Moreover, for chelation to be applicable in diseases of regional iron accumulation, the expected benefits of a chelator need to outweigh the risk of generating deficiencies of iron and other metals, both regionally and systemically.

In this chapter, FRDA was described as a model disorder in which intracellular iron misdistribution, rather than bulk iron accumulation, plays a key role in cell damage. Because misdistribution entails also (if not primarily) a deficiency in metabolic iron, only a cell iron-relocating agent or siderophore (such as DFP) can bring about the redistribution of accumulated iron, leading to restoration of functions affected by frataxin deficiency. Moreover, with this approach, the bulk of chelated iron can be recycled via its transfer to extracellular transferrin. The translation of these findings to the clinical setting indicated that agents such as DFP might be of therapeutic value in FRDA and NBIA, without seriously disturbing the hematological status of patients [90]. However, this proof of principle awaits more definitive clinical confirmation in well-controlled prospective long-term studies in which optimal regimens of DFP would need to be established to minimize possible body iron depletion. As for other neurodegenerative disorders associated with brain iron accumulation, it is not clear whether siderophores such as DFP, or chelators such as deferrioxamine and deferasirox, will prove to be effective at doses that are not hematologically deleterious and do not affect Zn or Cu status.

For future studies, the availability of bona fide animal models of the various neurodegenerative diseases might play an essential role in the design and screening of new siderophores and chelators. Modified siderophores with reduced systemic side effects and enhanced brain specificity are under development, such as an inactive glycosylated-DFP pro-drug, which crosses the BBB and, following β-glucosidase-induced cleavage of the attached carbohydrate, releases the active DFP [148]. In that respect, the possibility of using animal models such as the zebrafish [149], combined with the ability to visualize cellular labile iron redistribution by fluorescence microscopy [24], offers the unique opportunity of assessing the efficacy of novel chelators or siderophores in situ, in real time, and in a noninvasive manner. As for clinical applications, the adaptation of MRI for assessing iron accumulation in the brain [14, 60, 61] has proven invaluable for in situ assessment of chelation treatment in FRDA [3, 4] and NBIA [5, 106]. It is anticipated that these tools will improve the evaluation of the effects of chelators and siderophores on brain iron accumulation.

Acknowledgments Supported by grants from the Association Francaise contre les Myopathies (AFM), AFIRN (French-Israeli Organization for Research in Neuroscience), the Israel Science Foundation (ISF) and ApoPharma, Canada.

References

1. Zecca L, Youdim MBH, Riederer P, Connor JR, Crichton RR. Iron, brain ageing and neurodegenerative disorders. Nat Rev Neurosci. 2004;5(11):863–73.
2. Connor JR. Metal and oxidative damage in neurological disorders. New York: Plenum Press; 1997.
3. Gotz ME, Double KAY, Gerlach M, Youdim MBH, Riederere P. The relevance of iron in the pathogenesis of Parkinson's disease. Ann N Y Acad Sci. 2004;1012(1):193–208.
4. Honda K, Casadesus G, Petersen RB, Perry G, Smith MA. Oxidative stress and redox-active iron in Alzheimer's disease. Ann N Y Acad Sci. 2004;1012(1):179–82.
5. Johnson MA, Kuo YM, Westaway SK, Parker SM, Ching KH, Gitschier J, Hayflick SJ. Mitochondrial localization of human PANK2 and hypotheses of secondary iron accumulation in pantothenate kinase-associated neurodegeneration. Ann N Y Acad Sci. 2004;1012:282–98.
6. Pandolfo M. Friedreich ataxia. Semin Pediatr Neurol. 2003;10(3):163–72.
7. Simon D, Seznec H, Gansmuller A, et al. Friedreich ataxia mouse models with progressive cerebellar and sensory ataxia reveal autophagic neurodegeneration in dorsal root ganglia. J Neurosci. 2004;24(8):1987–95.
8. Hashimoto M, Takeda A, Hsu LJ, Takenouchi T, Masliah E. Role of cytochrome c as a stimulator of alpha-synuclein aggregation in Lewy body disease. J Biol Chem. 1999; 274(41):28849–52.
9. Kalinowski DS, Richardson DR. The evolution of iron chelators for the treatment of iron overload disease and cancer. Pharmacol Rev. 2005;57(4):547–83.
10. Wong A, Yang J, Cavadini P, et al. The Friedreich's ataxia mutation confers cellular sensitivity to oxidant stress which is rescued by chelators of iron and calcium and inhibitors of apoptosis. Hum Mol Genet. 1999;8(3):425–30.
11. Youdim MBH, Stephenson G, Shachar DB. Ironing iron out in Parkinson's disease and other neurodegenerative diseases with iron chelators: a lesson from 6-hydroxydopamine and iron chelators, desferal and VK-28. Ann N Y Acad Sci. 2004;1012(1):306–25.
12. Gaeta A, Hider RC. The crucial role of metal ions in neurodegeneration: the basis for a promising therapeutic strategy. Br J Pharmacol. 2005;146(8):1041–59.
13. Molina-Holgado F, Hider R, Gaeta A, Williams R, Francis P. Metals ions and neurodegeneration. Biometals. 2007;20(3):639–54.
14. Waldvogel D, van Gelderen P, Hallett M. Increased iron in the dentate nucleus of patients with Friedrich's ataxia. Ann Neurol. 1999;46(1):123–5.
15. Connor JR, Snyder BS, Arosio P, Loeffler DA, LeWitt P. A quantitative analysis of isoferritins in select regions of aged, parkinsonian, and Alzheimer's diseased brains. J Neurochem. 1995;65(2):717–24.

16. Rogers JT, Randall JD, Cahill CM, et al. An iron-responsive element type II in the 5′-untranslated region of the Alzheimer's amyloid precursor protein transcript. J Biol Chem. 2002;277(47):45518–28.
17. Wang L, Wang W, Zhao M, Ma L, Li M. Psychological stress induces dysregulation of iron metabolism in rat brain. Neuroscience. 2008;155(1):24–30.
18. Donovan A, Roy CN, Andrews NC. The ins and outs of iron homeostasis. Physiology (Bethesda). 2006;21:115–23.
19. Mackenzie EL, Iwasaki K, Tsuji Y. Intracellular iron transport and storage: From molecular mechanisms to health implications. Antiox Redox Signal. 2008;10(6):997–1030.
20. Epsztejn S, Kakhlon O, Glickstein H, Breuer W, Cabantchik ZI. Fluorescence analysis of the labile iron pool of mammalian cells. Anal Biochem. 1997;248(1):31–40.
21. Esposito BP, Epsztejn S, Breuer W, Cabantchik ZI. A review of fluorescence methods for assessing labile iron in cells and biological fluids. Anal Biochem. 2002;304(1):1–18.
22. Petrat F, de Groot H, Sustmann R, Rauen U. The chelatable iron pool in living cells: a methodically defined quantity. Biol Chem. 2002;383(3–4):489–502.
23. Gutteridge JMC, Halliwell B. Free radicals and antioxidants in the year 2000: a historical look to the future. Ann N Y Acad Sci. 2000;899:136–47.
24. Breuer W, Shvartsman M, Cabantchik ZI. Intracellular labile iron. Int J Biochem Cell Biol. 2008;40(3):350–4.
25. Kakhlon O, Cabantchik ZI. The labile iron pool: characterization, measurement, and participation in cellular processes. Free Radic Biol Med. 2002;33(8):1037–46.
26. Weiss G, Goodnough LT. Anemia of chronic disease. N Engl J Med. 2005;352(10):1011–23.
27. Hershko C, Link G, Konijn AM, Cabantchik ZI. Objectives and mechanism of iron chelation therapy. Ann N Y Acad Sci. 2005;1054(1):124–35.
28. Kushner JP, Porter JP, Olivieri NF. Secondary iron overload. Hematology. 2001;2001(1):47–61.
29. Delatycki MB, Camakaris J, Brooks H, et al. Direct evidence that mitochondrial iron accumulation occurs in Friedreich ataxia. Ann Neurol. 1999;45(5):673–5.
30. Gakh O, Park S, Liu G, et al. Mitochondrial iron detoxification is a primary function of frataxin that limits oxidative damage and preserves cell longevity. Hum Mol Genet. 2006;15(3):467–79.
31. Boultwood J, Pellagatti A, Nikpour M, et al. The role of the iron transporter ABCB7 in refractory anemia with ring sideroblasts. PLoS ONE. 2008;3(4):e1970.
32. Cazzola M. Sideroblastic anemias. Hematol Meet Rep. 2007;1:75–7.
33. Pondarre C, Campagna DR, Antiochos B, Sikorski L, Mulhern H, Fleming MD. Abcb7, the gene responsible for X-linked sideroblastic anemia with ataxia, is essential for hematopoiesis. Blood. 2007;109(8):3567–9.
34. Rouault TA, Tong WH. Iron-sulfur cluster biogenesis and human disease. Trends Genet. 2008;24(8):398–407.
35. Hayflick SJ. Neurodegeneration with brain iron accumulation: from genes to pathogenesis. Semin Pediatr Neurol. 2006;13(3):182–5.
36. Weiss G. Modification of iron regulation by the inflammatory response. Best Pract Res Clin Haematol. 2005;18(2):183–201.
37. Campuzano V, Montermini L, Molto MD, et al. Friedreich's ataxia: autosomal recessive disease caused by an intronic GAA triplet repeat expansion. Science. 1996;271(5254): 1423–7.
38. Delatycki MB, Williamson R, Forrest SM. Friedreich ataxia: an overview. J Med Genet. 2000;37(1):1–8.
39. Patel PI, Isaya G. Friedreich ataxia: from GAA triplet-repeat expansion to frataxin deficiency. Am J Hum Genet. 2001;69(1):15–24.
40. Clark RM, Bhaskar SS, Miyahara M, Dalgliesh GL, Bidichandani SI. Expansion of GAA trinucleotide repeats in mammals. Genomics. 2006;87(1):57–67.
41. Herman D, Jenssen K, Burnett R, Soragni E, Perlman SL, Gottesfeld JM. Histone deacetylase inhibitors reverse gene silencing in Friedreich's ataxia. Nat Chem Biol. 2006;2(10):551–8.

42. Karlberg T, Schagerlof U, Gakh O, et al. The structures of frataxin oligomers reveal the mechanism for the delivery and detoxification of iron. Structure. 2006;14(10):1535–46.
43. Schagerlof U, Elmlund H, Gakh O, Nordlund G, Herbert H, Lindahl M, Isaya G, Al-Karadaghi S. Structural basis of the iron storage function of frataxin from single-particle reconstruction of the iron-loaded oligomer. Biochemistry. 2008;47(17):4938–54.
44. Rotig A, de Lonlay P, Chretien D, et al. Aconitase and mitochondrial iron-sulphur protein deficiency in Friedreich ataxia. Nat Genet. 1997;17(2):215–7.
45. Culotta VC, Yang M, O'Halloran TV. Activation of superoxide dismutases: Putting the metal to the pedal. Biochim Biophys Acta Mol Cell Res. 2006;1763(7):747–58.
46. Tan G, Chen L-S, Lonnerdal B, Gellera C, Taroni FA, Cortopassi GA. Frataxin expression rescues mitochondrial dysfunctions in FRDA cells. Hum Mol Genet. 2001;10(19):2099–107.
47. Campanella A, Isaya G, O'Neill HA, et al. The expression of human mitochondrial ferritin rescues respiratory function in frataxin-deficient yeast. Hum Mol Genet. 2004;13(19):2279–88.
48. Cavadini P, Gellera C, Patel PI, Isaya G. Human frataxin maintains mitochondrial iron homeostasis in Saccharomyces cerevisiae. Hum Mol Genet. 2000;9(17)2523–30.
49. Sturm B, Bistrich U, Schranzhofer M, et al. Friedreich's ataxia, no changes in mitochondrial labile iron in human lymphoblasts and fibroblasts: a decrease in antioxidative capacity? J Biol Chem. 2005;280(8):6701–8.
50. Condo I, Ventura N, Malisan F, Tomassini B, Testi R. A pool of extramitochondrial frataxin that promotes cell survival. J Biol Chem. 2006;281(24):16750–6.
51. Lu C, Cortopassi G. Frataxin knockdown causes loss of cytoplasmic iron-sulfur cluster functions, redox alterations and induction of heme transcripts. Arch Biochem Biophys. 2007;457(1):111–22.
52. Stehling O, Elsasser H-P, Bruckel B, Muhlenhoff U, Lill R. Iron-sulfur protein maturation in human cells: evidence for a function of frataxin. Hum Mol Genet. 2004;13(23):3007–15.
53. Ristow M, Pfister MF, Yee AJ, et al. Frataxin activates mitochondrial energy conversion and oxidative phosphorylation. Proc Natl Acad Sci U S A. 2000;97(22):12239–43.
54. Meyer C, Carlqvist H, Vorgerd M, Schols L, Ostenson CG, Ristow M. Regular insulin secretory oscillations despite impaired ATP synthesis in Friedreich ataxia patients. Horm Metab Res. 2006;38(10):683–7.
55. Wilson RB, Roof DM. Respiratory deficiency due to loss of mitochondrial DNA in yeast lacking the frataxin homologue. Nat Genet. 1997;16(4):352–7.
56. Kakhlon O, Manning H, Breuer W, Melamed-Book N, Lu C, Cortopassi G, Munnich A, Cabantchik ZI. Cell functions impaired by frataxin deficiency are restored by drug-mediated relocation of iron. Blood. 2008;112:5219–27.
57. Cossee M, Puccio H, Gansmuller A, et al. Inactivation of the Friedreich ataxia mouse gene leads to early embryonic lethality without iron accumulation. Hum Mol Genet. 2000; 9(8):1219–26.
58. Puccio H, Simon D, Cossee M, et al. Mouse models for Friedreich ataxia exhibit cardiomyopathy, sensory nerve defect and Fe-S enzyme deficiency followed by intramitochondrial iron deposits. Nat Genet. 2001;27(2):181–6.
59. Wilson RB. Iron dysregulation in Friedreich ataxia. Semin Pediatr Neurol. 2006;13(3): 166–175.
60. Haacke EM, Cheng NYC, House MJ, et al. Imaging iron stores in the brain using magnetic resonance imaging. Magn Reson Imaging. 2005;23(1):1–25.
61. Schenck JF, Zimmerman EA. High-field magnetic resonance imaging of brain iron: birth of a biomarker? NMR Biomed. 2004;17(7):433–45.
62. Bulteau A-L, Dancis A, Gareil M, Montagne J-J, Camadro J-M, Lesuisse E. Oxidative stress and protease dysfunction in the yeast model of Friedreich ataxia. Free Radic Biol Med. 2007;42(10):1561–70.
63. Napier I, Ponka P, Richardson DR. Iron trafficking in the mitochondrion: novel pathways revealed by disease. Blood. 2005;105(5):1867–74.

64. Chantrel-Groussard K, Geromel V, Puccio H, et al. Disabled early recruitment of antioxidant defenses in Friedreich's ataxia. Hum Mol Genet. 2001;10(19):2061–7.
65. Chiueh CC. Iron overload, oxidative stress, and axonal dystrophy in brain disorders. Pediatr Neurol. 2001;25(2):138–47.
66. Li K, Besse EK, Ha D, Kovtunovych G, Rouault TA. Iron-dependent regulation of frataxin expression: implications for treatment of Friedreich ataxia. Hum Mol Genet. 2008;17(15): 2265–73.
67. Foury F, Talibi D. Mitochondrial control of iron homeostasis. A genome wide analysis of gene expression in a yeast frataxin-deficient strain. J Biol Chem. 2001;276(11):7762–8.
68. Hausmann A, Samans B, Lill R, Muhlenhoff U. Cellular and mitochondrial remodelling upon defects in iron-sulfur protein biogenesis. J Biol Chem. 2008;283(13):8318–30.
69. Kispal G, Csere P, Prohl C, Lill R. The mitochondrial proteins Atm1p and Nfs1p are essential for biogenesis of cytosolic Fe/S proteins. EMBO J. 1999;18(14):3981–9.
70. Whitnall M, Rahmanto YS, Sutak R, et al. The MCK mouse heart model of Friedreich's ataxia: alterations in iron-regulated proteins and cardiac hypertrophy are limited by iron chelation. Proc Natl Acad Sci U S A. 2008;105(28):9757–62.
71. Lill R, Muhlenhoff U. Maturation of iron-sulfur proteins in eukaryotes: mechanisms, connected processes, and diseases. Annu Rev Biochem. 2008;77(1):669–700.
72. Pierrel F, Cobine P, Winge D. Metal ion availability in mitochondria. Biometals. 2007;20(3): 675–82.
73. Neilands JB. Siderophores. Arch Biochem Biophys. 1993;302(1):1–3.
74. Kwiatkowski JL. Oral iron chelators. Pediatr Clin North Am. 2008;55(2):461–82.
75. Pietrangelo A. Iron chelation beyond transfusion iron overload. Am J Hematol. 2007; 82(S12):1142–6.
76. Porter JB. Concepts and goals in the management of transfusional iron overload. Am J Hematol. 2007;82(S12):1136–9.
77. Galanello R. Deferiprone in the treatment of transfusion-dependent thalassemia: a review and perspective. Ther Clin Risk Manag. 2007;3(5):795–805.
78. Whitnall M, Richardson DR. Iron: a new target for pharmacological intervention in neurodegenerative diseases. Semin Pediatr Neurol. 2006;13(3):186.
79. Lim CK, Kalinowski DS, Richardson DR. Protection against hydrogen peroxide-mediated cytotoxicity in Friedreich's ataxia fibroblasts using novel iron chelators of the 2-pyridylcarboxaldehyde isonicotinoyl hydrazone class. Mol Pharmacol. 2008;74(1):225–35.
80. Richardson DR. Novel chelators for central nervous system disorders that involve alterations in the metabolism of iron and other metal ions. Ann N Y Acad Sci. 2004;1012:326–41.
81. Jauslin ML, Meier T, Smith RAJ, Murphy MP. Mitochondria-targeted antioxidants protect Friedreich ataxia fibroblasts from endogenous oxidative stress more effectively than untargeted antioxidants. FASEB J. 2003;17(13):1972–4.
82. Cabantchik ZI, Glickstein H, Golenser J, Loyevsky M, Tsafack A. Iron chelators: mode of action as antimalarials. Acta Haematol. 1996;95(1):70–7.
83. Sohn Y-S, Breuer W, Munnich A, Cabantchik ZI. Redistribution of accumulated cell iron: a modality of chelation with therapeutic implications. Blood. 2008;111(3):1690–9.
84. Link G, Konijn AM, Breuer W, Cabantchik ZI, Hershko C. Exploring the "iron shuttle" hypothesis in chelation therapy: effects of combined deferoxamine and deferiprone treatment in hypertransfused rats with labeled iron stores and in iron-loaded rat heart cells in culture. J Lab Clin Med. 2001;138(2):130–8.
85. Zanninelli G, Glickstein H, Breuer W, et al. Chelation and mobilization of cellular iron by different classes of chelators. Mol Pharmacol. 1997;51(5):842–52.
86. Glickstein H, Ben El R, Shvartsman M, Cabantchik ZI. Intracellular labile iron pools as direct targets of iron chelators: a fluorescence study of chelator action in living cells. Blood. 2005;106(9):3242–50.

87. Glickstein H, El RB, Link G, et al. Action of chelators in iron-loaded cardiac cells: accessibility to intracellular labile iron and functional consequences. Blood. 2006;108 (9):3195–203.
88. Cohen AR, Galanello R, Piga A, Dipalma A, Vullo C, Tricta F. Safety profile of the oral iron chelator deferiprone: a multicentre study. Br J Haematol. 2000;108(2):305–1.
89. Breuer W, Ermers MJJ, Pootrakul P, Abramov A, Hershko C, Cabantchik ZI. Desferrioxamine-chelatable iron, a component of serum non-transferrin-bound iron, used for assessing chelation therapy. Blood. 2001;97(3):792–8.
90. Boddaert N, Le Quan Sang KH, Rotig A, et al. Selective iron chelation in Friedreich ataxia: biologic and clinical implications. Blood. 2007;110(1):401–8.
91. Evans RW, Sharma M, Ogwang W, Patel KJ, Bartlett AN, Kontoghiorghes GJ. The effect of alpha-ketohydroxypyridine chelators on transferrin saturation in vitro and in vivo. Drugs Today. 1992;28:19–23.
92. Bradbury MW. Transport of iron in the blood-brain-cerebrospinal fluid system. J Neurochem. 1997;69(2):443–54.
93. Moos T, Morgan EH. Evidence for low molecular weight, non-transferrin-bound iron in rat brain and cerebrospinal fluid. J Neurosci Res. 1998;54(4): 486–94.
94. Fredenburg AM, Sethi RK, Allen DD, Yokel RA. The pharmacokinetics and blood-brain barrier permeation of the chelators 1,2 dimethyl-, 1,2 diethyl-, and 1-[ethan-1'-ol]-2-methyl-3-hydroxypyridin-4-one in the rat. Toxicology. 1996;108(3):191–9.
95. Habgood MD, Liu ZD, Dehkordi LS, Khodr HH, Abbott J, Hider RC. Investigation into the correlation between the structure of hydroxypyridinones and blood-brain barrier permeability. Biochem Pharmacol. 1999;57(11):1305–10.
96. Waldmeier PC, Buchle AM, Steulet AF. Inhibition of catechol-*O*-methyltransferase (COMT) as well as tyrosine and tryptophan hydroxylase by the orally active iron chelator, 1,2-dimethyl-3-hydroxypyridin-4-one (L1, CP20), in rat brain in vivo. Biochem Pharmacol. 1993;45(12):2417–24.
97. Kontoghiorghes GJ. New concepts of iron and aluminium chelation therapy with oral L1 (deferiprone) and other chelators. A review. Analyst. 1995;120(3):845–51.
98. Singh S, Khodr H, Taylor MI, Hider RC. Therapeutic iron chelators and their potential side-effects. Biochem Soc Symp. 1995;61:127–37.
99. Di Prospero NA, Baker A, Jeffries N, Fischbeck KH. Neurological effects of high-dose idebenone in patients with Friedreich's ataxia: a randomised, placebo-controlled trial. Lancet Neurol. 2007;6(10):878–86.
100. Hausse AO, Aggoun Y, Bonnet D, et al. Idebenone and reduced cardiac hypertrophy in Friedreich's ataxia. Heart. 2002;87(4):346–9.
101. Rustin P, Rotig A, Munnich A, Sidi D. Heart hypertrophy and function are improved by idebenone in Friedreich's ataxia. Free Radic Res. 2002;36(4):467–9.
102. A study investigating the safety and tolerability of deferiprone in patients with Friedreich's ataxia. Clinical Trials.gov Identifier: NCT00530127. http://www.clinicaltrials.gov/ct2/show/NCT00530127?term=deferiprone&rank=2.
103. Gregory A, Hayflick SJ. Neurodegeneration with brain iron accumulation. Folia Neuropathol. 2005;43(4):286–96.
104. Hayflick SJ. Unraveling the Hallervorden-Spatz syndrome: pantothenate kinase-associated neurodegeneration is the name. Curr Opin Pediatr. 2003;15(6):572–577.
105. Zhou B, Westaway SK, Levinson B, Johnson MA, Gitschier J, Hayflick SJ. A novel pantothenate kinase gene (PANK2) is defective in Hallervorden-Spatz syndrome. Nat Genet. 2001;28(4):345–9.
106. Forni GLB, Cremonesi M, Abbruzzese L, Parodi G, Marchese RC. Regression of symptoms after selective iron chelation therapy in a case of neurodegeneration with brain iron accumulation. Mov Disord. 2008;23(6):904–7.

107. Cohen AR, Galanello R, Piga A, De Sanctis V, Tricta F. Safety and effectiveness of long-term therapy with the oral iron chelator deferiprone. Blood. 2003;102(5):1583–7.
108. Nick H. Iron chelation, quo vadis? Curr Opin Chem Biol. 2007;11(4):419–23.
109. Mariani R, Arosio C, Pelucchi S, et al. Iron chelation therapy in aceruloplasminaemia: study of a patient with a novel missense mutation. Gut. 2004;53(5):756–8.
110. Skidmore FM, Drago V, Foster P, Schmalfuss IM, Heilman KM, Streiff RR. Aceruloplasminaemia with progressive atrophy without brain iron overload: treatment with oral chelation. J Neurol Neurosurg Psychiatry. 2008;79(4):467–70.
111. Amit T, Avramovich-Tirosh Y, Youdim MBH, Mandel S. Targeting multiple Alzheimer's disease etiologies with multimodal neuroprotective and neurorestorative iron chelators. FASEB J. 2008;22(5):1296–305.
112. Wessling-Resnick M. Iron transport. Annu Rev Nutr. 2000;20(1):129–51.
113. Wessling-Resnick M. Iron imports. III. Transfer of iron from the mucosa into circulation. Am J Physiol Gastrointest Liver Physiol. 2006;290(1):G1–6.
114. Konijn AM, Glickstein H, Vaisman B, Meyron-Holtz EG, Slotki IN, Cabantchik ZI. The cellular labile iron pool and intracellular ferritin in K562 cells. Blood. 1999;94(6):2128–34.
115. Richardson DR, Mouralian C, Ponka P, Becker E. Development of potential iron chelators for the treatment of Friedreich's ataxia: ligands that mobilize mitochondrial iron. Bioch Biophys Acta Mol Basis Dis. 2001;1536(2–3):133–40.
116. Srichairatanakool S, Thephinlap C, Phisalaphong C, Porter JB, Fucharoen S. Curcumin contributes to in vitro removal of non-transferrin bound iron by deferiprone and desferrioxamine in thalassemic plasma. Med Chem. 2007;3(5):469–74.
117. Dexter DT, Carayon A, Javoy-Agid F, et al. Alterations in the levels of iron, ferritin and other trace metals in Parkinson's disease and other neurodegenerative diseases affecting the basal ganglia. Brain. 1991;114(4):1953–75.
118. Gordon N. Infantile neuroaxonal dystrophy (Seitelberger's disease). Dev Med Child Neurol. 2002;44(12):849–51.
119. Barnham KJ, Bush AI. Metals in Alzheimer's and Parkinson's diseases. Curr Opin Chem Biol. 2008;12(2):222–8.
120. Kostka M, Hogen T, Danzer KM, et al. Single particle characterization of iron-induced pore-forming {alpha}-synuclein oligomers. J Biol Chem. 2008;283(16):10992–1003.
121. Munch G, Luth HJ, Wong A, et al. Crosslinking of [alpha]-synuclein by advanced glycation endproducts: an early pathophysiological step in Lewy body formation? J Chem Neuroanat. 2000;20(3–4):253–7.
122. Ostrerova-Golts N, Petrucelli L, Hardy J, Lee JM, Farer M, Wolozin B. The A53T alpha-synuclein mutation increases iron-dependent aggregation and toxicity. J Neurosci. 2000;20(16):6048–54.
123. Uversky VN, Li J, Fink AL. Metal-triggered structural transformations, aggregation, and fibrillation of human alpha-synuclein. A possible molecular link between Parkinson's disease and heavy metal exposure. J Biol Chem. 2001;276(47):44284–96.
124. Good PF, Olanow CW, Perl DP. Neuromelanin-containing neurons of the substantia nigra accumulate iron and aluminum in Parkinson's disease: a LAMMA study. Brain Res. 1992;593(2):343–6.
125. Rogers JT, Leiter LM, McPhee J, et al. Translation of the Alzheimer amyloid precursor protein mRNA is up-regulated by interleukin-1 through 5-untranslated region sequences. J Biol Chem. 1999;274(10):6421–31.
126. House E, Joanna C, Ayesha K, Olga K, Guy B, Christopher E. Aluminium, iron, zinc and copper influence the in vitro formation of amyloid fibrils of Aß 42 in a manner which may have consequences for metal chelation therapy in Alzheimer's disease. J Alzheimers Dis. 2004;6(3):291–301.

127. Huang X, Atwood CS, Hartshorn MA, et al. The A beta peptide of Alzheimer's disease directly produces hydrogen peroxide through metal ion reduction. Biochemistry. 1999; 38(24):7609–16
128. Moalem S, Percy ME, Andrews DF, Kruck TP, Wong S, Dalton AJ, Mehta P, Fedor B, Warren AC. Are hereditary hemochromatosis mutations involved in Alzheimer disease? Am J Med Genet. 2000;93(1):58–66.
129. Roy CN, Andrews NC. Recent advances in disorders of iron metabolism: mutations, mechanisms and modifiers. Hum Mol Genet. 2001;10(20):2181–6.
130. Braak H, Braak E, Bohl J. Staging of Alzheimer-related cortical destruction. Eur Neurol. 1993;33(6):403–8.
131. Pulliam JF, Jennings CD, Kryscio RJ, Davis DG, Wilson D, Montine TJ, Schmitt FA, Markesbery WR. Association of HFE mutations with neurodegeneration and oxidative stress in Alzheimer's disease and correlation with APOE. Am J Med Genet. 2003;119B(1):48–53.
132. Rogers J, Mastroeni D, Leonard B, Joyce J, Grover A. Neuroinflammation in Alzheimer's disease and Parkinson's disease: are microglia pathogenic in either disorder? Int Rev Neurobiol. 2007;82:235–46.
133. Mandel S, Weinreb O, Amit T, Youdim MBH. Mechanism of neuroprotective action of the anti-Parkinson drug rasagiline and its derivatives. Brain Res Rev. 2005;48(2):379–87.
134. Zhang X, Haaf M, Todorish B, Rosstephan E, Schieremberg H, Surguladze N, Connor JR. Cytokine toxicity to oligodendrocyte precursors is mediated by iron. Glia. 2005;52:199–208.
135. Grunblatt E, Mandel S, Youdim MB. MPTP and 6-hydroxydopamine-induced neurodegeneration as models for Parkinson's disease: neuroprotective strategies. J Neurol. 2000; 247(suppl 2):II95–102.
136. Youdim MBH, Grunblatt E, Mandel S. The pivotal role of iron in NF-kappaB activation and nigrostriatal dopaminergic neurodegeneration: prospects for neuroprotection in Parkinson's disease with iron chelators. Ann N Y Acad Sci. 1999;890:7–25.
137. Mandel S, Amit T, Bar-Am O, Youdim MBH. Iron dysregulation in Alzheimer's disease: multimodal brain permeable iron chelating drugs, possessing neuroprotective-neurorescue and amyloid precursor protein-processing regulatory activities as therapeutic agents. Prog Neurobiol. 2007;82(6):348–60.
138. Jiang H, Luan Z, Wang J, Xie J. Neuroprotective effects of iron chelator desferal on dopaminergic neurons in the substantia nigra of rats with iron-overload. Neurochem Int. 2006;49(6):605–9.
139. U.S. patent application, Pub. No. US 2004/0101521 A1PNUA.
140. Blat D, Weiner L, Youdim MBH, Fridkin M. A novel iron-chelating derivative of the neuroprotective peptide NAPVSIPQ shows superior antioxidant and antineurodegenerative capabilities. J Med Chem. 2008;51(1):126–34.
141. Avramovich-Tirosh Y, Reznichenko L, Mit T, et al. Neurorescue activity, APP regulation and amyloid-beta peptide reduction by novel multi-functional brain permeable iron-chelating-antioxidants, M-30 and green tea polyphenol, EGCG. Curr Alzheimer Res. 2007;4(4):403–11.
142. Crapper McLachlan DR, Dalton AJ, Kruck TP, et al. Intramuscular desferrioxamine in patients with Alzheimer's disease. Lancet. 1991;337(8753):1304–8.
143. Miyajima H, Takahashi Y, Kamata T, Shimizu H, Sakai N, Gitlin JD. Use of desferrioxamine in the treatment of aceruloplasminemia. Ann Neurol. 1997;41(3):404–7.
144. Harris ZL, Klomp LW, Gitlin JD. Aceruloplasminemia: an inherited neurodegenerative disease with impairment of iron homeostasis. Am J Clin Nutr. 1998;67(5):972S–7.
145. Mitchell KM, Dotson AL, Cool KM, Chakrabarty A, Benedict SH, LeVine SM. Deferiprone, an orally deliverable iron chelator, ameliorates experimental autoimmune encephalomyelitis. Mult Scler. 2007;13(9):1118–26.
146. Bush AI, Tanzi RE. Therapeutics for Alzheimer's disease based on the metal hypothesis. Neurotherapeutics. 2008;5(3):421–32.
147. Rouault TA, Cooperman S. Brain iron metabolism. Semin Pediatr Neurol. 2006; 13(3):142–8.

148. Schugar H, Green DE, Bowen ML, Scott LE, Storr T, Bohmerle K, Thomas F, Allen DD, Lockman PR, Merkel M, Thompson KH, Orvig C. Combating Alzheimer's disease with multifunctional molecules designed for metal passivation. Angew Chem Int Ed Engl. 2007;46(10):1716–8.
149. Newman M, Musgrave FI, Lardelli M. Alzheimer disease: Amyloidogenesis, the presenilins and animal models. Biochim Biophys Acta Mol Basis Dis. 2007;1772(3):285.

Chapter 12
Oxidative Stress in Parkinson's Disease

Jacob Vaya, Yuval Aluf, and John P.M. Finberg

Abstract Oxidative (OS) and nitrative stress are thought to be major factors in the generation and maintenance of neurodegeneration in Parkinson's disease (PD). Generation and maintenance of the parkinsonian state may involve separate mechanisms or may be interrelated, because neurodegeneration in this disease condition usually starts late in life and continues at an accelerated rate, although some forms of PD resulting from gene mutations may appear early in life. In this review we present the major recent findings connecting OS and PD, but also review in brief the relevant physiological and pathological findings in PD for the uninitiated. The major enzyme systems connected with OS generation with possible relevance to PD are discussed, as well as available antioxidants. A major contention in the past has been that oxidative metabolism of dopamine by MAO creates a potentially damaging environment for the dopaminergic neuron, by production of hydrogen peroxide which in the presence of reactive iron can be further metabolized to reactive oxygen species by Fenton chemistry. We review here the factors controlling iron levels in the brain, in particular the emerging role of neuromelanin, and the possible interventions available. Recent findings based on staged appearance of synucleinopathy show that Lewy bodies appear initially in peripheral structures and migrate via connecting neuronal tracts to finally reside in upper brain areas. These findings have been interpreted as showing the involvement of an infective agent in PD, which, if true, introduces an additional factor to the simplified hypotheses based on dopamine turnover-induced OS. The recent demonstration of clinical neuroprotective action in PD of the MAO-B inhibitor rasagiline may provide support for the OS hypothesis, but on the other hand, alternative explanations for this finding are also available. We can anticipate much further activity in the direction of new therapeutics to control OS in PD as well as in other related conditions.

J. Vaya (✉)
Laboratory of Natural Medicinal Compounds Migal-Galilee Technology Center, and Tel Hai College, Kiryat Shmona, Israel
e-mail: vaya@migal.org.il

N. Gadoth and H.H. Göbel (eds.), *Oxidative Stress and Free Radical Damage in Neurology*, Oxidative Stress in Applied Basic Research and Clinical Practice,
DOI 10.1007/978-1-60327-514-9_12,

Keywords Parkinson's disease · Iron · Neuromelanin · Substantia nigra · Nitration · Dopamine · Glutathione · Lewy bodies · Antioxidants · Coenzyme Q10 · Genetic mutations · Mitochondria

Abbreviations

AGEs	Advanced glycated end products
CNS	Central nervous system
COMT	Catechol-O-methyltransferase
DA	Dopamine
DOPA	3,4-Dihydroxy-phenylalanine
GSH	Glutathione
LB	Lewy bodies
6-OHDA	6-Hydroxydopamine
MAO	Monoamine oxidase
MPTP	1-Methyl-4-phenyl-1,2,3,6-tetrahydropyridine
MPP+	1-Methyl-4-phenyl-2,3-dihydropyridium ion
OS	Oxidative stress
NM	Neuromelanin
PD	Parkinson disease
ROS	Reactive oxygen species
RNS	Reactive nitrogen species
TH	Tyrosine hydroxylase

1 Introduction to Parkinson's Disease

This chapter reviews the connection between oxidative stress (OS) and Parkinson's disease (PD). PD is diagnosed on the basis of neurological symptoms including bradykinesia, rest tremor, postural instability, muscle rigidity, flexed posture, and freezing of gait [1]. There are several parkinsonian states in addition to primary Parkinson's disease: secondary parkinsonism (e.g., caused by drugs, tumors, or toxins), parkinsonism-plus syndromes (with additional complicating neurodegenerative conditions), and heredodegenerative diseases (including Alzheimer's and Huntington's disease; see [1]). Many incorrect diagnoses are made on the basis of neurological signs only, and the completely correct diagnosis can only be made at autopsy, when the presence of the characteristic neuronal inclusion bodies (Lewy bodies, LB) is confirmed. The difficulty of diagnosis is a factor that must always be considered in reviewing the relationship between clinical findings and mechanisms. Most research on etiology of the disease is aimed at understanding the cause of primary parkinsonism, a condition in which 5–10% of the cases have been shown to have a genetic basis, with currently more than ten genes described as causing the disease, and the remaining 90% of cases without known genetic involvement, or other known cause, and referred to as idiopathic parkinsonism.

A major development in recent years has been the hypothesis that Parkinson's disease may in fact be caused by a prion-like infective agent. This proposal was based on the finding that LB exist in peripheral neuronal structures, for example, olfactory tract and intestinal nerve plexus, in the earliest stages of the disease, and show a gradual ascent within the central nervous system (CNS), reaching the cerebral cortex in the most advanced stages [2]. Even if some aspects of this theory are controversial, yet the now well known facts remain, that symptoms including anosmia and loss of cardiac sympathetic control precede the appearance of motor symptoms by several years, and that cortical involvement (dementia) occurs in the latest stages. The importance of this theory for the present discussion is that the primary factor causing neuronal degeneration may be the cause of subsequent biochemical dysfunction, such as OS, rather than the reverse. In addition, the importance of oxidative metabolism of dopamine (DA) in the causation of OS has been called into question because many nondopaminergic pathways also undergo degeneration. Be that as it may, we now describe the findings involving OS in PD, bearing in mind that idiopathic PD is almost certainly caused by varied pathology resulting in the same final clinical result, in a similar way to essential hypertension, in which various mechanisms can lead to the same end result of increased blood pressure and a host of other conditions.

1.1 The Dopaminergic System and Its Involvement in PD

The brain contains several dopaminergic pathways, mainly (1) the nigro-striatal pathway, (2) the mesolimbic/mesocortical pathway, and (3) the incerto-hypothalamic pathway. Of these, the nigro-striatal pathway is the one involved in movement, and mainly involved in PD, while the others are affected to only a minor degree. The nigro-striatal pathway stems from perikarya in the pars compacta of the substantia nigra (SNpc), with an axonal bundle bringing the dopaminergic nerve endings to the striatum. The neuronal cell bodies in the SNpc are large and contain all the essential components for DA production and storage. As a result, and an important fact that is often overlooked, the dopaminergic cell bodies continually produce DA, even though the enzymes and storage vesicles produced in the cell bodies are continually transported to the axon terminal. The cell body area therefore contains a large amount of free DA, some of which is oxidized by the nonenzymatic pathway (see below) with the production of melanin, subsequently combined with peptides to neuromelanin (NM), and other compounds. The occurrence of NM within these cells gives the appearance to the structure of a narrow black band on either side of the mesencephalon of the normal brain, which is absent from the parkinsonian brain [3].

DA released in the striatum from the axon terminals of the nigral neurons acts on receptors of the D1 and D2 family. The distribution of these receptors on striatal target neurons shows a marked selectivity; D1 receptors are expressed on dendritic processes of striato-nigral neurons (direct pathway), and D2 receptors are expressed on dendrites of striato-pallidal neurons (indirect pathway; Fig. 1); for further information, see [4]. All output neurons of the striatum are GABA-ergic, but their interaction with downstream targets in the basal ganglia confers on the striato-nigral neurons the property of activating thalamic input to the motor cortex, while striato-pallidal activity has an eventual inhibitory effect on the motor cortex. Loss of DA release in the striatum of the parkinsonian patient has marked effects on downstream neurons of the basal ganglia; in particular, the reduction of inhibitory D2-mediated input to the striato-pallidal neurons causes their increased firing, and increased inhibition of GABA-ergic input to the subthalamic nucleus, leading to an accelerated activity of the latter [5] (see Fig. 1). Despite the fact that loss of substantia nigra DA neurons is responsible for the characteristic motor symptoms of PD, this nucleus is only affected in stage 3 of the neurodegenerative process as defined by Braak et al. [2], with degeneration of raphe nucleus serotonergic neurons and other non-DA neurons, for example, vagal motor nucleus, preceding the DA neuron degeneration.

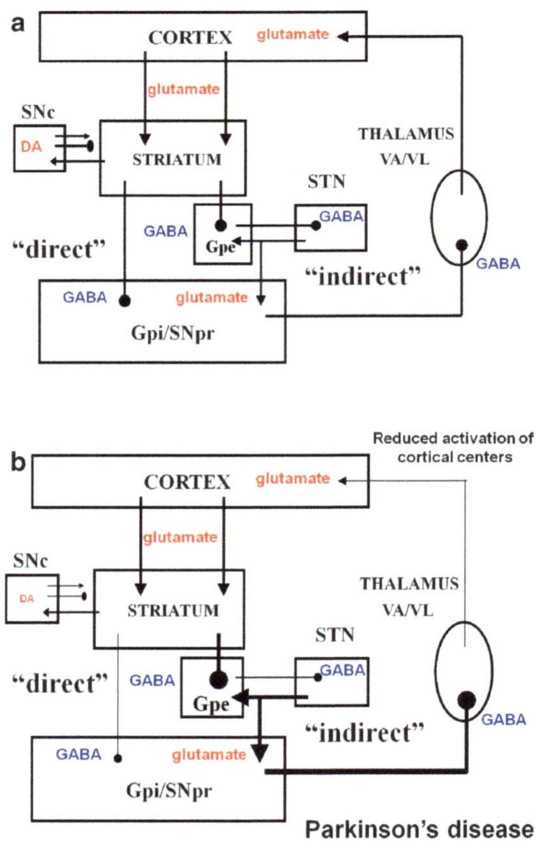

Fig. 1 (**a**) Major neurotransmitter pathways involved in control of motor activity in the normal brain: *arrows* represent excitatory pathway, *circles* represent inhibitory pathway. (**b**) Parkinsonian brain: line thickness of *arrows* represents neuronal activity level

Drugs used for treatment of PD include the immediate DA precursor, L-DOPA (3,4-dihydroxyalanine), direct DA agonists, and inhibitors of DA metabolizing enzymes (see following).

1.2 Dopamine Release and Metabolism

DA is synthesized throughout the neuron by the activity of two key enzymes, the first controlling one being tyrosine hydroxylase (TH) (expressed in all catecholaminergic neurons but not in others) and aromatic amino acid decarboxylase (expressed in several types of neurons and glial cells). DA is produced in the cytoplasmatic compartment of dopaminergic and noradrenergic neurons and is extensively metabolized therein by monoamine oxidase (MAO). DA molecules that escape the action of MAO are transported into storage vesicles by the vesicular monoaminergic transporter VMAT2. Following release into the synapse, DA is reabsorbed by the neuron via the high-affinity plasma membrane dopamine transporter (DAT), but can also be taken up and metabolized within other cells and neurons, specifically serotonergic and noradrenergic neurons and glial cells, which also express membrane transporters for monoamine neurotransmitters as well as MAO. Outside of catecholaminergic neurons, DA is *O*-methylated by the enzyme catechol-*O*-methyltransferase (COMT).

MAO is situated on the outer mitochondrial membrane. The breakdown of monamines by MAO can be described by the equation:

$$\text{R-CH}_2\text{-NH}_2 + \text{O}_2 + \text{H}_2\text{O} \xrightarrow{\text{MAO}} \text{R-CHO} + \text{NH}_3 + \text{H}_2\text{O}_2$$

Several important points of this reaction are the dependence on free oxygen, the initial production of an aldehyde, and the release of hydrogen peroxide as well as ammonia [6]. The aldehydes are metabolized by aldehyde-metabolizing enzymes, aldehyde reductase and aldehyde dehydrogenase, to carboxylic acids and glycols, and hydrogen peroxide is converted by catalase to water and oxygen. In the absence of such effective scavenger enzymes, however, potentially damaging reactive aldehydes and free radicals will accumulate.

In addition to the enzymatic metabolism of DA, the molecule is subject to nonenzymatic oxidation. As described in Sect. 1.3, DA is readily oxidized by free oxygen with the production of a quinone. The quinone structure enables formation of polymers such as melanin, as well as other reactive compounds such as salsolinol. An additional potentially toxic derivative of DA is 6-hydroxydopamine (6-OHDA), which has been identified in trace amounts in normal brain tissue.

The normal intracellular metabolism of DA is thus a potential source of OS, and because of the high concentration of DA and high rate of its turnover within neurons of the nigro-striatal pathway, these neurons are considered to be under a high level of OS threat in the resting state.

1.3 Nonenzymatic DA Metabolism

DA and DOPA are compounds containing a catechol ring (*ortho*-dihydroxyphenol). This reactive structure contributes significantly to their toxicity. Besides the enzymatic catabolism of DA by MAO and catechol-*O*-methyltransferase (COMT), this reactive species can yield cytotoxic materials such as quinones, 5-*S*-cysteinyl, and several catabolic DA autoxidation products. Free DA and DOPA, whether in the cytoplasm of the cell or in the synapse, can autoxidize at a high rate in the presence of oxygen. DA may also undergo oxidation via minor enzymatic pathways, including the enzymes prostaglandin H synthase, lipoxygenase, and tyrosinase [7].

It has been shown that DA cytotoxicity in neuroblastoma-derived cell lines and primary neuronal cultures is strongly mediated through quinone formation and is more powerful than cytotoxicity caused by other catecholamines [8]. Intrastriatal injection of DA to normal rats results in lesion formation in the corpus striatum in a dose-dependent manner that is mediated by quinone formation and protein modification [9].

The oxidation of DA and DOPA is enhanced in the presence of hydroxyl radicals [10], which may be formed as a by-product of the MAO metabolism (hydrogen peroxide) and in the presence of iron, copper, or manganese ions (Fenton reaction). This reaction in turn can facilitate the generation of quinone [11]. In the same note, treatment with an antioxidant such as glutathione (GSH) or ascorbate [12], an inducer of quinone reductase [13], or under DA depletion leads to neuronal protection through inhibition of quinone formation.

In addition to the production of hydrogen peroxide, the quinone (whether it originated from DA or DOPA) is a reactive molecule that can react easily with cysteine residues in the form of free amino acid or as a protein constituent. Conjunction of DA or DOPA quinone with the sulfhydryl group of cysteine forms a stable bond between the quinone and the amino acid. If this bond, usually 2- or 5-cysteinyl-DA/DOPA, is formed at the protein active site it can cause partial or complete deactivation of the protein, protein aggregation, and subsequent cell death [14]. Among the proteins that seem to be the physiological targets of quinones are alpha-synuclein [15], TH [16], and many others.

These reactive molecules can undergo further oxidation and cyclization to DA-chrome or DOPA-chrome, and with or without the addition of cysteine, polymerize to NM. NM is thought to be a mixture of eumelanin and pheomelanin [17] and is found only in dopaminergic and noradrenergic neurons of the CNS. Its definite composition is not totally elucidated; however, it contains melanic, aliphatic, and peptide residues, combined with metal depositions and lipids [18]. NM has been shown to be a strong metal chelator, with the ability to chelate iron and other metal ions, and it has been suggested that synthesis of NM could play a protective role within the cell by preventing the accumulation of toxic catechol derivatives integrating them into the NM polymer and removing metals such as iron (see following). Nevertheless, NM degradation caused by neuronal death or increased hydrogen peroxide concentrations [19] can lead to iron and quinone release, exacerbating neurodegeneration, and thus can play an important role in the pathogenesis of PD.

1.4 Genetic Forms of PD

For many years, PD was regarded as a sporadic disease with no genetic origin. However, during the past two decades following the research carried out on many family pedigrees, groups, and individuals from different origins and geographic backgrounds, the importance of inheritance and genetic factors has become more apparent than in the past. At present, it is common to attribute 5–10% of PD patients to a few monogenic mutations; however, there is growing evidence for the involvement of multiple genetic risk factors in idiopathic PD. To date, more than a dozen genetic loci have been implicated in rare forms of PD (usually referred to as PARK1, PARK2, etc.); among these are autosomal recessive and autosomal dominant forms. In six of these chromosomal loci, proteins have been identified by various groups around the world, and various mutations have been identified in different families and between family members suffering from PD. It is now apparent that the genetic component of PD plays a much more important role in the pathophysiology than was previously thought, and mutations in genes that were linked to familial PD are good candidates for involvement in the idiopathic form of PD. Four of the PARK genes, SNCA at PARK1, UCH-L1 at PARK5, PINK1 at PARK6, and LRRK2 at PARK8, have been implicated in sporadic PD [20].

1.5 Animal Models for PD

PD does not occur naturally in animals. To imitate the neurodegeneration of dopaminergic neurons or PD symptoms, several animal models have been developed and can be divided into two major categories: (1) administration of a neurotoxin systemically or through a direct injection to the CNS in order to deplete dopaminergic neurons, and (2) use of genetically modified animals related to disease pathology or to known genetic characteristics of the disease.

Neurotoxins: Neurotoxic agents have been used for simulating the neuropathological and neurological features of PD in laboratory animals. Thus, animals injected with the toxins develop parkinsonian symptoms such as hypokinesia, repetitive chewing movements, and catalepsia [21, 22]. Three dominant types of toxic materials are used. (i) General cellular toxins such as rotenone and the herbicides paraquat and epoxomicin, which are all naturally occurring toxins known to induce a high-affinity specific inhibition of mitochondrial complex I (rotenone and paraquat) or proteasomal inhibition. Rotenone, which is the most frequently used, inhibits the transfer of electrons from complex I to ubiquinone in the mitochondrial electron transfer chain. Rotenone is only mildly toxic for humans, but in rodents, particularly in rats, chronic infusion can induce a slowly progressing neurodegeneration of DA neurons. (ii) 1-Methyl-4-phenyl-1,2,3,6-tetrahydropyridine (MPTP) crosses the blood–brain barrier (BBB) easily after systemic administration and is converted into the active toxic metabolite 1-methyl-4-phenyl-2,3-dihydropyridium ion (MPP^+) by the enzyme monoamine oxidase-B (MAO-B). The metabolite MPP^+ is selectively taken up into dopaminergic neurons by the DA transporter (DAT) and irreversibly inhibits complex I of the mitochondrial respiratory chain, therefore producing selective degeneration of nigral DA cells in monkeys and mice, but it has little effect in rats. (iii) 6-OHDA induces degeneration of catecholaminergic cells in monkeys, rats, mice, and cats [23]. The relative specific toxicity of 6-OHDA for catecholamine neurons results from its uptake by DA and noradrenaline transporters into the cytosol and the ability of the toxin to act directly by inhibiting the mitochondrial respiratory chain at the level of complex I [24]. Beside its affect on the mitochondria, 6-OHDA is a major source of reactive oxygen species (ROS) and quinone, therefore resulting in a dramatic increase in OS and an elevated amount of modified proteins [25]. 6-OHDA inhibits the mitochondrial respiratory chain and generates ROS through its autoxidation and by microglial activation. ROS derived from NADPH oxidase act synergistically with 6-OHDA and are an early element of the 6-OHDA-induced cell death [22].

The route of administration of the toxins differs from one substance to another. Some are given systemically in a single bolus or chronically administrated, whereas others do not cross the BBB and thus are often administered directly to the brain tissue by intranigral, intrastriatal, or intracerebroventricular (i.c.v.t.) injections [14].

Genetically modified animals: There are principally two groups of genetic models that are based on the PD-linked genes: (i) models in which all cells of the organism are affected and (ii) knockout of dopaminergic neurons in a tissue-specific manner. In the first group, animal models are based on null mutations, an extra gene copy, or point mutations of genes located in different PARK loci (more details in the section about genetic factors and PD). Whether the mutation is a gain or loss of function, none of the genetic models based on PD-linked genes reproduces key symptoms of the disease, such as loss of dopaminergic neurons, but rather detect slight effects on the dopaminergic system, such as a small decrease in DAT binding and slightly reduced DA levels in the striatum [26]. Recently, a second group of genetically modified animals has been introduced, using conditional knockout strategies to disrupt the expression of genes in the dopaminergic system in a region- or neuron-specific manner. For this purpose, mice that express cre-recombinase (Cre) under the control of the DAT promoter are predominantly used to target postmitotic dopaminergic neurons in the midbrain. The target of Cre can also be wider populations of neurons available such as the TH promoter [27].

Overall, there are a number of models for the disease that reproduce either the symptoms or pathology and which can help to further explore the pathology, etiology, and symptoms of PD, contributing a great deal to pharmacological research. Still, despite the advantages in using the well-established or the newer models of PD, these models do not mimic the entire spectrum of clinical symptoms or pathology of human PD.

2 The Association Between PD and OS

Oxidative stress is a pathological factor that may contribute to neuronal death and PD progression. Analysis of postmortem brains from PD patients revealed enhanced OS in their SNpc with a decreased ratio of GSH/GSSG [28], augmented levels of iron [29], increased lipid peroxidation [30], elevated keto-protein formation [31], and DNA oxidation[32]. OS is defined as a disturbance in the balance between the formation and accumulation of ROS and reactive nitrogen species (RNS) to their removal by the organ via its defense systems, in the favor of the former. Such imbalance can result in oxidation of major macromolecules of the body, which can change their composition, structure, function, and activity, leading to impairment of membrane fluidity and transport of ions and nutrients. Formation of chlorinated, hydroxylated, and nitrated proteins may damage their function. Oxidized nucleic acids can be generated in DNA with the formation of mutations. OS is associated with many factors that are thought to be involved in the pathogenesis of PD. Mitochondrial damage may result from OS and, vice versa, mitochondrial impairment may enhance ROS/RNS release to the cytosol. Inflammation is always associated with high ROS/RNS generation (nitric oxide NO, hypochlorous acid HOCl, superoxide anion, hydrogen peroxide, etc.) as direct defense species and as signaling elements to induce a series of protecting genes. OS can contribute to protein misfolding and, when linked to an inability of the ubiquitin-proteasome system to degrade and remove them, such damage can result in the accumulation and aggregation of these abnormal proteins and induce what is termed proteolytic stress [33]. Misfolding and aggregation of α-synuclein have been described in association with familial PD [34, 35]. OS is linked to DOPA and DA enzymatic and nonenzymatic metabolism with the formation of DOPA and DA quinones, a reactive species that can further react with endogenous compounds such as cysteine to form cytotoxic agents. OS is also associated with MAO activity, as described in Sect. 1.2, as well as in other enzyme activities present in the brain such as TH or tyrosinase, both of which hydroxylate tyrosine to DOPA, or the enzyme heme oxygenase-1 (HO-1) responsible for the degradation of heme into biliveridin, carbon monoxide (CO), and iron. All these enzyme activities can be involved in OS, which could lead to selective neuronal death in the substantia nigra and to LB formation. The question whether OS is the initial stage and the cause or the consequence of neurodegeneration in PD remains open. The aim of the present chapter is to review the present knowledge related to such an association between OS and risk factors considered to contribute to neuronal death in the SN.

2.1 Biomarkers of OS in PD

The early clinical diagnosis of motor symptoms in PD patients is made when about 60% of the dopaminergic neurons of the substantia nigra have been lost. Early diagnosis of PD in the human is important for early treatment, for monitoring the progression of the disease, and for following up the effectiveness of therapeutic interventions. Ideal biomarkers of PD should be specific to the disease, sensitive to identification of the disease as early as possible, before the other clinical symptoms arises, have high detection accuracy (with a very low percentage of false-negative and false-positive results), and be noninvasive, easy to use, and not expensive. Such markers do not exist. Several biomarkers for PD are associated with OS and can be applied during a patient's lifespan (in opposition to postmortem indicators) are under development. Methods such as imaging techniques and clinical and genetic tests are outside the scope of this review.

Degeneration of dopaminergic neurons in the brain and the formation of the LB are the hallmark events of PD. OS is believed to contribute at least in part to these two incidents, and biomarkers that might be generated during the pathways of their formation are under investigation. It was assumed that, as a consequence of dopaminergic cell death, NM will be present not only in the brain but also in the plasma. Antibodies specific to NM have been found in the plasma of patients with PD, but it is not clear yet if the NM detected in plasma of PD patients is specific to the disease or common to other neurological disorders or even may arise from skin diseases. In other studies, superoxide dismutase (SOD) activity was shown to decrease in the blood of advanced PD patients vs. control subjects and a negative correlation was anticipated between this activity and disease duration [36]. The level of oxidized pyrimidins and altered purine nucleotides in plasma of PD patients vs. controls was also shown to be elevated together with a higher incidence of DNA strand breaks, both of which can be generated via OS pathways [37]. Münch et al. identified advanced glycated end products (AGEs) in SN of patients with incidental Lewy body disease, which is considered an early form of PD [38]. The patients examined showed no clinical signs of PD, which suggests that AGEs are present in a very early stage of PD before symptoms are detected. AGEs are reactive ketones or aldehydes formed during the reaction of carbohydrates with the free amino groups in protein or during unsaturated fatty acid oxido-degradations. These intermediates are reactive species that can further lead to protein cross-linking and to protein dysfunction. The AGE-cross-linked proteins were detected by specific polyclonal antibodies raised against different AGE-protein adducts.

2.2 *Dopamine Metabolism: Evidence for an Association Between DA Metabolism and OS*

An association has been shown in various animal models and in brains taken from deceased PD patients between neuronal destruction in the brain and the generation of a massive OS. The cause or time course of the OS is still elusive; however, various studies have pointed to a role of DA as one of the major factors influencing redox balance, ROS production, and OS. The high rate of oxidative metabolism of DA by MAO generating ammonia, hydrogen peroxide, and a highly reactive aldehyde metabolite (3,4-dihydroxyphenylacetaldehyde, DOPAL) in conjunction with the spontaneous autoxidation of DA to form quinones or semiquinones are major factors accelerating neurodegeneration.

DA quinones are cytotoxic because of their interaction with the sulfhydryl group of cysteine forming various bioactive molecules, predominantly 5-cysteinyl-DA [39]. Because cysteine sulfhydryl groups are in many cases part of the active site of functional proteins, covalent modification of cysteine residues by quinones to form 5-cysteinyl-DA results in irreversible alteration or inhibition of the protein's function. Indeed, DA quinone covalently binds to key molecules in dopaminergic neurons including TH [16], DAT, and parkin protein [40], consequently inactivating these molecules [39].

In neuronal cell cultures, DA has been used as an inhibitor of brain mitochondria [41], as a precursor to 6-OHDA formation [42], and, as mentioned earlier, as a major source of quinone and semiquinone formation. Intrastriatal injections of DA caused dose-dependent increases in neuronal degradation and in the amount of quinoproteins found in the rat brain [12]. In a hemi-parkinsonian rat model of PD, treatment with L-DOPA, which caused an excess amount of DA outside the synaptic vesicles, caused increased DA turnover and quinoprotein formation in the damaged side and may exert neurodegenerative effect on the dopaminergic nerve terminal. The quinone formation and subsequent dopaminergic neuronal damage in vivo and in vitro can be slowed down or prevented through pretreatment with antioxidants, such as SOD and GSH [43].

Amphetamine, methamphetamine: These substances and their analogues have been shown to release DA from dopaminergic nerve terminals by several different mechanisms, such as the reversal of the DA uptake transporter (DAT), or through interfering with the storage of DA in the vesicular stores. It has been shown that chronic administration or high doses of amphetamines can be extremely neurotoxic and cause a dramatic neurodegeneration, a reduction in the amount of TH-positive neurons, and a reduction in the amount of active DA in the brain [44]. Several hypotheses have been proposed to explain the neurotoxic effect of amphetamines, including adenosine triphosphate (ATP) depletion, mitochondrial inhibition, and toxin-induced OS. In many studies of chronic amphetamine administration, a clear increase in OS was observed and tissue levels of lipid peroxidation and protein oxidation were elevated. Even after a single large dose of amphetamine, an increase in hydroxyl radical was observed, and an increase in malonaldialdehyde (MDA) was seen 7 days after administration [45]. This increase in OS was abolished when animals were pretreated with the antioxidants *N*-acetylcysteine (NAC) and α-phenyl-*N*-tertbutylnitrone (PBN). In other experiments, animals were pretreated with antioxidants, such as ascorbate and vitamin E, before amphetamine, which resulted in reduced OS and fewer neuronal lesions [46].

Antipsychotic drugs: These drugs are widely used in the treatment of psychotic disorders, but their long-term use may be accompanied by a number of unwanted side effects including tardive dyskinesia (TD), a movement disorder characterized by involuntary movements especially of the mouth and face, but also of other parts of the body. The pathophysiology of this disorder is still elusive; however, it is probably associated with a specific loss of dopaminergic receptor responsiveness, resulting in D2 receptor hypersensitivity, or increased DA turnover [47]; both these theories invoke an increased OS and excess ROS formation. The classic antipsychotic drugs (such as haloperidol), are mainly D2 dopaminergic receptor antagonists and have two different mechanisms by which they may increase OS. (i) By blocking presynaptic D2 receptors, which have an inhibitory effect on DA release they induce an increase in DA turnover in an attempt to overcome the postsynaptic receptor blockade. This increased turnover can yield an augmented amount of ROS production and quinone formation, thus increasing OS. (ii) The haloperidol pyridinium ion (HP^+) metabolite, which is derived from haloperidol, is a potent inhibitor of mitochondrial complex I and can interfere with electron transport at both complexes I and II, causing an interruption in mitochondrial activity and an increase in the resulting ROS production [48]. This situation is therefore another example of increased neuronal activity leading to OS and neuronal damage.

2.3 *Mitochondrial Impairment and PD*

The first link between mitochondrial dysfunction and PD was the finding of decreased activity of complex I in the SN of parkinsonian brain [49] with a decrease in complex I protein level. A recent study that strongly implicates a defect in complex I with PD was carried out using mice with a knockout of respiratory chain proteins in their dopaminergic neurons. These animals show progressive impairment of motor function together with formation of intraneuronal inclusions and dopaminergic cell death [50]. Inhibition of complex I by toxins such as MPTP, paraquat, or rotenone caused dopaminergic cell death or induced the formation of intracellular filamentous inclusions with α-synuclein protein as in the LB [51]. The critical role of α-synuclein, the major component of LB in PD, in mitochondrial impairment, and specifically in complex I deficiency, was evaluated in experiments with mice lacking the gene for α-synuclein. These mice were shown to be resistant to the toxic effects of the complex I inhibitor MPTP and did not show the pattern of DA neurodegeneration characteristic of MPTP-induced mitochondrial complex I inhibition [52]. These findings indicate that in the absence of α-synuclein protein, the neurotoxin MPTP cannot inhibit complex I. To further associate such observations with OS, Giasson et al. were able to show extensive accumulation of nitrated α-synuclein within the inclusion bodies of PD, dementia with LB, and in the major filamentous building blocks of these inclusion bodies. These nitrations occur at the tyrosine aromatic ring residue of the α-synuclein protein and could be formed through different pathways such as the generation of NO, which in the presence of superoxide anion (O_2^-) may form highly reactive peroxynitrite ($ONOO^-$) [53]. Oxidation of α-synuclein by either nitration or other ROS/RNS alters its protein structure and changes its physical properties, including decreased solubility, which may have an important role in the aggregation of α-synuclein in sporadic PD. Neurotoxin-induced inhibition of complex I induces OS generation in a dose-dependent manner in isolated brain mitochondria, as detected by augmentation of hydrogen peroxide level [54]. Injection of MPTP to mice causes oxidation of cardiolipin (phospholipids that bind the cytochrome *c* in the inner mitochondrial membrane) in ventral midbrain with the formation of phospholipid hydroperoxide, an established marker of OS [54], which was also obtained in mitochondria isolated from brain of mice treated with complex I inhibitors. Interestingly, the formation of oxidized phospholipids as a result of complex I inhibition was reduced by addition of an antioxidant known to convert superoxide anion to hydrogen peroxide (similar to SOD). These results suggest that O_2^- is generated and contributes to the cardiolipin oxidation as a result of complex I inhibition by MPTP. The same authors showed that the inhibition of complex I is linked to the release of cytochrome *c* from the inner mitochondrial membrane (where it is bound to phospholipids) to the intermembrane space as soluble cytochrome *c*. It is concluded that with the mitochondrial impairment, ROS are generated that oxidize cardiolipin and cause the release of cytochrome *c* from the inner mitochondrial membrane. Advanced postmortem analysis of proteins from mitochondrial preparations obtained from frontal cortex of PD patients revealed a significantly increased level of keto-proteins in the complex I protein fraction compared with age-matched controls. This enhanced oxidation was mostly localized in a fraction of hydrophobic proteins thought to form the catalytic core of complex I. When proteins of complex I were exposed to exogenous ROS such as hydrogen peroxide to reproduce the type of damage to complex I proteins observed in the samples from PD patients, such an oxidative pattern was not formed but did form when NADH was used to transfer the electrons, suggesting that the ROS involved in the oxidation of complex I proteins are formed within the mitochondria and not from outside [55].

In summary, mitochondrial impairment may result in complex I defect, increased free radical generation, and damage to macromolecules such as α-synuclein protein, which further accelerate mitochondrial dysfunction and free radical generation. This augmented superoxide anion O_2^- leakage, hydrogen peroxide accumulation, and protein nitration in conjunction with ubiquitin proteasomal dysfunction induces proteolytic stress, which results in protein misfolding, oxidation, and aggregation. The foregoing alteration at the molecular level results in tissue damage, alteration in mitochondria membrane permeability, damage to the respiratory chain, decreased ATP production, reduced membrane potential, formation of LB, apoptosis, and neuronal death (Fig. 2).

2.4 Genetic Alterations

In concordance with the topic of this chapter, we only cover the loci in which the proteins are already identified and which may be involved in alteration in redox balance, changing OS status, and increased neurodegeneration or, in contrast, have neuroprotective properties.

α-Synuclein: This protein is a major component of the typical LB in sporadic forms of PD. The discovery of a mutation in the α-synuclein molecule in a group of Greek and Sicilian families [35] and its connection to the familial forms of PD initiated years of research for the function of this protein, and the mechanism by

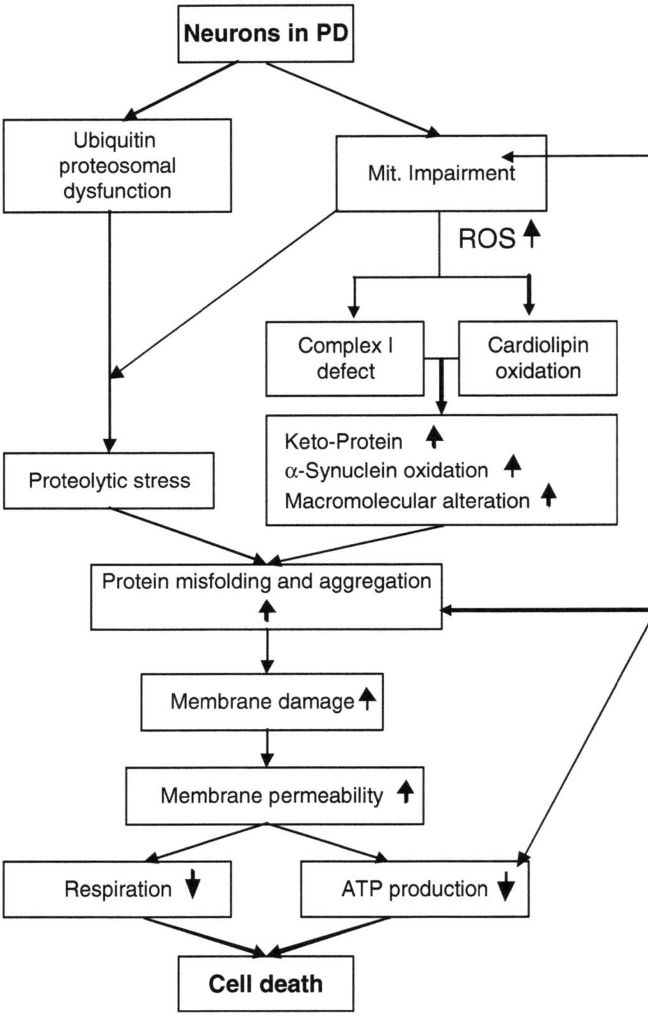

Fig. 2 Mitochondrial impairment and oxidative stress (OS). Parkinson's disease (PD) is associated with mitochondrial dysfunction, which may result from increased reactive oxygen species (ROS) generation, damaging complex I, decreasing its activity level, and oxidizing mitochondrial cardiolipin. The result of increasing oxidative stress is formation of keto-protein in the mitochondria, oxidation of α-synuclein, and damage to other macromolecules, including mitochondrial DNA. When mitochondrial impairment is associated with ubiquitin proteosomal dysfunction, it leads to proteolytic stress, which is expressed as a higher level of protein misfolding and aggregation. Protein aggregation further enhances ROS generation and increases mitochondrial injury and tissue damage. The mitochondrial membrane permeability alters, and thus respiratory level and ATP production decrease, leading to cell death

which it aggregates to form soluble oligomers, soluble protofibrils, and insoluble fibrils (a major component in LB); the precise function of α-synuclein, however, is still unknown.

Parkin: Parkin is an ubiquitin E3 ligase encoded by the PARK2 gene and is involved in the proteosomal degradation of target proteins. Mutations in the parkin gene are the major causes of recessive, early-onset Parkinsonism. When the normal function of the ligase is disrupted, several known mutations related to PD cause considerable slowing down of degradation of the putative parkin substrates, and a massive increase in protein accumulation, a process that is thought to cause neuronal degradation in the substantia nigra and locus coeruleus of humans.

Parkin is very sensitive to cell oxidative/nitrative homeostasis; increased oxidative DA modifications (quinone formation) or stress-induced phosphorylation of parkin can inhibit the ligase activity and reduce its solubility [56, 57]. However, the neuroprotective properties of parkin against a variety of insults, including ROS production and quinone formation, have been shown in various cells and animal models. Overexpression of parkin protected against α-synuclein-induced damage in primary neuronal cultures, *Drosophila*, and rat brains [58]; similarly, lenti-parkin expression protected rat brain from the neurotoxicity of 6-OHDA. The neuroprotective mechanism of parkin is unknown; however, recent work showed that parkin overexpression attenuates tyrosinase-induced neurotoxicity mediated by activation of JNK signaling and subsequent cleavage of caspase 3 [59]. In parkin-knockout *Drosophila*, mitochondrial dysfunction and shorter longevity was observed in mutant flies, which had increased susceptibility to OS [60].

As mentioned previously, mutations in the parkin gene and overexpression of mutant forms of parkin in NT-2 or SK-N-MC cell lines increases oxidative damage to proteins and lipids, specifically, keto-protein and lipid peroxidation [61]. The precise mechanism of this parkin-mediated OS is unknown.

UCHL1: The UCHL1 gene encodes for the ubiquitin carboxyl-terminal hydrolase-L1 protein, which is found in neurons throughout the brain. This protein, which is part of the ubiquitin-proteosome system, is involved in α-synuclein degradation and has two enzymatic activities: (1) hydrolase removes and recycles ubiquitin molecules from degraded proteins and (2) ligase links together ubiquitin molecules for tagging proteins for disposal. Although the protein's main role is still elusive, opposite functions have been implicated in a few known mutations in the UCHL1 gene; the S18Y polymorphism reduces the ligase activity of the UCHL1 enzyme but has little effect on the hydrolase activity, therefore increasing the rate of protein degradation and inhibiting α-synuclein accumulation. This mutation is inversely associated with sporadic PD and is more common in Chinese and Japanese populations. Another mutation found in the UCHL1 gene is I93M, which decreases the hydrolase activity, therefore interfering with the ubiquitination of proteins, resulting in protein accumulation and possible cell damage. In vitro experiments in which downregulation of the UCHL1 gene was performed using antisense cDNA resulted in increased sensitivity to oxygen deprivation stress [62]. A clue to the relationship between UCHL1 and OS has been shown in the S18Y mutant cell line in which the neuroprotective role of this mutation against MPP^+ and $H_2O_2^-$ induced OS has been shown to be mediated through its antioxidant function [63]; however, the mechanism is still unknown.

PTEN-induced kinase 1 (PINK1): PINK1 is a widely expressed protein with a serine–threonine kinase and a mitochondrial targeting motif whose function is still unclear. Mutations in PINK1 account for about 1–2% of early-onset PD. Inhibition of the Drosophila homologue of PINK1 using RNAi technique resulted in a massive dopaminergic neurodegeneration, a process that could be prevented by expressing human PINK1 or through the addition of antioxidants [45]. This neuroprotective treatment suggests that PINK1 may play a functional role in preventing neuronal loss through OS. Although the mechanism for this neuroprotection remains to be determined, a potential hint may be the mitochondrial localization of *PINK1*. As already mentioned, mitochondria are the main source of reactive oxygen and nitrogen species, and PD-associated PINK1 mutants may disrupt homeostasis of mitochondria, resulting in OS and neuronal death.

Work from Pridgeon et al. identified a mitochondrial heat shock protein [Hsp75, or tumor necrosis factor (TNF) receptor-associated protein 1, TRAP1] as a substrate for PINK1. Phosphorylation of TRAP1 by PINK1 protected cells against OS-induced apoptosis by suppressing cytochrome *c* release from mitochondria. PINK1 depletion causes increased cytochrome *c* release from mitochondria and sensitizes cells to OS-induced cell death [64].

DJ1: DJ-1 is a homodimeric protein of 189 amino acid monomers that is found both in the cytoplasm and in the mitochondria and protects cells from various types of OS. Under OS conditions, DJ-1 is upregulated and modified by oxidation of cysteine and methionine; the modified form is then shifted to the mitochondria. This system is a postulated defense mechanism; because of the natural ROS-producing environment of the mitochondria, the accumulation of modified DJ-1 may prevent mitochondrial injury or decrease mitochondrial ROS production, thereby protecting the cell from damage.

In isolated cells, knocking down DJ-1 expression with small interfering RNA (siRNA) resulted in susceptibility to OS and endoplasmic reticulum stress. Recent analyses of DJ-1 knockout mice have revealed nigrostriatal dopaminergic dysfunction, motor deficits, and hypersensitivity to MPTP and to OS. Although the mechanism of protection is not apparent, it is known that the regulation of DJ-1 is OS dependent, and increasing stress (rotenone, 6-OHDA) immediately upregulates DJ-1 expression, a process prevented by pretreatment with antioxidants [43].

LRRK: The predicted product of the leucine-rich repeat kinase 2 (LRRK2) gene is the large protein, dardarin (PARK8), which is highly expressed in regions of the mammalian brain containing dopaminergic neurons such as the caudate-putamen, frontal cerebral cortex, and substantia nigra. Mutations in LRRK2 are the most common known cause of late-onset PD. Not much work has been done concerning the protective role of LRRK2 against OS; however, it has been shown that wild-type LRRK2 but not Y1699C or G2019S mutants attenuated H_2O_2-induced cell death. Moreover, the presence of the LRRK2 wild-type gene product but not the Y1699C or G2019S mutant gene product conferred protection against H_2O_2-induced cell death through activation of the extracellular signal-regulated kinase (ERK) pathway mediated by its kinase domain in HEK293 and SH-SY5Y cells [65].

2.5 *Iron and OS*

Iron is transported into cells, and across the BBB, by binding to transferrin receptor. Within cells that store iron, the metal ion is bound to ferritin in a nonreactive form. In the brain, iron levels vary among different areas, with highest levels in the basal ganglia [66]. Oligodendrocytes and microglia have the highest levels of iron and transferrin per cell, as revealed by Perl's stain [67]. Brain iron levels are maintained constant, even in the face of severe dietary restriction [68], mainly as a result of binding to ferritin. Other proteins that contain bound iron include the enzymes MAO and TH [6, 69], in which iron acts as a cofactor. Free ferrous iron (Fe^{2+}) is a potent catalyst for a variety of oxidative reactions, and its concentration is maintained at very low levels within cells. In particular, Fe^{2+} catalyzes the Fenton reaction, in which hydrogen peroxide is converted to hydroxyl anion, and to hydroxyl free radical, which is among the most active ROS species. The potential toxicity of iron in the brain has been adequately demonstrated in earlier studies in which a dopaminergic lesion was created by direct infusion of iron to the brain of rats [70–72]. In most tissues, and in most parts of the brain, ferrous iron is rapidly hydrolyzed to ferric iron by the enzyme ferroxinase and stored in ferritin as Fe^{3+} [73]. In areas of the brain that contain pigmented cells, an additional way in which iron is bound in a nonreactive form is by binding to NM, which is found in SNpc and the locus coeruleus and which strongly chelates iron. The Fe content of SN increases with age, as does the NM content; however, NM levels outstrip Fe levels after the fourth decade [74].

A number of studies have found high Fe levels in parkinsonian SNpc, as well as in the globus pallidus and dentate gyrus, although the levels in the substantia nigra pars reticulata (SNpr) are not increased [75, 76]. In addition, the ratio of Fe^{2+} to Fe^{3+} is increased in PD [77]. Increased Fe-containing pigments are also found in basal ganglia areas in other neurodegenerative diseases, including Hallervorden–Spatz disease, striato-nigral degeneration, and Huntington's and Alzheimer's diseases [78].

Gerlach et al. [79] described the possible sources of the increased Fe levels in parkinsonian SNpc. First, passage of Fe across the BBB could be increased. Evidence exists for a localized increase in BBB permeability in the mesencephalon of PD patients [80]. Second, brain Fe uptake could be selectively increased in a particular brain area if transferrin receptor levels increased in that area; in fact, the opposite is the case for parkinsonian SNpc [81, 82], although levels of lactoferrin receptors are increased [83]. A third possibility could be transfer of Fe from areas of higher iron content to SN cells by axonal transport along connecting fiber tracts, but no evidence exists for such a possibility [84, 85]. The fourth possible cause is release of Fe from intracellular storage sites or reduction in iron binding by proteins such as NM and ferritin.

Iron storage molecules: The intracellular distribution of Fe is regulated by the iron regulatory proteins, IRP1 and IRP2 [86, 87]. These proteins respond to altered cytoplasmic free Fe levels by increasing or decreasing the expression of ferritin and transferrin receptor. Targeted deletion of these proteins in mice leads to accumulation of cytosolic iron in axons and cell bodies, followed by severe neurodegeneration and loss of cell bodies, particularly in SN [88, 89]. Ferritin molecule exists in H (heavy) and L (light) forms [85]. Several studies have assessed ferritin levels in parkinsonian SN tissue, with variable results, because of technical problems including the specificity of antibodies used and the type of ferritin (H or L) against which they were directed (see [85] for review). The expression level of H-ferritin protein is low in neurons of SNpc, whereas high levels are seen in oligodendrocytes and microglia. L-Ferritin is absent from neurons but strongly expressed in oligodendrocytes. It is, therefore, apparent that determination of whole-tissue ferritin levels will yield minimal information on changes in iron storage in the target dopaminergic neurons of SNpc. Levels of ferritin mRNA also do not change in parkinsonian SN [82, 84]. The lack of change in transferrin receptors and ferritin in PD SNpc is suggestive of a role of NM in binding the increased Fe in the PD brain. Despite the lack of clear correlation between presence or absence of PD and changes in ferritin levels, it is clear that alterations in the iron-binding capacity of this important molecule, for example, genetically based polymorphism, could exert significant effects on the tissue levels of reactive iron. Three different insertional mutations have been described in the ferritin molecule, and disease forms including parkinsonism have been identified in patients who carry mutated genes [85, 90, 91].

In normal SNpc, iron is mainly bound to NM in dopaminergic neurons, and to ferritin in glial cells [74, 92]. NM is a complex of iron–melanin and peptides [93] with additional lipid material [74]. Natural NM obtained from SNpc has a tenfold greater ability to bind Fe than synthetic DA melanin [94]. Iron binds to high- and low-affinity sites on NM [95]. The metal ions are bound in clusters of high-spin, pseudo-octahedral iron (III) ions with an oxo-hydroxo bridge [74, 96]. The role of NM in physiological and pathophysiological function is much disputed and poorly understood. Because NM is largely absent from the brain at birth, but accumulates with age, it may play a passive role in normal aging by inactivating free surplus metal ions. However, several workers have proposed an important, active role for NM in PD. In healthy brain, NM has reserve capacity for Fe binding, but NM from the PD brain has decreased ability to bind iron [97, 98] so that reactive Fe levels in the parkinsonian brain may be disproportionately high [93, 95, 99]. NM levels in individual DA neurons of SNpc are decreased in PD [100], so that more Fe in parkinsonian SNc may be bound in a low-affinity, reactive form and therefore will increase the oxidative environment [101]. Parkinsonian NM is also qualitatively different from normal NM. Faucheux et al. [102] showed that redox activity of NM aggregates from PD brain positively correlated with severity of neuronal loss. In addition, however, NM released from degenerating neurons is broken down by H_2O_2 with the release of reactive iron and generation of toxic free radicals [93]. Alterations in the Fe-binding properties of NM could therefore be an initiating factor in the dopaminergic neuron death in parkinsonian SN. Recent work [19] has thrown a new light on the importance of NM in neuroprotection and neurodegeneration, because metal ions other than iron, namely, lead, aluminum, mercury, zinc, and others, are also concentrated in pigment areas in SN and other brain areas. This finding shows the importance of NM as a slowly turning over storage site for potentially toxic metals. Initial damage to this metal ion store could precipitate a cycle of ongoing cell destruction. In addition to its iron-chelating ability, human NM when added to isolated neuronal cells in culture reduces the toxic effect of an OS (Fenton reagent), supporting the concept of a neuroprotective role for human NM [103].

Further research is necessary to evaluate the multiple factors regulating intracellular Fe storage and release, but it would appear that many possibilities exist whereby reactive Fe levels can be altered by aberrant protein expression or degradation, as well as by altered NM characteristics.

Iron chelators as neuroprotectants: Considering the abundant evidence of increased toxic iron levels in parkinsonian SN, iron chelator treatment may be considered as a potential strategy in neuroprotective therapy of the early parkinsonian patient. Effective iron chelators must be capable of crossing the BBB, and must not damage the physiological Fe pool, but should reduce excessive free Fe levels. A number of drugs have been described, although none are currently in clinical use in PD as such; future developments, however, are predicted in this area.

Desferrioxamine (desferral) is a highly effective iron chelator that has been studied in animal experiments, but must be given directly to the brain, for example, by intracerebroventricular injection (i.c.v.), because it does not cross the BBB. Given this way, desferral substantially blocked the DA depletion caused by administered 6-OHDA [104]; however, dopaminergic cells were not counted in this study, and so a positive index of neuroprotection was not obtained. Similarly, desferral blocked dopaminergic toxicity induced by systemic iron overload caused by administration of iron–dextran complex [45]. More recently, a BBB-permeable drug, VK-28 (5-[4-(2-hydroxyethyl) piperazine-1-ylmethyl]-quinoline-8-ol), was found to significantly protect against 6-OHDA-induced dopaminergic toxicity in low dosage, by systemic or i.c.v. administration [105]. An additional strategy of development of effective iron chelator drugs combined with an additional neuroprotective or other CNS active molecule has been developed by Youdim and coworkers [106]. One such drug (M30) combines the iron-chelating moiety of VK-28 with the neuroprotective MAO-B inhibitor rasagiline and effectively antagonized lactacystin-induced dopaminergic damage in the mouse in vivo [107]; however, as this drug is also an effective MAO inhibitor, this result cannot be solely attributed to iron chelation. Green tea polyphenols are known iron chelators that also have neuroprotective action against a variety of dopaminergic neurotoxins, but the relationship between the two actions is at present only hypothetical [108]. Chelation of iron by the antibiotic cloquinol, or increased binding of tissue iron by overexpression of ferritin, significantly reduced MPTP toxicity in mice [109]. As in many similar studies, these findings indicate that iron chelation can protect against dopaminergic neurotoxins, but such animal models do not reproduce the etiology of most PD cases.

2.6 Inflammation and PD

A number of factors are thought to be involved in the progression of PD, among which inflammation is believed to play an important role. Microglia are the immune system cells resident in the CNS that provide permanent close watch of the extracellular environment. It is known that the immune system and its inflammatory response are not the initial cause of the disease, but rather a consequence of damage, that is, tissue or cellular modifications in the CNS, such as neurodegeneration. Such threats activate the microglia, which respond through morphological changes in which the cells are converted to an amoeboid state with enlarged cytoplasmic processes capable of phagocytosis, migrate to the injured environment, alter gene expression, and release of inflammatory mediators such as cytokines, chemokines, ROS, and RNS. Naturally, the primary role of these microglial cells is to stand up to the "threat", whether an external danger such as a virus or a toxin, or even an internal hazard such as cell debris or malignant cells is involved, but recent evidence suggests that the immune response can have either protective or harmful effects, depending on the pathological condition, its development, and its cause. Indeed, excessive, chronic, or unregulated microglial activation may be harmful to neurons.

The phagocytic activity of microglia is beneficial during neuronal development and in injury because of the effectiveness of this process to remove cellular debris and injured cells, but dysregulation or excessive activation and as a consequence excessive ROS formation can lead to neuronal oxidative burden [110].

The involvement of inflammation in neuronal death in PD has been observed post mortem in which an increase in the expression of the cyclooxygenase (COX) enzyme and of inflammatory mediators has been shown in the injured striatum. However, most of the data available have come from models of the disease, such as the lipopolysaccharide (LPS), 6-OHDA, and MPTP models. In fact, the addition of LPS, which activates microglial cells, and induces the expression of cytokines in a pure dopaminergic culture, did not change the cell viability, although the addition of LPS to a culture consisting of both neurons and microglia caused an increase in neuronal death [111]. In vivo experiments demonstrated that a single intranigral injection of LPS caused no damage to GABAergic neurons but strongly injured dopaminergic neurons [112]. Antiinflammatory therapies have provided a strong neuroprotective role in different kinds of illness and pathologies, and the use of nonsteroidal antiinflammatory drugs (NSAIDs), such as aspirin and other COX inhibitors, is known to confer at least partial protection against the neurodegeneration seen in the MPTP and 6-OHDA models of PD [113]. Moreover, in a large clinical trial, it was shown that users of ibuprofen (a common NSAID) had a 35% lower risk for PD; however, no other NSAID had the same results [114]. The involvement of the immune system and its activation in PD has been seen clearly post mortem and in vitro and in vivo models, but the precise mechanism by which activated microglia can worsen dopaminergic neuronal degradation is still a mystery. Microglial activation can occur through many different pathways including interleukin-4 (IL-4), glucocorticoids, aggregated α-synuclein, and LPS.

This microglial activation can accelerate neurodegeneration through various mechanisms, but central to all is the involvement of OS, either through the release of superoxide anions derived from NADPH oxidase, increased peroxynitrite production, or the nitration of α-synuclein [115].

2.7 Enzyme Alterations in PD

Heme oxygenase-1 (HO-1): HO-1, a protein expressed in brain and other tissues, is responsible for the catabolism of heme to biliverdin, free iron, and CO. HO-1 expression was significantly higher in PD substantia nigra than that of control subjects [116]. HO-1 is an inducible enzyme upregulated by several inducers including hydrogen peroxide and cytokines [117]. HO-1 confers cytoprotection by the enhancement of the breakdown of pro-oxidant heme to form the antioxidants biliverdin and bilirubin [118], but under certain conditions the heme-derived end products, CO and especially iron, can exacerbate OS in cells and damage protein, unsaturated fatty acids, and DNA [119]. In a recent publication it was shown that HO-1 upregulation in astrocytes suppressed cholesterol level and augmented oxysterol formation by enhanced OS [120]. The mechanism mediating these HO-1 effects was further investigated, demonstrating that the 30% increase in cholesterol biosynthesis was catalyzed by the CO released; CO and iron stimulated the cholesterol efflux, whereas iron (Fe^{2+}) alone induced OS with a threefold increase in oxysterols formation[121].

Tyrosine hydroxylase (TH): TH, necessary for the biosynthesis of DA, is selectively localized within dopaminergic neurons. A marked decrease (10–20% of controls) was found in the activity of TH in the striatum of PD patients [122]. Riederer et al. [123] found TH activity to be decreased also in the adrenal medulla in PD, indicating the general impairment of the catecholamine system. Decreased activity of DA beta-hydroxylase (DBH) was also found for noradrenaline synthesis and phenylethanolamine *N*-methyltransferase (PNMT) for adrenaline synthesis in PD brains. In these brains, enzyme immunoassay showed lower TH activity and rediced protien content. Although both TH protein and TH activity in the striatum were markedly decreased in PD brains as compared with those of the control brains, the molecular activity (activity per enzyme protein) was significantly increased. The increase in the molecular activity of residual TH in PD brains suggests that the remaining neurons compensate for the neuronal loss by increasing their DA release, to bring normal amounts of DA to the postsynaptic receptors, and therefore increase their tyrosine hydroxylation level. This overactivity of TH may further lead to increase in OS as a result of quinone formation.

Tyrosinase: A key enzyme in the synthesis of melanin in skin and hair, tyrosinase has also been proposed to contribute to the formation of NM. Tyrosinase mRNA has been detected in human substantia nigra, but presence of the protein in brain is controversial [124]. In dopaminergic cell cultures (SH-SY5Y), expression of tyrosinase induced OS, which was reduced by simultaneous expression of parkin [59].

Recent work showed that the exposed tyrosine side chains of α-synuclein are reactive centers that can be modified by tyrosinase, leading to α-synuclein aggregates, and the reactivity of α-synuclein as a tyrosinase substrate depends on whether it is free or membrane bound [125]. Because of the conflicting data on whether tyrosinase is present in the human brain, few data exist to connect tyrosinase to PD. Greggio et al. [124] could not find any association between occurrence of tyrosinase polymorphisms and PD but demonstrated a role for tyrosinase in dopaminergic neurotoxicity.

Inducible nitric oxide synthase: iNOS is one of the NO-synthesizing enzymes that have been shown to induce dopaminergic neuronal loss. NO can react vigorously with superoxide anion to form a powerful RNS that may further degrade into other reactive intermediates. In iNOS-deficient mice, the toxic effect of MPTP treatment on dopaminergic neurons is eliminated [126], and the pharmacologic inhibition of iNOS reduced the neuronal death caused by LPS both in vitro and in vivo [127]. In the cerebrospinal fluid of PD patients, a marked increase in the concentration of nitrite was observed, but with the increase in the metabolite, there was an increase in the density of glial cells expressing iNOS in the substantia nigra of PD patients, suggesting the involvement of NO as well as iNOS induction in the pathogenesis of the disease [128].

2.8 Endogenous Antioxidant Alteration During PD

Glutathione (GSH): GSH is the major antioxidant in organs, and the ratio between GSH and the oxidized disulfide form (GSH/GSSG) has a major role in preserving the oxido-redux homeostasis (the thiol balance). GSH has a role in cells, in preventing proteins from oxidation, in maintaining intramolecular and intermolecular cysteine–cystine balance, in keeping unsaturated fatty acids from oxidation to hydroperoxide and 2′-deoxyguanosine nucleic acid in DNA from oxidizing to the 8-oxo derivative. Glutathione also functions as a substrate for many enzymes including glutathione peroxidase, reductase, and thioredoxine by donating or accepting hydrogen atoms.

It is well documented that patients with PD have a decreased total GSH level in their brain without increased GSSG level [129]. The suppressed glutathione concentration is specific to the substantia nigra of PD and does not occur in other brain regions (cerebral cortex, globus pallidus, putamen) or in other neurodegenerative illnesses such as in multiple system atrophy. Furthermore, decreased GSH levels occur in the brain of patients with incidental LB disease before any change occurs in the complex I activity or in iron metabolism, which suggests that GSH depletion is an early event in PD progression [130]. In a small group of nine PD patients, administration of GSH daily for a month resulted in improved symptoms of PD [131]. Intracarotid injection of labeled GSH to rats resulted in only 0.5% of the injected GSH appearing in the brain extract, which demonstrates the inefficiency of the GSH penetration through the BBB. Zeevalk et al. tested the possibility of improving GSH concentration in brain by administration of GSH ethyl ester derivative (GEE); subcutaneous administration of GEE to rats (up to 50 mg/kg/day) for 28 days failed to increase brain tissue levels of GSH, but when GEE was delivered directly to the left cerebral ventricle, the GSH level in the brain increased significantly [132]. GEE also provided partial protection from loss of striatal DA by MPP^+ coadministration, but when given before MPP^+ administration, it completely protected DA neurons against MPP^+ toxicity. A similar approach of using a modified drug to overcome the obstacle of crossing the BBB was made by modifying cysteine to *N*-acetyl cysteine (NAC). NAC is a powerful thiol antioxidant that, when given systemically, passes the BBB and releases cysteine in the brain, elevating GSH [7]. NAC increased dopaminergic neurons survival against MPTP toxicity and, following subcutaneous administration, it induced about 30% reduction of the dopaminergic lesion [133].

Coenzyme Q-10: Coenzyme Q-10 (Q_{10}, ubiquinone) is a highly lipophilic endogenous compound of 1,4-quinone structure with ten isoprene units attached to the quinone ring. Q_{10} acts as an electron acceptor in mitochondrial complexes I and II during the electron transport chain, forming the Q_{10} reduced form (quinol). The quinol form, as many other phenols, can also act as an antioxidant by donating two electrons to ROS or RNS such as to phospholipid hydroperoxide, or by regenerating oxidized vitamin E, thus preventing free radical propagation reactions, while the quinol is converted to its oxidized Q_{10} (quinone) form. The link between Q_{10} to PD was established by observations that (a) inhibition of mitochondrial complex I by MPTP caused parkinsonism in animals and humans [134] (see also Sect. 2.3. Mitochondria Impairment and PD), (b) Q_{10} levels and complex I activity decreased whereas oxidized Q_{10} concentrations increased in blood of PD patients compared to age-matched controls [135–137], and (c) supplementation of mice and monkeys with Q_{10} before MPTP treatment significantly attenuated the loss of nigral dopaminergic neurons [138]. These observations encouraged the conduction of clinical trials to elevate Q_{10} concentration in blood as a treatment for PD progression in human, although efficacy in animal models has not predicted efficacy in humans [139]. The major aim of the clinical experiments conducted so far has been to assess the safety and tolerability of Q_{10} supplementation at various doses (200–3,000 mg/day, together or without vitamin E), rather than to test the efficacy of the therapeutic treatments. Results show that patients receiving Q_{10} had elevated Q_{10} concentration in their blood and that administration of Q_{10} is safe. In some of the trials UPDRS(Unified Parkinson's Disease Rating Scale) was shown to improve [140] whereas in others no or only mild improvement was observed in PD patients who are already on standard (selegiline HCl + pergolide mesylate or levodopa + decarboxylase inhibitor) anti-parkinsonian therapy [141, 142]. It is unclear if the beneficial effects observed following supplementation of Q_{10} in PD patients in some of the clinical trials are caused by the antioxidant property of Q_{10} or to other Q_{10} actions such as its ability to enhance mitochondrial function. Supplementation of Q_{10} at any dosage evaluated did not delay the need for initiating symptomatic therapy [140]. In a one year phase II trial, a does of 2,400 mg/day Q10 with 1.200 IU vitamin E supplement to PD patients was not effective in neuroprotection (The NINDS NET-PD Investigation, 2007).

Melatonin: *N*-Methoxytryptamine is formed mostly by the pineal gland and released into the circulation with the ability to easily cross cell and organelle membranes. Melatonin regulates many physiological functions related to the circadian rhythms such as the reproductive system and sleep/wake cycle, plays a role in the immune, cardiovascular, and digestive systems and is an important antioxidant in the brain. Melatonin has been shown to scavenge ROS and RNS [143] and to induce antioxidant enzyme activities such as those of SOD, catalase, and glutathione peroxidase [144]. It can donate an electron and be converted into a nitrogen-centered radical, which may further scavenge superoxide and convert it into a stable compound [145]. In PD patients, pineal activity and melatonin circulation were reduced [146]. In MPP^+-mediated parkinsonism in rats, melatonin reduced lipid peroxidation, protected nigral dopaminergic neurons [147], DNA fragmentation, and protein misfolding; in 6-OHDA-mediated parkinsonism, similar indoleamine compounds also prevented the PD-like behavioral changes [147]. Many derivatives of melatonin have been synthesized as potential drugs to overcome some of the disadvantages associated with melatonin, mainly its short circulation half-life resulting from rapid catabolism and lack of selectivity at the target site [148].

3 Reduction of OS in Brain as a Methodology to Delay or Stop PD Progression

Antioxidants: Antioxidants are compounds that, when present at low concentrations, are capable of delaying or preventing the oxidation and destruction of endogenous macromolecules such as proteins, unsaturated fatty acids, and DNA. Cumulative evidence emphasizes the role of OS in PD development and progression (see Sect. 2), which raised the hope that antioxidants may lower such risk. Major food antioxidants present in fruits and vegetables are polyphenols of flavonoid and nonflavonoid structure. Prospective cohort studies (PCS) were conducted correlating the effect of consumption of a diet rich in polyphenols on PD progression, such as that of Checkoway et al., which showed that consumption of green tea was associated with reduced risk for PD [149]. It was assumed that this protective effect may result from to the ability of tea polyphenols to scavenge singlet oxygen, superoxide, hydroxyl radicals, and peroxides. Similarly, various correlations were carried out between consumption of polyphenols from other sources, such as extracts of blueberries or *Ginkgo biloba*, vs. parkinsonian symptom development. The conclusions from most of such experiments were that a diet rich in polyphenols such as catechin, epi-catechin, anthocyanins, quercetin, and kaempferol had a beneficial effect on preventing development of neurological diseases including PD. PCS usually aim to prove or disprove a certain hypothesis without providing solid proof to the causal factor and thus are usually followed by clinical trials. The PCS type of research investigating the effects of diet or their extracts is outside the scope of the present review, which focused only on clinical trials testing specific defined agents.

Such clinical trials, when carried out using vitamin E, coenzyme Q-10, and glutathione, showed that vitamin E was not effective in slowing the progression of PD or in its prevention. Supplementation of vitamin E alone or in combination with selegiline was not found to reduce the probability of requiring levodopa therapy [150]. In two large cohort studies covering 120,000 participants, the associations between risk of PD and use of vitamin E, vitamin C, carotenoids, or vitamin supplementation was examined [127]. No association was found between vitamin E or vitamin C intake and PD development. Meta-analysis studies on the effect of vitamin C, vitamin E, and β-carotene covering the period 1966–2005 on the risk of PD development revealed that β-carotene or vitamin C had no beneficial effects on the risk of developing PD, whereas a diet rich in vitamin E, as opposed to pure vitamin E, has some beneficial effect [151].

Resveratrol: Resveratrol is a well-known, nonflavonoid antioxidant present in grapes and red wine, and some of the protecting effects of red wine on the development of cardiovascular diseases have been attributed to the presence of resveratrol [152]. Mice treated with MPTP displayed severe neuronal loss whereas administration of MPTP in combination with resveratrol significantly protected mice from MPTP-induced motor coordination impairment and neuronal loss [43].

Curcumin: A nonflavonoid polyphenol present in turmeric, curcumin possesses antiinflammatory and antioxidant effects. Administration of curcumin to a mouse model of GSH depletion restores the cellular GSH pool, significantly delays protein oxidation, and preserves mitochondrial complex I activity caused by glutathione depletion [153]. Systemic administration of curcumin and its metabolite tetrahydrocurcumin reversed the MPTP-induced depletion of DA, which was presumed to be the result of inhibition of the enzyme MAO-B [154]. In a 6-OHDA model of PD, rats pretreated with curcumin exhibited protection of the number of TH-positive cells in the SN and of DA levels in the striatum. A similar effect was observed with naringenin but not with quercetin or fisetin [155]. There are insufficient human clinical trials evaluating the beneficial effects of pure exogenous antioxidants, other than vitamin E, vitamin C, and β-carotene, on PD development; such clinical trials are urgently required.

MAO-B inhibitors: The propargylamine MAO-B inhibitors selegiline and rasagiline have the potential to reduce OS by reducing DA oxidative deamination, and both drugs possess neuroprotective effects in preclinical studies, both in vivo and in vitro. The in vitro neuroprotective effect of the compounds is exerted at concentrations below those normally required for MAO inhibition and has been attributed to an intrinsic property of the drug molecules to improve cell resistance to a variety of stressors by increased expression of Bcl2, bcl-xL, SOD, and catalase, and by other mechanisms. When administered to patients, the drugs are used at doses that effectively inhibit MAO-B and so could potentially induce neuroprotection by (a) reducing DA catabolism and (b) increasing levels of DA, which induces release of neurotrophic factors. A previous multicenter study was not able to conclude that selegiline had a neuroprotective effect in humans [150], but very recently rasagiline has been demonstrated to reduce the rate of disease progression in human PD patients [156]. At present, however, it is impossible to conclude whether the indirect antioxidant effect of rasagiline is responsible for its neuroprotective effect in PD.

References

1. Fahn S. Levodopa in the treatment of Parkinson's disease. J Neural Transm Suppl. 2006;71:1–15.
2. Braak H, Rub U, Gai WP, Del Tredici K. Idiopathic Parkinson's disease: possible routes by which vulnerable neuronal types may be subject to neuroinvasion by an unknown pathogen. J Neural Transm. 2003;110:517–36.
3. Tretiakoff C. Contribution a l'etude de l'anatomie pathologique du locus niger de Soemmering avec quelques deductions relatif a la pathogenie des troubles de tonus musculaire et de la maladie de Parkinson. Thesis. Paris: University of Paris; 1919.
4. Standaert DG, Young AB. Treatment of central nervous system degenerative disorders. In: Brunton LL, Lazo JS, Parker KL (eds) Goodman and Gilman's the pharmacological basis of therapeutics. New York: McGraw-Hill; 2006. p. 527–45.
5. Wichman T, DeLong MR. Basal ganglia discharge abnormalities in Parkinson's disease. J Neural Transm Suppl. 2006;70:21–5.
6. Youdim MBH, Finberg JPM, Tipton KF. Amine oxidases. In: Trendlenburg U, Weiner N (eds) Handbook of experimental pharmacology. Berlin: Springer Verlag; 1988. p. 119–92.
7. Asanuma M, Miyazaki I, Diaz-Corrales FJ, Ogawa N. Quinone formation as dopaminergic neuron-specific oxidative stress in the pathogenesis of sporadic Parkinson's disease and neurotoxin-induced parkinsonism. Acta Med Okayama. 2004;58:221–33.
8. Graham DG. Oxidative pathways for catecholamines in the genesis of neuromelanin and cytotoxic quinones. Mol Pharmacol. 1978;14:633–43.
9. Stokes AH, Hastings TG, Vrana KE. Cytotoxic and genotoxic potential of dopamine. J Neurosci Res. 1999;55:659–65.
10. Nappi AJ, Vass E, Prota G, Memoli S. The effects of hydroxyl radical attack on dopa, dopamine, 6-hydroxydopa, and 6-hydroxydopamine. Pigment Cell Res. 1995;8:283–93.
11. Serra PA, Esposito G, Enrico P, et al. Manganese increases L-DOPA auto-oxidation in the striatum of the freely moving rat: potential implications to L-DOPA long-term therapy of Parkinson's disease. Br J Pharmacol. 2000;130:937–45.
12. Hastings TG, Lewis DA, Zigmond MJ. Role of oxidation in the neurotoxic effects of intrastriatal dopamine injections. Proc Natl Acad Sci USA. 1996;93:1956–61.
13. Miyazaki I, Asanuma M, Diaz-Corrales FJ, et al. Methamphetamine-induced dopaminergic neurotoxicity is regulated by quinone-formation-related molecules. FASEB J. 2006; 20:571–3.
14. Asanuma M, Miyazaki I, Ogawa N. Dopamine- or L-DOPA-induced neurotoxicity: the role of dopamine quinone formation and tyrosinase in a model of Parkinson's disease. Neurotox Res. 2003;5:165–76.
15. Bisaglia M, Mammi S, Bubacco L. Kinetic and structural analysis of the early oxidation products of dopamine: analysis of the interactions with alpha-synuclein. J Biol Chem. 2007;282:15597–605.
16. Kuhn DM, Arthur RE Jr, Thomas DM, Elferink LA. Tyrosine hydroxylase is inactivated by catechol-quinones and converted to a redox-cycling quinoprotein: possible relevance to Parkinson's disease. J Neurochem. 1999;73:1309–17.
17. Odh G, Carstam R, Paulson J, Wittbjer A, Rosengren E, Rorsman H. Neuromelanin of the human substantia nigra: a mixed-type melanin. J Neurochem. 1994;62:2030–6.
18. Fedorow H, Tribl F, Halliday G, Gerlach M, Riederer P, Double KL. Neuromelanin in human dopamine neurons: comparison with peripheral melanins and relevance to Parkinson's disease. Prog Neurobiol. 2005;75:109–24.
19. Zecca L, Bellei C, Costi P, et al. New melanic pigments in the human brain that accumulate in aging and block environmental toxic metals. Proc Natl Acad Sci USA. 2008;105: 17567–72.
20. Belin AC, Westerlund M. Parkinson's disease: a genetic perspective. FEBS J. 2008; 275:1377–83.

21. Rodriguez Diaz M, Abdala P, Barroso-Chinea P, Obeso J, Gonzalez-Hernandez T. Motor behavioural changes after intracerebroventricular injection of 6-hydroxydopamine in the rat: an animal model of Parkinson's disease. Behav Brain Res. 2001;122:79–92.
22. Rodriguez-Pallares J, Parga JA, Munoz A, Rey P, Guerra MJ, Labandeira-Garcia JL. Mechanism of 6-hydroxydopamine neurotoxicity: the role of NADPH oxidase and microglial activation in 6-hydroxydopamine-induced degeneration of dopaminergic neurons. J Neurochem. 2007;103:145–56.
23. Shimohama S, Sawada H, Kitamura Y, Taniguchi T. Disease model: Parkinson's disease. Trends Mol Med. 2003;9:360–5.
24. Glinka Y, Tipton KF, Youdim MB. Nature of inhibition of mitochondrial respiratory complex I by 6-hydroxydopamine. J Neurochem. 1996;66:2004–10.
25. Soto-Otero R, Mendez-Alvarez E, Hermida-Ameijeiras A, Munoz-Patino AM, Labandeira-Garcia JL. Autoxidation and neurotoxicity of 6-hydroxydopamine in the presence of some antioxidants: potential implication in relation to the pathogenesis of Parkinson's disease. J Neurochem. 2000;74:1605–12.
26. Fleming SM, Fernagut PO, Chesselet MF. Genetic mouse models of parkinsonism: strengths and limitations. NeuroRx. 2005;2:495–503.
27. Terzioglu M, Galter D. Parkinson's disease: genetic versus toxin-induced rodent models. FEBS J. 2008;275:1384–91.
28. Sofic E, Lange KW, Jellinger K, Riederer P. Reduced and oxidized glutathione in the substantia nigra of patients with Parkinson's disease. Neurosci Lett. 1992;142:128–30.
29. Good PF, Olanow CW, Perl DP. Neuromelanin-containing neurons of the substantia nigra accumulate iron and aluminum in Parkinson's disease: a LAMMA study. Brain Res. 1992;593:343–6.
30. Dexter DT, Carter CJ, Wells FR, et al. Basal lipid peroxidation in substantia nigra is increased in Parkinson's disease. J Neurochem. 1989;52:381–9.
31. Alam ZI, Daniel SE, Lees AJ, Marsden DC, Jenner P, Halliwell B. A generalised increase in protein carbonyls in the brain in Parkinson's but not incidental Lewy body disease. J Neurochem. 1997;69:1326–9.
32. Alam ZI, Jenner A, Daniel SE, et al. Oxidative DNA damage in the parkinsonian brain: an apparent selective increase in 8-hydroxyguanine levels in substantia nigra. J Neurochem. 1997;69:1196–203.
33. Olanow CW. The pathogenesis of cell death in Parkinson's disease. Mov Disord. 2007; 22Suppl 17:S335–42.
34. Chartier-Harlin MC, Kachergus J, Roumier C, et al. Alpha-synuclein locus duplication as a cause of familial Parkinson's disease. Lancet. 2004;364:1167–9.
35. Polymeropoulos MH, Lavedan C, Leroy E, et al. Mutation in the alpha-synuclein gene identified in families with Parkinson's disease. Science. 1997;276(5321):2045–7.
36. Bostantjopoulou S, Kyriazis G, Katsarou Z, Kiosseoglou G, Kazis A, Mentenopoulos G. Superoxide dismutase activity in early and advanced Parkinson's disease. Funct Neurol. 1997;12:63–8.
37. Migliore L, Scarpato R, Coppede F, Petrozzi L, Bonuccelli U, Rodilla V. Chromosome and oxidative damage biomarkers in lymphocytes of Parkinson's disease patients. Int J Hyg Environ Health. 2001;204:61–6.
38. Munch G, Luth HJ, Wong A, et al. Crosslinking of alpha-synuclein by advanced glycation endproducts: an early pathophysiological step in Lewy body formation? J Chem Neuroanat. 2000;20:253–7.
39. Miyazaki I, Asanuma M, Hozumi H, Miyoshi K, Sogawa N. Protective effects of metallothionein against dopamine quinone-induced dopaminergic neurotoxicity. FEBS Lett. 2007;581:5003–8.
40. LaVoie MJ, Ostaszewski BL, Weihofen A, Schlossmacher MG, Selkoe DJ. Dopamine covalently modifies and functionally inactivates parkin. Nat Med. 2005;11:1214–21.

41. Jana S, Maiti AK, Bagh MB, et al. Dopamine but not 3,4-dihydroxy phenylacetic acid (DOPAC) inhibits brain respiratory chain activity by autoxidation and mitochondria catalyzed oxidation to quinone products: implications in Parkinson's disease. Brain Res. 2007;1139:195–200.
42. Linert W, Herlinger E, Jamèson RF, Kienzl E, Jellinger K, Youdim MB. Dopamine, 6-hydroxydopamine, iron, and dioxygen: their mutual interactions and possible implication in the development of Parkinson's disease. Biochim Biophys Acta. 1996;1316: 160–8.
43. Asanuma M, Miyazaki I, Diaz-Corrales FJ, Miyoshi K, Ogawa N, Murata M. Preventing effects of a novel anti-parkinsonian agent zonisamide on dopamine quinone formation. Neurosci Res. 2008;60:106–13.
44. Afanas'ev II, Anderzhanova EA, Kudrin VS, Rayevsky KS. Effects of amphetamine and sydnocarb on dopamine release and free radical generation in rat striatum. Pharmacol Biochem Behav. 2001;69:653–8.
45. Jiang H, Luan Z, Wang J, Xie J. Neuroprotective effects of iron chelator Desferal on dopaminergic neurons in the substantia nigra of rats with iron-overload. Neurochem Int. 2006;49:605–9.
46. Lotharius J, O'Malley KL. Role of mitochondrial dysfunction and dopamine-dependent oxidative stress in amphetamine-induced toxicity. Ann Neurol. 2001;49:79–89.
47. Paulson GW. Historical comments on tardive dyskinesia: a neurologist's perspective. J Clin Psychiatry. 2005;66:260–4.
48. Reinke A, Martins MR, Lima MS, Moreira JC, Dal-Pizzol F, Quevedo J. Haloperidol and clozapine, but not olanzapine, induces oxidative stress in rat brain. Neurosci Lett. 2004;372:157–60.
49. Schapira AH, Cooper JM, Dexter D, Jenner P, Clark JB, Marsden CD. Mitochondrial complex I. deficiency in Parkinson's disease. Lancet. 1989;1:1269.
50. Ekstrand MI, Terzioglu M, Galter D, et al. Progressive parkinsonism in mice with respiratory-chain-deficient dopamine neurons. Proc Natl Acad Sci USA. 2007;104:1325–30.
51. Sherer TB, Kim JH, Betarbet R, Greenamyre JT. Subcutaneous rotenone exposure causes highly selective dopaminergic degeneration and alpha-synuclein aggregation. Exp Neurol. 2003;179:9–16.
52. Dauer W, Kholodilov N, Vila M, et al. Resistance of alpha -synuclein null mice to the parkinsonian neurotoxin MPTP. Proc Natl Acad Sci USA. 2002;99:14524–9.
53. Giasson BI, Duda JE, Murray IV, et al. Oxidative damage linked to neurodegeneration by selective alpha-synuclein nitration in synucleinopathy lesions. Science. 2000;290:985–9.
54. Perier C, Tieu K, Guegan C, et al. Complex I deficiency primes Bax-dependent neuronal apoptosis through mitochondrial oxidative damage. Proc Natl Acad Sci USA. 2005;102: 19126–31.
55. Keeney PM, Xie J, Capaldi RA, Bennett JP Jr. Parkinson's disease brain mitochondrial complex I has oxidatively damaged subunits and is functionally impaired and misassembled. J Neurosci. 2006;26:5256–64.
56. Avraham E, Rott R, Liani E, Szargel R, Engelender S. Phosphorylation of Parkin by the cyclin-dependent kinase 5 at the linker region modulates its ubiquitin-ligase activity and aggregation. J Biol Chem. 2007;282:12842–50.
57. Wong ES, Tan JM, Wang C, et al. Relative sensitivity of parkin and other cysteine-containing enzymes to stress-induced solubility alterations. J Biol Chem. 2007;282:12310–8.
58. Petrucelli L, O'Farrell C, Lockhart PJ, et al. Parkin protects against the toxicity associated with mutant alpha-synuclein: proteasome dysfunction selectively affects catecholaminergic neurons. Neuron. 2002;36:1007–19.
59. Hasegawa T, Treis A, Patenge N, Fiesel FC, Springer W, Kahle PJ. Parkin protects against tyrosinase-mediated dopamine neurotoxicity by suppressing stress-activated protein kinase pathways. J Neurochem. 2008;105:1700–15.

60. Pesah Y, Pham T, Burgess H, et al. Drosophila parkin mutants have decreased mass and cell size and increased sensitivity to oxygen radical stress. Development. 2004;131:2183–94.
61. Danielson SR, Andersen JK. Oxidative and nitrative protein modifications in Parkinson's disease. Free Radic Biol Med. 2008;44:1787–94.
62. Shen H, Sikorska M, Leblanc J, Walker PR, Liu QY. Oxidative stress regulated expression of ubiquitin carboxyl-terminal hydrolase-L1: role in cell survival. Apoptosis. 2006;11:1049–59.
63. Kyratzi E, Pavlaki M, Stefanis L. The S18Y polymorphic variant of UCH-L1 confers an antioxidant function to neuronal cells. Hum Mol Genet. 2008;17:2160–71.
64. Pridgeon JW, Olzmann JA, Chin LS, Li L. PINK1 Protects against oxidative stress by phosphorylating mitochondrial chaperone TRAP1. PLoS Biol. 2007;5:e172.
65. Hung SY, Liou HC, Kang KH, Wu RM, Wen CC, Fu WM. Overexpression of heme oxygenase-1 protects dopaminergic neurons against 1-methyl-4-phenylpyridinium-induced neurotoxicity. Mol Pharmacol. 2008;74:1564–75.
66. Hallgren B, Sourander P. The effect of age on the non-haem iron in the human brain. J Neurochem. 1958;3:41–51.
67. Connor JB. Cellular and regional maintenance of iron homeostasis in the brain: normal and diseased states. In: Riederer P, Youdim MBH (eds) Iron in central nervous system disorders. New York: Springer-Verlag; 1993:1–18.
68. Yehuda S, Youdim MBH. Brain iron: neurochemical and behavioural aspects. London: Taylor & Francis; 1988.
69. Kaushik P, Gorin F, Vali S. Dynamics of tyrosine hydroxylase mediated regulation of dopamine synthesis. J Comput Neurosci. 2007;22:147–60.
70. Wesemann W, Blaschke S, Clement HW, et al. Iron and neurotoxin intoxication: comparative in vitro and in vivo studies. In: Riederer P, Youdim MBH (eds) Iron in central nervous system disorders. New York: Springer; 1993. p. 79–86.
71. Ben-Shachar D, Youdim MB. Intranigral iron injection induces behavioral and biochemical "parkinsonism" in rats. J Neurochem. 1991;57:2133–5.
72. Arendash GW, Olanow CW, Sengstock GJ. Intranigral iron infusion in rats: a progressive model for excess nigral iron levels in Parkinson's disease? In: Riederer P, Youdim MBH (eds) Iron in central nervous system disorders. New York: Springer; 1993. p. 87–101.
73. Galazka-Friedman J, Bauminger ER, Koziorowski D, Friedman A. Mossbauer spectroscopy and ELISA studies reveal differences between Parkinson's disease and control substantia nigra. Biochim Biophys Acta. 2004;1688:130–6.
74. Zecca L, Gallorini M, Schunemann V, et al. Iron, neuromelanin and ferritin content in the substantia nigra of normal subjects at different ages: consequences for iron storage and neurodegenerative processes. J Neurochem. 2001;76:1766–73.
75. Riederer P, Sofic E, Rausch WD, et al. Transition metals, ferritin, glutathione, and ascorbic acid in parkinsonian brains. J Neurochem. 1989;52:515–20.
76. Dexter DT, Wells FR, Lees AJ, et al. Increased nigral iron content and alterations in other metal ions occurring in brain in Parkinson's disease. J Neurochem. 1989;52:1830–6.
77. Minotti G, Aust SD. The requirement for iron (III) in the initiation of lipid peroxidation by iron (II) and hydrogen peroxide. J Biol Chem. 1987;262:1098–104.
78. Jellinger K, Kienzl E. Iron deposits in brain disorders. In: Riederer P, Youdim MBH (eds) Iron in central nervous system disorders. New York: Springer; 1993. p. 19–36.
79. Gerlach M, Double KL, Youdim MB, Riederer P. Potential sources of increased iron in the substantia nigra of parkinsonian patients. J Neural Transm Suppl. 2006;70:133–42.
80. Kortekaas R, Leenders KL, van Oostrom JC, et al. Blood-brain barrier dysfunction in parkinsonian midbrain in vivo. Ann Neurol. 2005;57:176–9.
81. Faucheux BA, Hauw JJ, Agid Y, Hirsch EC. The density of [^{125}I]-transferrin binding sites on perikarya of melanized neurons of the substantia nigra is decreased in Parkinson's disease. Brain Res. 1997;749:170–4.

82. DC Mash, Singer J, Pablo J, Basile M, Bruce J, Weiner WJ. Iron storage and transport markers in Parkinson's disease and MPTP-treated mice. In: Riederer P, Youdim MBH (eds) Iron in central nervous system disorders. New York: Springer; 1993. p. 103–16.
83. Faucheux BA, Nillesse N, Damier P, et al. Expression of lactoferrin receptors is increased in the mesencephalon of patients with Parkinson disease. Proc Natl Acad Sci USA. 1995;92:9603–7.
84. Hirsch EC. Altered regulation of iron transport and storage in Parkinson's disease. J Neural Transm Suppl. 2006;71:201–4.
85. Snyder AM, Connor JR. Iron, the substantia nigra and related neurological disorders. Biochim Biophys. Acta 2009;1790(7):606–614.
86. Rouault TA. Post-transcriptional regulation of human iron metabolism by iron regulatory proteins. Blood Cells Mol Dis. 2002;29:309–14.
87. Hanson ES, Rawlins ML, Leibold EA. Oxygen and iron regulation of iron regulatory protein-2. J Biol Chem. 2003;278:40337–42.
88. LaVaute T, Smith S, Cooperman S, et al. Targeted deletion of the gene encoding iron regulatory protein-2 causes misregulation of iron metabolism and neurodegenerative disease in mice. Nat Genet. 2001;27:209–14.
89. Smith SR, Cooperman S, Lavaute T, et al. Severity of neurodegeneration correlates with compromise of iron metabolism in mice with iron regulatory protein deficiencies. Ann N Y Acad Sci. 2004;1012:65–83.
90. Curtis AR, Fey C, Morris CM, et al. Mutation in the gene encoding ferritin light polypeptide causes dominant adult-onset basal ganglia disease. Nat Genet. 2001;28:350–4.
91. Vidal R, Ghetti B, Takao M, et al. Intracellular ferritin accumulation in neural and extra-neural tissue characterizes a neurodegenerative disease associated with a mutation in the ferritin light polypeptide gene. J Neuropathol Exp Neurol. 2004;63:363–80.
92. Zecca L, Stroppolo A, Gatti A, et al. The role of iron and copper molecules in the neuronal vulnerability of locus coeruleus and substantia nigra during aging. Proc Natl Acad Sci USA. 2004;101:9843–8.
93. Zecca L, Casella L, Albertini A, et al. Neuromelanin can protect against iron-mediated oxidative damage in system modeling iron overload of brain aging and Parkinson's disease. J Neurochem. 2008;106:1866–75.
94. Double KL. Functional effects of neuromelanin and synthetic melanin in model systems. J Neural Transm. 2006;113:751–56.
95. Double KL, Gerlach M, Schunemann V, et al. Iron-binding characteristics of neuromelanin of the human substantia nigra. Biochem Pharmacol. 2003;66:489–94.
96. Aime S, Bergamasco B, Biglino D, et al. EPR investigations of the iron domain in neuromelanin. Biochim Biophys Acta. 1997;1361:49–58.
97. Aime S, Bergamasco B, Casu M, et al. Isolation and ^{13}C-NMR characterization of an insoluble proteinaceous fraction from substantia nigra of patients with Parkinson's disease. Mov Disord. 2000;15:977–81.
98. Lopiano L, Chiesa M, Digilio G, et al. Q-band EPR investigations of neuromelanin in control and Parkinson's disease patients. Biochim Biophys Acta. 2000;1500:306–12.
99. Gotz ME, Double K, Gerlach M, Youdim MB, Riederer P. The relevance of iron in the pathogenesis of Parkinson's disease. Ann N Y Acad Sci. 2004;1012:193–208.
100. Kastner A, Hirsch EC, Lejeune O, Javoy-Agid F, Rascol O, Agid Y. Is the vulnerability of neurons in the substantia nigra of patients with Parkinson's disease related to their neuromelanin content? J Neurochem. 1992;59:1080–9.
101. Double KL, Halliday GM. New face of neuromelanin. J Neural Transm Suppl. 2006;70: 119–23.
102. Faucheux BA, Martin ME, Beaumont C, Hauw JJ, Agid Y, Hirsch EC. Neuromelanin associated redox-active iron is increased in the substantia nigra of patients with Parkinson's disease. J Neurochem. 2003;86:1142–8.
103. Li J, Scheller C, Koutsilieri E, et al. Differential effects of human neuromelanin and synthetic dopamine melanin on neuronal and glial cells. J Neurochem. 2005;95:599–608.

104. Ben-Shachar D, Eshel G, Finberg JP, Youdim MB. The iron chelator desferrioxamine (Desferal) retards 6-hydroxydopamine-induced degeneration of nigrostriatal dopamine neurons. J Neurochem. 1991;56:1441–4.
105. Youdim MB, Stephenson G, Ben Shachar D. Ironing iron out in Parkinson's disease and other neurodegenerative diseases with iron chelators: a lesson from 6-hydroxydopamine and iron chelators, desferal and VK-28. Ann N Y Acad Sci. 2004;1012:306–25.
106. Youdim MB, Buccafusco JJ. Multi-functional drugs for various CNS targets in the treatment of neurodegenerative disorders. Trends Pharmacol Sci. 2005;26:27–35.
107. Zhu W, Xie W, Pan T, et al. Prevention and restoration of lactacystin-induced nigrostriatal dopamine neuron degeneration by novel brain-permeable iron chelators. FASEB J. 2007; 21:3835–44.
108. Mandel S, Weinreb O, Reznichenko L, Kalfon L, Amit T. Green tea catechins as brain-permeable, non toxic iron chelators to "iron out iron" from the brain. J Neural Transm Suppl. 2006;71:249–57.
109. Kaur D, Yantiri F, Rajagopalan S, et al. Genetic or pharmacological iron chelation prevents MPTP-induced neurotoxicity in vivo: a novel therapy for Parkinson's disease. Neuron. 2003;37:899–909.
110. Meredith GE, Sonsalla PK, Chesselet MF. Animal models of Parkinson's disease progression. Acta Neuropathol. 2008;115:385–98.
111. Hartmann A, Hunot S, Hirsch EC. Inflammation and dopaminergic neuronal loss in Parkinson's disease: a complex matter. Exp Neurol. 2003;184:561–4.
112. Herrera AJ, Castano A, Venero JL, Cano J, Machado A. The single intranigral injection of LPS as a new model for studying the selective effects of inflammatory reactions on dopaminergic system. Neurobiol Dis. 2000;7:429–47.
113. Villoslada P, Moreno B, Melero I, et al. Immunotherapy for neurological diseases. Clin Immunol. 2008;128:294–305.
114. Chen H, Zhang SM, Hernan MA, et al. Nonsteroidal anti-inflammatory drugs and the risk of Parkinson disease. Arch Neurol. 2003;60:1059–64.
115. McGeer PL, McGeer EG. Glial reactions in Parkinson's disease. Mov Disord. 2008;23:474–83.
116. Schipper HM, Liberman A, Stopa EG. Neural heme oxygenase-1 expression in idiopathic Parkinson's disease. Exp Neurol. 1998;150:60–8.
117. Schipper HM. Heme oxygenase expression in human central nervous system disorders. Free Radic Biol Med. 2004;37:1995–2011.
118. Llesuy SF, Tomaro ML. Heme oxygenase and oxidative stress. Evidence of involvement of bilirubin as physiological protector against oxidative damage. Biochim Biophys Acta. 1994;1223:9–14.
119. Piantadosi CA, Carraway MS, Suliman HB. Carbon monoxide, oxidative stress, and mitochondrial permeability pore transition. Free Radic Biol Med. 2006;40:1332–9.
120. Vaya J, Song W, Khatib S, Geng G, Schipper HM. Effects of heme oxygenase-1 expression on sterol homeostasis in rat astroglia. Free Radic Biol Med. 2007;42:864–71.
121. Hascalovici JR, Song W, Vaya J, et al. Impact of heme oxygenase-1 on cholesterol synthesis, cholesterol efflux and oxysterol formation in cultured astroglia. J Neurochem. 2009; 108:72–81.
122. Nagatsu T, Sawada M. Biochemistry of postmortem brains in Parkinson's disease: historical overview and future prospects. J Neural Transm Suppl. 2007;72:113–20.
123. Riederer P, Rausch WD, Birkmayer W, Jellinger K, Seemann D. CNS modulation of adrenal tyrosine hydroxylase in Parkinson's disease and metabolic encephalopathies. J Neural Transm Suppl. 1978;14:121–31.
124. Greggio E, Bergantino E, Carter D, et al. Tyrosinase exacerbates dopamine toxicity but is not genetically associated with Parkinson's disease. J Neurochem. 2005;93:246–56.
125. Tessari I, Bisaglia M, Valle F, et al. The reaction of alpha-synuclein with tyrosinase: possible implications for Parkinson disease. J Biol Chem. 2008;283:16808–17.

126. Okuno T, Nakatsuji Y, Kumanogoh A, et al. Loss of dopaminergic neurons by the induction of inducible nitric oxide synthase and cyclooxygenase-2 via CD 40: relevance to Parkinson's disease. J Neurosci Res. 2005;81:874–82.
127. Gao HM, Jiang J, Wilson B, Zhang W, Hong JS, Liu B. Microglial activation-mediated delayed and progressive degeneration of rat nigral dopaminergic neurons: relevance to Parkinson's disease. J Neurochem. 2002;81:1285–97.
128. Chen J, Tang XQ, Zhi JL, et al. Curcumin protects PC12 cells against 1-methyl-4-phenylpyridinium ion-induced apoptosis by bcl-2-mitochondria-ROS-iNOS pathway. Apoptosis. 2006;11:943–53.
129. Sian J, Dexter DT, Lees AJ, et al. Alterations in glutathione levels in Parkinson's disease and other neurodegenerative disorders affecting basal ganglia. Ann Neurol. 1994;36:348–55.
130. Jenner P, Dexter DT, Sian J, Schapira AH, Marsden CD. Oxidative stress as a cause of nigral cell death in Parkinson's disease and incidental Lewy body disease. The Royal Kings and Queens Parkinson's Disease Research Group. Ann Neurol. 1992;32 Suppl:S82–7.
131. Sechi G, Deledda MG, Bua G, et al. Reduced intravenous glutathione in the treatment of early Parkinson's disease. Prog Neuropsychopharmacol Biol Psychiatry. 1996;20:1159–70.
132. Zeevalk GD, Manzino L, Sonsalla PK, Bernard LP. Characterization of intracellular elevation of glutathione (GSH) with glutathione monoethyl ester and GSH in brain and neuronal cultures: relevance to Parkinson's disease. Exp Neurol. 2007;203:512–20.
133. Munoz AM, Rey P, Soto-Otero R, Guerra MJ, Labandeira-Garcia JL. Systemic administration of N-acetylcysteine protects dopaminergic neurons against 6-hydroxydopamine-induced degeneration. J Neurosci Res. 2004;76:551–62.
134. Langston JW, Forno LS, Rebert CS, Irwin I. Selective nigral toxicity after systemic administration of 1-methyl-4-phenyl-1,2,5,6-tetrahydropyrine (MPTP) in the squirrel monkey. Brain Res. 1984;292:390–4.
135. Matsubara T, Azuma T, Yoshida S. Serum coenzyme Q10 level in Parkinson syndrome. In: Folkers K, Littarru GP, Yamagami T (eds) Biochemical and clinical aspects of coenzyme Q. New York: Elsevier; 1991. p. 159–66.
136. Shults CW, Haas RH, Passov D, Beal MF. Coenzyme Q10 levels correlate with the activities of complexes I and II/III in mitochondria from parkinsonian and nonparkinsonian subjects. Ann Neurol. 1997;42:261–4.
137. Sohmiya M, Tanaka M, Tak NW, et al. Redox status of plasma coenzyme Q10 indicates elevated systemic oxidative stress in Parkinson's disease. J Neurol Sci. 2004;223:161–6.
138. Horvath TL, Diano S, Leranth C, et al. Coenzyme Q induces nigral mitochondrial uncoupling and prevents dopamine cell loss in a primate model of Parkinson's disease. Endocrinology. 2003;144:2757–60.
139. Galpern WR, Cudkowicz ME. Coenzyme Q treatment of neurodegenerative diseases of aging. Mitochondrion. 2007;7 Suppl:S146–53.
140. Shults CW, Oakes D, Kieburtz K, et al. Effects of coenzyme Q10 in early Parkinson disease: evidence of slowing of the functional decline. Arch Neurol. 2002;59:1541–50.
141. Horstink MW, van Engelen BG. The effect of coenzyme Q10 therapy in Parkinson's disease could be symptomatic. Arch Neurol.2003;60:1170–2; author reply 2–3.
142. Muller T, Buttner T, Gholipour AF, Kuhn W. Coenzyme Q10 supplementation provides mild symptomatic benefit in patients with Parkinson's disease. Neurosci Lett. 2003;341:201–4.
143. Reiter RJ, Acuna-Castroviejo D, Tan DX, Burkhardt S. Free radical-mediated molecular damage. Mechanisms for the protective actions of melatonin in the central nervous system. Ann N Y Acad Sci. 2001;939:200–15.
144. Antunes F, Barclay LR, Ingold KU, et al. On the antioxidant activity of melatonin. Free Radic Biol Med. 1999;26:117–28.
145. Winterbourn CC, Kettle AJ. Radical–radical reactions of superoxide: a potential route to toxicity. Biochem Biophys Res Commun. 2003;305:729–36.
146. Sandyk R. Pineal melatonin functions: possible relevance to Parkinson's disease. Int J Neurosci. 1990;50:37–53.

147. Jin BK, Shin DY, Jeong MY, et al. Melatonin protects nigral dopaminergic neurons from 1-methyl-4-phenylpyridinium (MPP+) neurotoxicity in rats. Neurosci Lett. 1998;245:61–4.
148. Suzen S. Recent developments of melatonin related antioxidant compounds. Comb Chem High Throughput Screen. 2006;9:409–19.
149. Checkoway H, Powers K, Smith-Weller T, Franklin GM, Longstreth WT, Jr., Swanson PD. Parkinson's disease risks associated with cigarette smoking, alcohol consumption, and caffeine intake. Am J Epidemiol. 2002;155:732–8.
150. Group TPS. Effects of tocopherol and deprenyl on the progression of disability in early Parkinson's disease. N Engl J Med. 1993;328:176–83.
151. Etminan M, Gill SS, Samii A. Intake of vitamin E, vitamin C, and carotenoids and the risk of Parkinson's disease: a meta-analysis. Lancet Neurol. 2005;4:362–5.
152. Sun AY, Simonyi A, Sun GY. The "French Paradox" and beyond: neuroprotective effects of polyphenols. Free Radic Biol Med. 2002;32:314–8.
153. Jagatha B, Mythri RB, Vali S, Bharath MM. Curcumin treatment alleviates the effects of glutathione depletion in vitro and in vivo: therapeutic implications for Parkinson's disease explained via in silico studies. Free Radic Biol Med. 2008;44:907–17.
154. Rajeswari A, Sabesan M. Inhibition of monoamine oxidase-B by the polyphenolic compound, curcumin and its metabolite tetrahydrocurcumin, in a model of Parkinson's disease induced by MPTP neurodegeneration in mice. Inflammopharmacology. 2008;16:96–9.
155. Zbarsky V, Datla KP, Parkar S, Rai DK, Aruoma OI, Dexter DT. Neuroprotective properties of the natural phenolic antioxidants curcumin and naringenin but not quercetin and fisetin in a 6-OHDA model of Parkinson's disease. Free Radic Res. 2005;39:1119–25.
156. Olanow CW, Hauser RA, Jankovic J, et al. A randomized, double-blind, placebo-controlled, delayed start study to assess rasagiline as a disease modifying therapy in Parkinson's disease (the ADAGIO study): rationale, design, and baseline characteristics. Mov Disord. 2008;23:2194–201.
157. The NINDS NET-PD Investigators, 2007. A Randomized, Double Blind, Futility Clinical Trial of Coenzyme Q10 and GPI-1485 in Early Parkinson's Disease, Neurology 68(!), 20-28.

Chapter 13
Tetrahydrobiopterin Deficiency

Mary Kay Koenig and Ian J. Butler

Abstract Tetrahydrobiopterin has been well described as an essential cofactor for the synthesis of the neurotransmitters dopamine, norepinephrine, and serotonin. A delicate balance of tetrahydrobiopterin is also crucial for the maintenance of the oxidative environment within cells through its involvement in the nitric oxide synthase system. Failure of this system results in the production of large amounts of free radicals that serve to impair the mitochondrial electron transport chain and ultimately decrease cellular production of ATP. Tetrahydrobiopterin deficiency has been linked to many neurological disorders including atypical phenylketonuria, dystonia, Parkinson's disease, Alzheimer's disease, depression, and schizophrenia.

Keywords Tetrahydrobiopterin · Nitric oxide · Neurotransmitter metabolites · Phenylketonuria

1 Introduction

1.1 Chemical Structure and Synthetic Pathways

Tetrahydrobiopterin belongs to a group of chemical compounds known as the pteridines [1]. Pteridines consist of fused pyrazine and pyrimidine rings with an amino group in position 2 and an oxy group in position 4 [1]. Substitution at position 6 with a

I.J. Butler (✉)
Department of Pediatrics, Division of Child and Adolescent Neurology,
The University of Texas Health Science Center, Medical School at Houston,
6431 Fannin Street, MSB 3.153, Houston, TX 77030, USA
e-mail: Ian.J.Butler@uth.tmc.edu

N. Gadoth and H.H. Göbel (eds.), *Oxidative Stress and Free Radical Damage in Neurology*, Oxidative Stress in Applied Basic Research and Clinical Practice,
DOI 10.1007/978-1-60327-514-9_13,

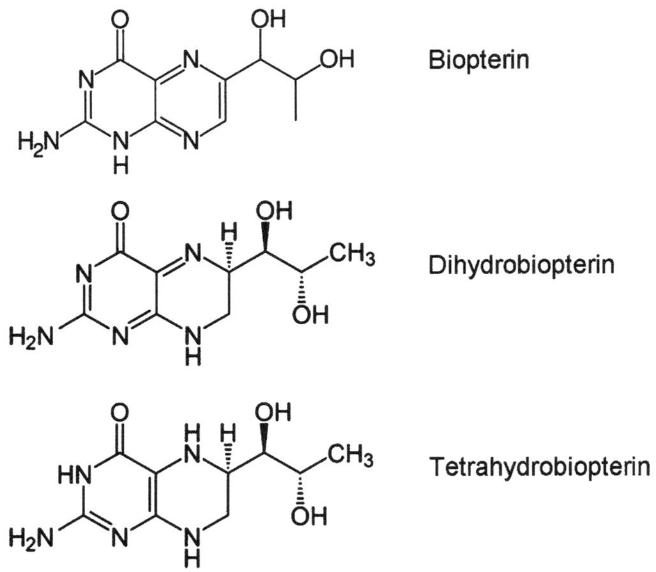

Fig. 1 Oxidation states of biopterin

dihydroxyl propyl side chain gives biopterin [1]. Physiologically, biopterin exists in either a fully oxidized, dihydro, or tetrahydro state (Fig. 1) [1, 2]. The tetrahydro state, that is, tetrahydrobiopterin, is the only form that exhibits biological activity [1].

Tetrahydrobiopterin is synthesized from guanosine triphosphate (GTP). Its synthesis involves three enzymes: GTP cyclohydrolase (rate limiting), 6-pyruvoyl tetrahydropterin synthase, and sepiapterin reductase [1, 3, 4]. GTP cyclohydrolase activity, and therefore tetrahydrobiopterin synthesis, are increased in the presence of bacterial lipopolysaccharides, tumor necrosis factor-alpha, interleukin-1 beta, and interferon-gamma [1, 5, 6].

During physiological reactions, tetrahydrobiopterin becomes oxidized [1] to its biopterin or dihydrobiopterin state. Regeneration of the tetrahydro molecule needs to occur to ensure a continuous supply of tetrahydrobiopterin for cellular activity [4]. Regeneration is facilitated by the action of pterin-4a carbinolamine dehydratase, dihydrofolate reductase [7], and dihydropteridine reductase [1, 4, 7].

1.2 Neurotransmitter Synthesis

Neurotransmitters are chemicals found in the central nervous system that relay, amplify, and modulate signals between neurons and other cells. Tetrahydrobiopterin is involved in the synthesis of three widely distributed neurotransmitters: dopamine, norepinephrine, and serotonin. Dopamine has many functions in the brain, including roles in behavior, cognition, movement, motivation, reward, sleep, mood, attention, learning, and regulation of hormone secretion. Norepinephrine is also involved in maintenance of mood and attention. Additionally, norepinephrine is involved in activating the sympathetic nervous system during stress. In the brain, serotonin functions to modulate anger, aggression, body temperature, mood, sleep, sexuality, appetite, and metabolism.

Tetrahydrobiopterin functions as a cofactor for the aromatic ring hydroxylases to form dopamine, norepinephrine, and serotonin [1]. Phenylalanine hydroxylase converts phenylalanine to tyrosine [8]; tyrosine hydroxylase converts tyrosine to L-dihydroxyphenylalanine (L-DOPA) [9]; and tryptophan hydroxylase converts tryptophan to 5-hydroxytryptophan [10]. L-DOPA and 5-hydroxytryptophan are decarboxylated by an aromatic amino acid decarboxylase to form dopamine and serotonin [1]. Tetrahydrobiopterin donates two electrons to each reaction allowing the aromatic ring hydroxylases to incorporate a single oxygen atom from molecular oxygen (O_2) into substrate, while reducing the other oxygen atom to water (H_2O) [7]. The donation of electrons results in the reduction of tetrahydrobiopterin to dihydrobiopterin. Tetrahydrobiopterin must then be regenerated for biological activity. Norepinephrine is formed from dopamine by beta-oxidation.

1.3 Nitric Oxide Synthase

Under physiological conditions, nitric oxide synthase converts reduced nicotinamide adenine dinucleotide phosphate (NADPH), oxygen, and L-arginine to nicotinamide adenine dinucleotide phosphate ($NADP^+$), nitric oxide, water, and citrulline [11, 12]. Tetrahydrobiopterin is a cofactor for all isoforms of the nitric oxide synthase reaction [1, 11, 13–17]. Nitric oxide synthase exists as a homodimer [7, 11]. Tetrahydrobiopterin stabilizes the dimeric structure and increases the affinity of nitric oxide synthase for arginine [1, 7, 12, 18, 19]. In the absence of tetrahydrobiopterin, nitric oxide synthase is unable to bind arginine, and nitric oxide formation is impaired [1, 20]. In addition to its role as an allosteric regulator, tetrahydrobiopterin acts as a single electron donor during nitric oxide formation [1, 12, 19], resulting in the reduction of tetrahydrobiopterin to trihydrobiopterin ($BH3^\bullet$).

Each monomeric subunit of nitric oxide synthase contains a C-terminal reductase domain and an N-terminal oxidase domain [7]. The C-terminal domain contains a binding site for NADPH. The N-terminal domain contains a prosthetic heme group and binding sites for tetrahydrobiopterin, molecular oxygen, and L-arginine [7]. Flavins in the N-terminal oxidase domain mediate the electron transfer from NADPH (C-terminal) to the heme center (N-terminal) where they reduce and activate molecular oxygen [7]. L-Arginine is hydroxylated to N-hydroxyl-L-arginine via the reduction of tetrahydrobiopterin to the trihydro state of $BH3^\bullet$ [7]. Tetrahydrobiopterin is then regenerated by the action of pterin-4a carbinolamine dehydratase, dihydrofolate reductase [7], and dihydropteridine reductase [1, 4, 7]. *N*-Hydroxyl-L-arginine is further oxidized to L-citrulline and nitric oxide [7].

Nitric oxide synthase is a potent factor in neuronal modulation, blood circulation, and the immune response [11]. Nitric oxide synthase has also been demonstrated to mediate ischemic-reperfusion injuries [21]. Lipopolysaccharides and interferon-gamma stimulate the formation of both tetrahydrobiopterin and nitric oxide synthase through an upregulation of GTP cyclohydrolase activity [1].

2 Free Radical Formation

During physiological processes, the involvement of tetrahydrobiopterin in the amino acid hydroxylase reactions does not typically produce free radicals. The physiological nitric oxide synthase reaction produces a small amount of $BH3^{\bullet}$ and superoxide (O_2^-) free radicals. The superoxide is quickly scavenged by the free radical scavenger glutathione [1, 15, 22, 23], and the $BH3^{\bullet}$ is regenerated to tetrahydrobiopterin [11] by pterin-4a carbinolamine dehydratase, dihydrofolate reductase [7], and dihydropteridine reductase [1, 4, 7].

The nitric oxide synthase reaction is said to be "coupled" when arginine hydroxylation is coupled to nitric oxide formation. When coupled, nitric oxide synthase catalyzes the formation of nitric oxide, L-citrulline, $NADP^+$, and water from L-arginine, NADPH, and oxygen [22]. The state of nitric oxide synthase coupling is determined largely by the availability of tetrahydrobiopterin and L-arginine [1, 12, 22, 24–26]. Depletion of either tetrahydrobiopterin or L-arginine "uncouples" the NOS reaction, that is, NADPH oxidation and oxygen reduction become independent of arginine hydroxylation and nitric oxide formation [7]. Uncoupled nitric oxide synthase continues to oxidize NADPH, but now the reaction also catalyzes the synthesis of reactive oxygen species such as superoxide (O_2^-), hydrogen peroxide (H_2O_2), and peroxynitrite ($ONOO^-$) [1, 7, 12, 15, 18, 22, 26, 27].

If a cell has an intermediate amount of tetrahydrobiopterin, one subunit of the nitric oxide synthase dimer will contain a tetrahydrobiopterin molecule and the other will not. In this situation, the monomer with tetrahydrobiopterin will produce nitric oxide and the one without tetrahydrobiopterin will produce superoxide [12, 15]. The nitric oxide and superoxide will react to form peroxynitrite [12, 15]. In the presence of glutathione, peroxynitrite is converted to *S*-nitrosoglutathione [12, 15]. When not scavenged by glutathione, peroxynitrite further uncouples the nitric oxide reaction by releasing zinc from the zinc-thiolate cluster of nitric oxide synthase and by forming disulfide bonds between the monomers [7]. Peroxynitrite also attacks cytochrome *c* oxidase, complex IV of the mitochondrial electron transport chain [28–31]. Impairment of the mitochondrial electron transport chain leads to a deficiency of adenosine triphosphate (ATP), increased mitochondrial free radical production, proton leak, loss of mitochondrial membrane potential, opening of the mitochondrial permeability transition pore, and the release of apoptotic inducing factors [29]. Mitochondrial permeability transition pore opening allows the exchange of solute and small proteins between the mitochondrial matrix and the cytosol, leading to swelling of the mitochondria and ultimately to rupture of the outer mitochondrial membrane [29].

In addition to nitric oxide synthase uncoupling, a variety of other cellular reactive oxygen-generating systems can stimulate the nitroso-redox imbalance [7]. Tetrahydrobiopterin itself is prone to oxidation to bihydrobiopterin by a number of reactive species such as the hydroxyl radical and various reactive nitrogen species [1]. Dihydrobiopterin inhibits the amino acid hydroxylases and uncouples the nitric oxide synthase reaction.

Initially, cellular antioxidant defense mechanisms will limit the degree of oxidative damage to cells, but sustained generation of reactive oxygen and nitrogen species will eventually deplete free radical scavenging systems, such as glutathione [12, 29]. Regardless of the initiating source, once the process of free radical production, tetrahydrobiopterin oxidation, tetrahydrobiopterin depletion, and nitric oxide synthase uncoupling is generated, the process is perpetuated [7].

3 Disease States

3.1 Atypical Phenylketonuria

Butler et al. described a new variant of phenylketonuria in 1975 that was unresponsive to a low phenylalanine diet. A deficiency of dihydropteridine reductase was reported [32, 33]. Mutations in dihydropteridine reductase prevent the normal recycling of tetrahydrobiopterin from dihydrobiopterin. Initial reports described symptoms as being related to impaired neurotransmitter synthesis [34], but lack of dihydropteridine reductase also results in diminished recycling of dihydrobiopterin and increased free radical production. It is therefore likely that increased free radical formation also plays a role in symptomatology. Symptom onset is usually between 4 and 6 months of age. Patients develop a typical neurological picture with truncal hypotonia, limb hypertonia, tremors, lethargy, irritability, hypokinesia, swallowing difficulties, hypersalivation, abnormal movements, seizures, and temperature instability [1, 12, 34]. Without treatment, the disorder is progressive, and children rarely survive into adulthood. Treatment is with exogenous oral tetrahydrobiopterin, 5-hydroxytryptophan, and levodopa. If begun early in the disease course, treatment is curative [34].

Several other autosomal recessive disorders of tetrahydrobiopterin synthesis/regeneration have since been described. All display phenotypic similarities to dihydropteridine reductase deficiency and have similar prognoses and treatment recommendations [34] (Table 1).

Table 1 Autosomal recessive forms of tetrahydrobiopterin deficiency

Dihydropteridine reductase deficiency (DHPR)
Guanosine triphosphate (GTP) cyclohydrolase I (GTPCH)
6-Pyruvoyl-tetrahydropterin synthase (PTPS)
Pterin-4a-carbinolamine dehydratase (PCD)
Sepiapterin reductase (SR)

3.2 *Dopa-Responsive Dystonia*

Dopa-responsive dystonia or Segawa disease is a disorder of tetrahydrobiopterin synthesis. It results from mutations in the GTP cyclohydrolase gene and therefore decreased tetrahydrobiopterin production [35]. The disorder can be inherited in either an autosomal recessive or autosomal dominant fashion, and intrafamilial variability in expression is common [36]. Symptom onset ranges from early childhood (6 years) to adulthood. Symptoms initially are dominated by gait disturbances with a diurnal fluctuation [37]. Symptoms classically worsen throughout the day and may progress to include tremor and fixed dystonic posturing. Intellectual and cognitive functions are preserved, but patients may demonstrate a variety of psychiatric symptoms including anxiety, depression, and obsessive-compulsive disorder [36].

Patients generally demonstrate a dramatic and sustained response to low doses of oral levodopa [35], but the potential impact of tetrahydrobiopterin cofactor deficiency on other hydroxylation steps suggests the involvement of 5-hydroxytryptophan in the pathogenesis of this disorder [36, 38]. Patients may benefit from administration of drugs targeted to increase the availability of 5-hydrotryptophan (pyridoxine, selective serotonin re-uptake inhibitors) [36, 38].

3.3 *Neurodegenerative Diseases*

In recent years, oxidative stress has been postulated to play an important role in the pathogenesis of several neurodegenerative diseases [1, 39, 40]. Decreased cerebrospinal fluid (CSF) concentrations of tetrahydrobiopterin have been demonstrated in both Parkinson's [41] and Alzheimer's disease [42].

It has been proposed that Parkinson's disease begins with a loss of glutathione synthase from the substantia nigra. The diminished glutathione synthesis leads to a decreased ability to scavenge reactive oxygen and reactive nitrogen species, resulting in a state of oxidative stress. Free radicals oxidize tetrahydrobiopterin, causing a partial tetrahydrobiopterin depletion state and uncoupling of nitric oxide synthase. Nitric oxide synthase uncoupling leads to further tetrahydrobiopterin catabolism and further increases oxidative stress. As the tetrahydrobiopterin deficiency becomes more pronounced, impairment of dopamine and serotonin metabolism occurs [1].

It has also been postulated that the combination of tetrahydrobiopterin and dopamine within cells is the initiating factor. In vivo studies have demonstrated that excess tetrahydrobiopterin can facilitate production of dopamine quinone products [28, 39, 43]. The dopamine quinone products increase the vulnerability of cells to oxidative stress and induce dopaminergic cell apoptosis [40, 44]. Tetrahydrobiopterin-exposed dopaminergic neurons in vivo have demonstrated loss of mitochondrial membrane potential and increased levels of oxidized proteins [45].

3.4 *Neuropsychiatric Conditions*

Tetrahydrobiopterin has also been postulated to play a role in the pathophysiology of both depression and schizophrenia. Tetrahydrobiopterin is a coenzyme of the amino acid hydroxylases and therefore is known to be required for the synthesis of dopamine and serotonin [46]. Tetrahydrobiopterin deficiency has been demonstrated in schizophrenic patients [47]. Electroconvulsive therapy, used for refractory depression, increases the activity of GTP cyclohydrolase 1 and therefore increases concentrations of tetrahydrobiopterin [46].

3.5 *Cardiovascular Disease*

Via the nitric oxide synthase system, tetrahydrobiopterin synthesis and function play an important role in diseases such as hypertension, atherosclerosis, diabetes mellitus, and cardiac hypertrophy/failure [7]. Nitric oxide synthase-deficient mice have been shown to develop hypertension, insulin resistance, hyperlipidemia, and increased ischemia-reperfusion injuries [18].

3.6 *Pain*

Tetrahydrobiopterin has also been described as an intrinsic regulator of pain sensitivity and chronicity [48].

4 Summary and Conclusions

The role of tetrahydrobiopterin in physiological systems is ubiquitous. Tetrahydrobiopterin has been known for many years to play a role in the synthesis of neurotransmitters (serotonin, norepinephrine, and dopamine), but only recently has its importance in the oxidative environment of cells been elucidated. A delicate balance of tetrahydrobiopterin appears to be required for proper functioning of the nitric oxide synthase system, and any imbalance in this system results in catastrophic biochemical events inside the cell. This oxidative environment is implicated to be involved in the pathogenesis of a variety of disorders including atypical phenylketonuria, dopa-responsive dystonia, Alzheimer dementia, Parkinson's disease, depression, schizophrenia, cardiovascular disorders, and chronic pain.

References

1. Foxton RH, Land JM, Heales SJR. Tetrahydrobiopterin availability in Parkinson's and Alzheimer's disease; potential pathogenic mechanisms. Neurochem Res. 2007;32:751–6.
2. Heales SJ, Blair JA, Meinschad C, Ziegler I. Inhibition of monocyte luminol-dependent chemiluminescence by tetrahydrobiopterin, and the free radical oxidation of tetrahydrobiopterin, dihydrobiopterin, and dihydroneopterin. Cell Biochem Funct. 1988;6:191–5.
3. Koshimura K, Murakami Y, Tanaka J, Kato Y. The role of 6R-tetrahydrobiopterin in the nervous system. Prog Neurobiol. 2000;61:415–38.
4. Thony B, Auerback G, Blau N. Tetrahydrobiopterin biosynthesis, regeneration and functions. Biochem J. 2000;347:1–16.
5. Gross SS, Levi R. Tetrahydrobiopterin synthesis. An absolute requirement for cytokine-induced nitric oxide generation by vascular smooth muscle. J Biol Chem. 1992;267:25722–9.
6. Dal Pra I, Chiarini A, Nemeth EF, et al. Roles of Ca^{2+} and the Ca^{2+}-sensing receptor (CASR) in the expression of inducible NOS (nitric oxide synthase)-2 and its BH4 (tetrahydrobiopterin)-dependent activation in cytokine-stimulated adult human astrocytes. J Cell Biochem. 2005;96:428–38.
7. Moens AL, Kass DA. Therapeutic potential of tetrahydrobiopterin for treating vascular and cardiac disease. J Cardiovasc Phamacol. 2007;50:238–46.
8. Kaufman S. Studies on the mechanism of the enzymatic conversion of pheylalanine to tyrosine. J Biol Chem. 1959;234:2677–82.
9. Nagatsu T, Levitt M, Udenfriend S. Tyrosine hydroxylase: the initial step in norepinephrine biosynthesis. J Biol Chem. 1964;239:2910–7.
10. Lovenberg W, Jequier E, Sjoerdsma. Tryptophan hydroxylation: measurement in pineal gland, brainstem, and carcinoid tumor. Science. 1967;155:217–9.
11. Bec N, Gorren AFC, Mayer B, et al. The role of tetrahydrobiopterin in the activation of oxygen by nitric-oxide synthase. J Inorg Biochem. 2000;81:207–11.
12. Lam AAJ, Hyland K, Heales SJR. Tetrahydrobiopterin availability, nitric oxide metabolism and glutathione status in the *hph*-1 mouse; implications for the pathogenesis and treatment of tetrahydrobiopterin deficiency states. J Inherit Metab Dis. 2007;30:256–62.
13. Kwon NS, Nathan CF, Stuehr DJ. Reduced biopterin as a cofactor in the generation of nitrogen oxides by murine macrophages. J Biol Chem. 1989;264:20496–501.
14. Dunbar AY, Jenkins GJ, Jianmongkol S, Nakatsuka M, Lowe ER, Lau M, Osawa Y. Tetrahydrobiopterin protects against guanabenz-mediated inhibition of neuronal nitric-oxide synthase in vitro and in vivo. Drug Metab Dispos. 2006;34:1448–56.
15. Gorren AC, Mayer B. Tetrahydrobiopterin in nitric oxide synthesis: a novel biological role for pteridines. Curr Drug Metab. 2002;3:133–57.
16. Heales SJR, Canevari L, Brand MP, et al. Cerebrospinal fluid nitrite plus nitrate correlates with tetrahydrobiopterin concentration. J Inherit Metab Dis. 1999;22:221–3.
17. Tayeh MA, Marletta MA. Macrophage oxidation of L-arginine to nitric oxide, nitrite, and nitrate. Tetrahydrobiopterin is required as a cofactor. J Biol Chem. 1989;264:19654–8.
18. Vasquez-Vivar J, Kalyanaraman B, Martasek P. The role of tetrahydrobiopterin in superoxide generation from eNOS: enzymology and physiological implications. Free Radic Res. 2003; 37:121–127.
19. Wei CC, Wang ZQ, Wang Q, et al. Rapid kinetic studies link tetrahydrobiopterin radical formation to heme-dioxy reduction and arginine hydroxylation in inducible nitric-oxide synthase. J Biol Chem. 2001;276:315–9.
20. Rafferty SP, Boyington JC, Kulansky R, et al. Stoichiometric arginine binding in the oxygenase domain of inducible nitric oxide synthase requires a single molecule of tetrahydrobiopterin per dimer. Biochem Biophys Res Commun. 1999;257:344–7.
21. Hara Y, Teramoto K, Ishidate K, et al. Cytoprotective function of tetrahydrobiopterin in rat liver ischemia/reperfusion injury. Surgery (St. Louis). 2006;139:377–84.

22. Berka V, Wang LH, Tsai AL. Oxygen-induced radical intermediates in the nNOS oxygenase domain regulated by L-arginine, tetrahydrobiopterin, and thiol. Biochemistry. 2008;47: 405–20.
23. Brown JM, Yamamoto BK. Effects of amphetamines on mitochondrial function: role of free radicals and oxidative stress. Phamacol Ther. 2003;99:45–53.
24. Brand MP, Helaes SJR, Land JM, et al. Tetrahydrobiopterin deficiency and nitric oxide metabolism in the *hph*-1 mouse. J Inherit Metab Dis. 1995;18:33–9.
25. Cosentino F, Barker JE, Brand MP, et al. Reactive oxygen species mediate endothelium-dependent relaxations in tetrahydrobiopterin-deficient mice. Arterioscler Thromb Vasc Biol. 2001;21:496–502.
26. Bendall JK, Alp NJ, Warrick N, et al. Stoichiometric relationships between endothelial tetrahydrobiopterin, endothelial NO synthase (eNOS) activity, and eNOS coupling in vivo: insights from transgenic mice with endothelial targeted GTP cyclohydrolase 1 and eNOS overexpression. Circ Res. 2005;97:864–71.
27. Pou S, Pou WS, Bredt DS, et al. Generation of superoxide by purified brain nitric oxide synthase. J Biol Chem. 1992;267:24173–6.
28. Choi HJ, Lee SY, Cho Y, et al. Tetrahydrobiopterin causes mitochondrial dysfunction in dopaminergic cells: implications for Parkinson's disease. Neurochem Int. 2006;48:255–62.
29. Heales SJR, Bolanos JP. Impairment of brain mitochondrial function by reactive nitrogen species: the role of glutathione in dictating susceptibility. Neurochem Int. 2002;40:469–74.
30. Bolanos JP, Peuchen S, Heales SJR, et al. Nitric oxide mediated inhibition of the mitochondrial electron transport chain in cultured astrocytes. J Neurochem. 1994;63:910–6.
31. Bolanos JP, Heales SJR, Land JM, et al. Effect of peroxynitrite on the mitochondrial respiratory chain; differential effects of neurons and astrocytes in primary culture. J Neurochem. 1995;64:1965–72.
32. Butler IJ, Holtzman NA, Kaufman S, et al. Phenylketonuria due to deficiency of dihydropteridine reductase. Pediatr Res. 1975;9:348.
33. Kaufman S, Holtzman NA, Milstien S, et al. Phenylketonuria due to a deficiency of dihydropteridine reductase. N Engl J Med. 1975;293:785–90.
34. Ponzone A, Spada M, Ferraris S, et al. Dihydropteridine reductase deficiency in man: from biology to treatment. Med Res Rev. 2004;24:127–50.
35. Segawa M, Nomura Y, Nishiyama N. Autosomal dominant guanosine triphosphate cyclohydrolase I deficiency (Segawa disease). Ann Neurol. 2003;54:S32–45.
36. Hahn H, Trant MR, Brownstein MJ, et al. Neurologic and psychiatric manifestations in a family with a mutation in exon 2 of the guanosine triphosphate-cyclohydrolase gene. Arch Neurol. 2001;58:749–55.
37. Segawa M, Nomura Y. Hereditary progressive dystonia with marked diurnal fluctuation. Pathophysiological importance of the age of onset. Adv Neurol. 1993;60:568–76.
38. Assman B, Kohler M, Hoffman GF, et al. Selective decrease in central nervous system serotonin turnover in children with dopa-nonresponsive dystonia. Pediatr Res. 2002;52:91–4.
39. Chae SW, Bang YJ, Kim KM, et al. Role of cyclooxygenase-2 in tetrahydrobiopterin-induced dopamine oxidation. Biochem Biophys Res Commun. 2007;359:735–41.
40. Kim ST, Choi JH, Chang JW, et al. Immobilization stress causes increases in tetrahydrobiopterin, dopamine, and neuromelanin and oxidative damage in the nigrostriatal system. J Neurochem. 2005;95:89–98.
41. Fujishiro K, Hagihara M, Takahashi A, Nagatsu T. Concentrations of neopterin and biopterin in the cerebrospinal fluid of patients with Parkinson's disease. Biochem Med Metab Biol. 1990;44:97–100.
42. Kay AD, Milstien S, Kaufman S, et al. Cerebrospinal fluid biopterin is decreased in Alzheimer's disease. Arch Neurol. 1986;43:996–9.
43. Choi JH, Jang YJ, Kim JH, et al. Tetrahydrobiopterin is released from and causes preferential death of catecholaminergic cells by oxidative stress. Mol Pharmacol. 2000;58:633–40.

44. Choi HJ, Kim SW, Lee SY, et al. Dopamine-dependent cytotoxicity of tetrahydrobiopterin: a possible mechanism for selective neurodegeneration in Parkinson's disease. J Neurochem. 2003;86:143–52.
45. Lee SY, Moon Y, Choi DH, et al. Particular vulnerability of rat mesencephalic dopaminergic neurons to tetrahydrobiopterin: relevance to Parkinson's disease. Neurobiol Dis. 2007;24: 112–20.
46. Miura H, Kitagami T, Ozaki N. Suppressive effect of paroxetine, a selective serotonin uptake inhibitor, on tetrahydrobiopterin levels and dopamine as well as serotonin turnover in the mesoprefrontal system of mice. Synapse. 2007;61:698–706.
47. Richardson MA, Read LL, Clelland CLT, et al. Evidence for a tetrahydrobiopterin deficit in schizophrenia. Neuropsychobiology. 2005;52:190–201.
48. Tegeder I, Costigan M, Griffin RS, et al. GTP cyclohydrolase and tetrahydrobiopterin regulate pain sensitivity and persistence. Nat Med. 2006;12:1269–77.

Chapter 14
Radicals Attack the Ear The Toll: A Loss of Hearing

Haim Sohmer and Cahtia Adelman

Abstract Transduction of sound stimuli into nerve impulses in the cochlea begins with an initial passive mechanical stage, the energy for which comes from the sound stimulus itself. In the normal cochlea, this passive event initiates active components of transduction, the energy for which comes from energy metabolism, and involves creation and maintenance of electrochemical gradients. These gradients must be maintained to enable continued transduction. This metabolic activity leads to the release of reactive oxygen species (ROS) and other free radicals, which, in the course of transduction of conventional sound intensities, is balanced by the endogenous antioxidant compounds present in the cochlear tissues. However, during acoustic overstimulation, the elevated levels of the highly reactive ROS molecules produced exceed the inherent levels of antioxidants present in the tissue, and this can lead to structural damage in the cochlea, with cell death and hearing loss. It has been suggested that such mechanisms are also involved in the hearing loss caused by ototoxic drugs and by aging (presbyacusis). Several strategies can be used to reduce the resulting hearing loss, whether induced by noise exposure, by ototoxic drugs, or by aging. These approaches include induction of elevated levels of endogenous antioxidants in the cochlea, before the noise exposure. Exogenous antioxidant drugs can also be administered before the exposure. An additional therapeutic strategy involves "rescue" by administering antioxidant drugs after the noise exposure. These therapeutic strategies are reviewed and evaluated.

Keywords Reactive oxygen species (ROS) · Free radicals · Antioxidants · Noise · Hearing loss · Transduction · Ototoxic · Antibiotics · Antineoplastic · Aging · Cochlea

---

H. Sohmer (✉)
Department of Physiology, Hebrew University-Hadassah Medical School,
POB 12272, Jerusalem 91120, Israel
e-mail: haims@ekmd.huji.ac.il

N. Gadoth and H.H. Göbel (eds.), *Oxidative Stress and Free Radical Damage in Neurology*, Oxidative Stress in Applied Basic Research and Clinical Practice,
DOI 10.1007/978-1-60327-514-9_14,

1 Introduction

One of the major causes of free radical-induced hearing loss is exposure to noise. The industrial revolution brought with it many positive benefits including better health, longevity, improvements in quality of life, availability of goods and their distribution. On the other hand, one of the negative aspects of this revolution is the exposure to noise levels rarely encountered by our ancestors. Noise exposure today is an almost unavoidable and nevertheless undesirable universal pollutant in industry, in military activity, and even in modern mechanized farming. In addition, youth voluntarily expose themselves to increased levels and duration of music provided by personal music players using insert earphones. Although recent studies have not found a significant increase in noise-induced hearing loss (NIHL) in youth [1], it is possible that future epidemiological studies will yield evidence for an increase in hearing loss in today's youth and perhaps for earlier and more widespread prevalence of presbyacusis (the decrease in hearing accompanying normal aging) compared to previous generations.

2 Noise-Induced Hearing Loss

Depending on the intensity and duration of the noise exposure, the resulting hearing loss can be temporary (temporary threshold shift, TTS) or permanent (permanent threshold shift, PTS). The debilitating effects of the presence of such a hearing loss are often aggravated by the addition of tinnitus, which is a sensation of sound (buzzing, ringing, etc.) in the absence of external sound. The NIHL, evaluated functionally and structurally, may result from damage to delicate inner ear structures caused directly by the excessive sound-induced mechanical vibrations [2, 3] and/or indirectly by the synthesis and release of highly reactive, potentially harmful molecules (free radicals) resulting from the noise exposure [4]. These molecules can cause cell damage by reacting with vital cellular components [5, 6] and may eventually end in cell death. Henceforth, we refer to these molecules using the general terms reactive oxygen species (ROS) and free radicals. These entities include, for example, molecules bearing oxygen with unpaired electrons such as superoxide, hydrogen peroxide, and hydroxyl radicals that can cause oxidative damage to various cell components such as membranes, proteins, enzymes, and DNA. ROS have also been implicated in the hearing loss resulting from ototoxic drugs [7] and aging. Permanent hearing loss, whether caused by excessive mechanical vibrations (e.g., during noise exposure) or by elevated ROS levels (e.g., induced by noise or by ototoxic drugs), will involve structural damage. The hearing loss caused by ototoxic drugs may also be accompanied by vestibular lesions (see Table 1), because the vestibular end organs (three semicircular canals, the receptors for angular acceleration, and two otolith organs, the receptors for linear acceleration) are also parts of the inner ear, and the fluid channels (scalae) of the cochlea are continuous with those of the vestibular

Table 1 Antibiotic and antineoplastic ototoxic drugs that are cochleotoxic (and/or vestibulotoxic)

Cochleotoxic antibiotics
Erythromycin
Gentamicin[a]
Streptomycin[a]
Dihydrostreptomycin[a]
Tobramycin[a]
Netilmicin
Amikacin
Neomycin[a]
Kanamycin[a]
Vancomycin
Capreomycin
Antineoplastic drugs (chemotherapy)
Cisplatin[a]
Carboplatin

[a]Indicates vestibulotoxicity

receptors. For example, gentamicin is toxic not only to the cochlea but also to the vestibular end organs [8]. On the other hand, exposure to intense noise is usually not accompanied by overt vestibular disorders [9], unless some other lesion facilitates access of acoustic energy to the vestibular end organs [10, 11].

As opposed to reptiles and birds, in mammals the loss of hair cells is not followed by initiation of new hair cells [12], and thus it is highly important to prevent or treat damage caused by ROS. Attempts are being made to induce hair cell regeneration in mammals also [13]. In the present review, the relationships between each of the factors – exposure to noise and ototoxic drugs and aging, release of damaging molecules, and hearing loss – is discussed.

3 Cochlear Transduction

3.1 Passive Mechanics

As a prerequisite for such a discussion, the mechanisms of cochlear transduction must first be reviewed, because the cochlea is the site of both auditory transduction and the production of the damaging molecules. Sound stimuli initiate cochlear transduction by first producing a mechanical event that activates the outer hair cells (OHCs). This event is entirely passive, that is, the energy for this initial mechanical event comes from the sound stimulus itself and not from metabolism. This initial mechanical event was studied by von Bekesy [14] in cadavers, using stroboscopic measuring techniques, and therefore at high stimulus intensities. Based on his analyses and those of others, and as a result of inner ear anatomy (a completely bony enclosed cavity containing incompressible fluid, with only two windows), it seems that stapes footplate volume displacements (the magnitude of which is a function of sound intensity) in the oval window are accompanied by opposite phase vibrations of the round window [15, 16]. This displacement induces pressure differences across the basilar membrane, passively displacing it. Because of gradients in mass and stiffness along the length of the basilar membrane, the displacements begin at the base and propagate toward the apex. At the higher stimulus frequencies, maximal displacement is seen at the base, whereas with lower frequencies, maximal displacements are more apical, which gives the appearance of a mechanical traveling wave propagating along the basilar membrane from the base toward the apex. This passive basilar membrane traveling wave activates the OHCs. It has been assumed that the basilar membrane traveling wave described by Bekesy in cadavers at high stimulus intensities is also the effective mechanical event in the normal cochlea. However, at low stimulus intensities, the OHCs may be directly activated by the fluid pressures (condensations/rarefactions) induced by the vibrations of the stapes footplate [17, 18].

3.2 Active Mechanics

In any case, this initial passive mechanical event (whether by fluid pressures directly or via the traveling wave along the basilar membrane) activates the OHCs by opening ion channels [19–21]. The resulting ion current modulates the electrical resting potential of the OHCs [22]; this is therefore referred to as the stage of mechano-electrical transduction. These electrical changes in the OHC activate a specific motor protein called prestin, situated in the lateral wall of these cells [23], that induces changes in the length of the OHCs and thus motility at the frequency of the sound stimulus [24]. This is the electro-mechanical stage of transduction, also referred to as the cochlear amplifier. These changes in length of a group of adjacent OHCs deliver active mechanical feedback to the basilar membrane, enhancing its displacement [25–28]. The energy for this active mechanical component is derived from the metabolism, which requires adequate blood flow to the cochlea, especially to the stria vascularis and the organ of Corti. When the initial mechanical passive event induces the opening of OHC ion channels, the magnitude of the resulting ion current into the OHCs will depend on the intensity of the stimulus and on the electrochemical gradients across the hair cell membrane, derived from energy metabolism. These gradients include the K^+ and Na^+ ion concentrations in the perilymph (high Na^+, low K^+) and endolymph (high K^+, low Na^+) [29], the ion concentrations in the hair cells, the OHC and inner hair cell (IHC) resting potentials (mean OHC, −83 mV; mean IHC, −37 mV) [22], and the magnitude of the extracellular resting potential of the endolymph (about +80 mV) called the endocochlear potential [30]. Because the mean resting potential of the OHCs is −83 mV [22], there is an electrical potential difference of about 163 mV across that part of the OHC membrane containing the hair bundle and exposed to endolymph. With respect to the IHCs, the potential difference is about 117 mV. Figure 1 is a diagram showing

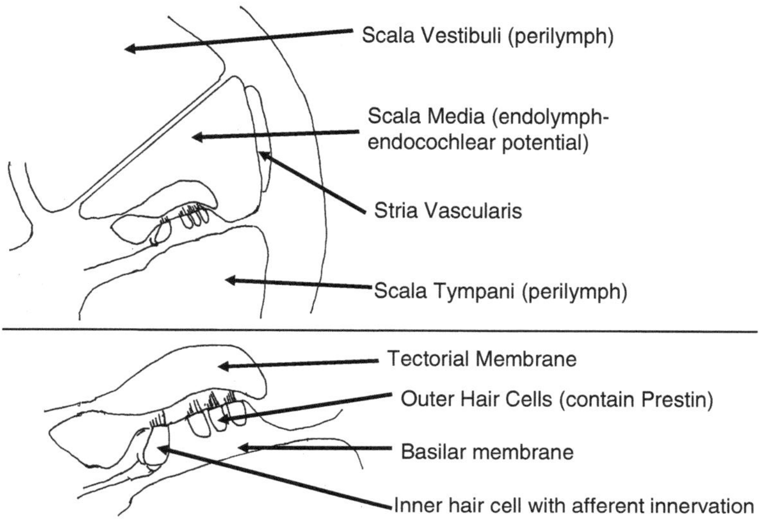

Fig. 1 A diagram of a cross section of a turn of the cochlea (*above*) and of the organ of Corti (*below*) indicating the electrochemical energy sources and gradients (endocochlear positive potential in scala media, with high K^+ and low Na^+ levels; hair cell negative resting potential, with high K^+ and low Na^+ intracellular levels), derived from energy metabolism, which are required for transduction (see text for details). During acoustic overstimulation, the elevated metabolic demands needed to maintain these gradients probably lead to formation of excessive reactive oxygen species (ROS) and free radicals

the electrochemical energy sources and gradients that must be continuously maintained so that auditory transduction can continue. The energy for the maintenance of these gradients comes from metabolism. The high K^+ concentration, coupled with the positive potential (endocochlear) in the endolymph in scala media, is generated by active transport electrogenic ion pumps situated in the highly vascular lateral wall of scala media, the stria vascularis, moving ions against their electrochemical gradient. It seems that the electrogenicity is even greater because, when the activity of these ion pumps is depressed, the electrical potential measured in scala media is about -40 mV [31]. Therefore, it is likely that the pumps are generating a positive potential of about 120 mV, so that the actual (net) potential recorded in the scala media is $+80$ mV. Maintenance of this potential is essential for normal hearing [28].

The OHCs in turn activate the IHCs, which then excite the auditory nerve fibers [32] by means of the synaptic transmitter glutamate [33].

In trying to assess the mechanisms of hearing loss resulting from exposure to noise and ototoxic drugs, it would not be ethical to conduct the relevant experiments directly on human subjects. Therefore, most such studies have been conducted on animals, with the inherent problems in applying the results to the human situation. Such experiments have involved exposing animals to intense sound stimuli, application of relevant drugs, agents, and manipulations, coupled with electrophysiological and histopathological assessments of the resulting NIHL and damage.

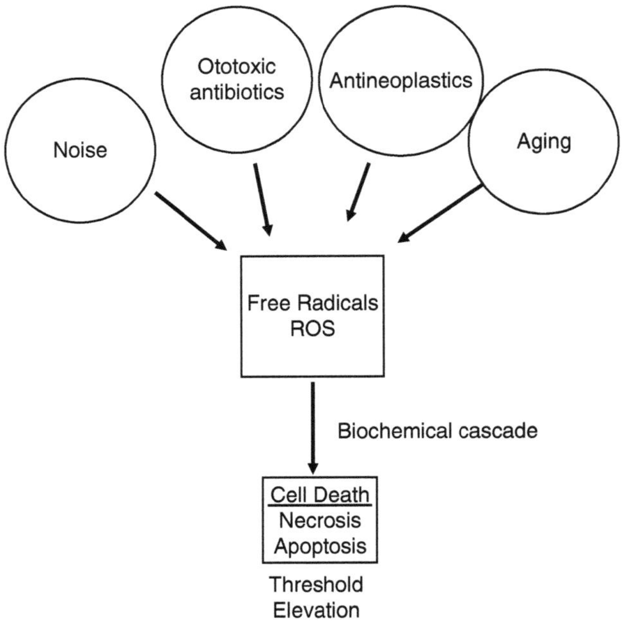

Fig. 2 The cascade of events that can lead to hearing loss as a result of noise exposure, aminoglycoside antibiotics, antineoplastic drugs, and probably aging: this includes induction of elevated levels of ROS and free radicals in the inner ear, which can cause cell death by necrosis or apoptosis, accompanied by hearing loss

4 ROS, Free Radicals, and Antioxidants

In all tissues and organs, regular ongoing levels of metabolism produce concentrations of ROS that are balanced and neutralized by intrinsic antioxidant defense mechanisms; this is also true when the cochlea transduces conventional sound levels into nerve impulses in the cochlea. The metabolism required for this transduction is accompanied by the generation of a level of ROS that is adequately balanced by the natural levels of endogenous antioxidant compounds in the tissue, including vitamins, glutathione, catalases, heat shock proteins, and their relevant enzymes. However, during acoustic overstimulation, the levels of ROS produced exceed the inherent levels of antioxidants in the cochlear tissues (Fig. 2). Elevated levels of ROS have been demonstrated as early as 1–2 h after the end of the noise exposure [34], and these become progressively elevated, reaching a maximum 7–10 days after the exposure, and then decline [4].

5 Ototoxic Drugs and ROS

The functional and structural damage seen in the ear as a result of acoustic overstimulation is similar to that found following administration of ototoxic drugs such as certain antibiotics (e.g., gentamicin, kanamycin) and antineoplastic agents (e.g., cisplatin). Therefore, it has been suggested [7] that these hearing loss-inducing factors share a common pathological mechanism, namely, that each of these agents (perhaps also aging) cause hearing loss by somehow inducing the generation of excessive ROS levels (see Fig. 2). This idea is supported by the finding that adding an antioxidant drug to organ of Corti cultures can attenuate the degree of hair cell loss following antibiotic (gentamicin) application [35].

6 Aging and ROS

It has also been suggested that oxidative stress may be part of the aging process (see Fig. 2), leading to age-related hearing loss (presbyacusis). This stress may be caused by the reductions in circulation accompanying aging [36]. This stress has been assessed in aged mice in which an important enzyme (Cu/Zn superoxide dismutase), which removes the ROS hydrogen peroxide, thereby preventing oxidative damage, was deficient (knockout). These animals had significant auditory nerve brainstem evoked response (ABR) threshold elevations and greater hair cell and neuronal loss than control mice, providing evidence that presbyacusis may be caused by ROS-induced hair cell loss [37, 38]. Other studies in aging mice have shown an elevation in levels of markers of ROS, coupled with a reduction in antioxidant molecules over the lifespan [39]. These findings have led to the suggestion that the progression of aging involves a continuous reduction in antioxidants, rendering the ear more susceptible to oxidative imbalance. Even though this suggestion was not supported by experiments showing that overexpression of the same antioxidant enzyme in other mice did not protect them from age-related hearing loss [38] and from NIHL [40], it is nevertheless generally assumed that free radicals are involved in presbyacusis.

7 Elevated ROS

With respect to acoustic overstimulation, the mechanism by means of which noise can lead to elevated levels of ROS is easily understood. This mechanism has been demonstrated [4, 34] accompanying the higher levels of metabolic activity required to continue the transduction of sound into nerve impulses while at the same time to maintain the electrochemical gradients required for this transduction. However, the mechanism whereby generation of excessive levels of ROS in the ear result from administration of ototoxic drugs such as aminoglycoside antibiotics and the antineoplastic drug cisplatin is not as directly obvious. Evidence for the involvement of ROS in antibiotic (gentamicin and kanamycin) ototoxicity has come from several experimental approaches, including the demonstration of elevated ROS in inner ear explants following aminoglycoside application [41] and that gentamicin in vivo, acting with iron, can indirectly catalyze the formation of ROS [42]. This point is supported by studies showing that antioxidant therapy (salicylate) can reduce gentamicin ototoxicity in guinea pigs [43] and in humans [44]. With respect to cisplatin ototoxicity (see

Fig. 2), ROS have been demonstrated in the inner ear following cisplatin application [45] along with a reduction in antioxidants in the cochlea in rats [46]. Administration of an antioxidant drug in rats was able to reduce the cisplatin-induced auditory threshold shift [45] and OHC loss [46]. Carboplatin is a second-generation ototoxic antineoplastic drug that seems less ototoxic than cisplatin. However, interestingly, this drug, in the chinchilla, affects mainly the IHCs [47, 48] and thereby causes hearing loss. A by-product of this finding is support for the conclusion that the otoacoustic emissions (a product of the cochlear amplifier) are generated by the OHCs [49].

How do aminoglycoside antibiotics induce elevations of ROS levels in the cochlea? It has been suggested that these antibiotics cause ototoxicity by overstimulating *N*-methyl-D-aspartate (NMDA) receptors, leading to glutamate excitotoxicity, because it has been shown that NMDA antagonists are able to reduce the hearing loss and hair cell loss following aminoglycoside administration [7].

Alternatively, ROS may not be directly involved in ototoxicity, and the similar patterns of functional and structural damage observed following noise exposure and administration of ototoxic drugs may be caused by similar final stages of the complex pathophysiological cascade of biochemical processes induced by each of these agents that triggers necrotic and apoptotic cell death in the final stage. It may be possible to influence these later stages by the use of different drugs. For example, leupeptin, an inhibitor of several protease enzymes involved in the breakdown of cellular components during apoptosis, was able to protect hair cells in culture from gentamicin [50] and from neomycin [51] ototoxicity. In vivo, leupeptin infused into scala tympani of the basal turn was able to protect the hair cells from noise-induced structural damage [52].

On the other hand, it is also possible that a part of the NIHL observed following acoustic overstimulation may be a direct result of excessive mechanical vibrations of delicate inner ear structures (e.g., hair cells) induced by the sound stimuli [53]. This result may occur mainly during more intense acoustic stimulation, such as that following impulse noise (firearms).

8 NIHL Caused by Active Mechanics

Studies in this laboratory have provided evidence that the NIHL produced by a continuous broadband noise at 113 dB SPL is more likely caused by the active mechanical vibrations (OHC motility-cochlear amplifier) that induce the active displacements of the basilar membrane. The evidence for this came from studies that showed that when drugs which reversibly depress OHC motility and therefore reduce the magnitude of the active components of vibration are administered at appropriate times before the noise exposure, the degree of NIHL is smaller; that is, each of these drugs, following a single injection alone, just before the noise exposure had a protective effect on the degree of NIHL [54, 55]. The first drug, salicylate, competitively interacts with the anion-binding sites on the prestin molecule, blocking the motility of the OHCs and thereby depressing the cochlear amplifier. The second drug, furosemide, a loop diuretic, depresses the endocochlear potential so that the electrical gradient for mechano-electrical transduction and for electromechanical transduction (the cochlear amplifier) is reduced. Besides these pharmacologic techniques for temporarily depressing the cochlear amplifier during noise exposure and thereby obtaining protection, there is evidence that the ear is equipped with a natural intrinsic physiological noise protection mechanism acting on the cochlear amplifier (OHCs). This intrinsic mechanism is mediated by the efferent pathway to the cochlea, and it reduces the electromotility of the OHCs [56, 57]. Based on these results, one may suggest two possible mechanisms for the involvement of the active mechanics in the NIHL. First, it is possible that the NIHL is a direct result of the excessive active mechanical vibrations of the basilar membrane and the hair cells, that is, direct mechanical damage, so that reduction of the active displacements by the drugs leads to less NIHL. On the other hand, it is also possible that reduction of the active vibration components leads to a reduction of the metabolic demands on the cochlear tissues, with reduced production of ROS, leading to reduced NIHL. Further research is required to differentiate between these two possibilities: noise causing a direct mechanical lesion as a result of excessive active vibrations or a lesion resulting from the production of high levels of ROS.

The degree of the NIHL is assessed functionally by determining the audiometric threshold of the subject (subjective evaluation) or, more objectively, by recording the auditory evoked potentials, such as the ABR, assessing its threshold. The NIHL can also be evaluated objectively by recording otoacoustic emissions generated by the OHCs, which reflect the activity of the cochlear amplifier and thus serve as a clinical "handle" to evaluate the activity of the cochlear amplifier. These objective measures can be used in human subjects and in experimental animals: they are used to evaluate sensorineural hearing loss, which is caused by a structural or functional lesion of the inner ear, including the sensory receptor hair cells and the synapse between the IHC and the primary sensory neurons. In fact, most, if not all, cases of sensory neural loss are the result of damage in one way or another to some component of the cochlear amplifier.

9 ROS and Hearing Loss

What is the mechanism whereby elevated levels of ROS, no matter what agent initiates them, causes hearing loss? As highly reactive molecules, ROS can lead to structural damage as a result of their reactions with lipids and proteins, lesions to DNA, and finally induce cell death by passive necrosis and by apoptosis (programmed cell death activated by caspases). The cell death cascade can continue for days after cessation of the noise exposure [6, 58].

Following intense noise exposure, cochlear blood flow is reduced, along with a decrease in the partial pressure of oxygen in the perilymph and a depression of cochlear responses [59, 60]. This reduction in oxygen availability leads to the generation of superoxide (ROS). Following reperfusion, the levels of superoxide can be even greater than those induced by the initial ischemia.

10 Temporary Threshold Shifts and Excitotoxicity

The hearing loss following a noise exposure may also involve excitotoxicty at the IHC afferent dendrite synapse because of excessive release of glutamate onto the postsynaptic dendrites [61]. The first sign of this is swelling of the dendrites beneath the IHCs from increased permeability of the postsynaptic membrane, followed by influx of water and ions, resulting in disruption of the synapse and loss of function. After several days, synaptic repair is apparent together with partial or total recovery of cochlear potentials. Therefore, this mechanism has been implicated in TTS [62–64]. Additional mechanisms which have been suggested as leading to TTS involve uncoupling of the stereocilia extensions of the OHCs (near the tips of the stereocilia are the transduction channels) from their insertion sites in the overlying tectorial membrane [65] and inactivation of transduction channels in the stereocilia themselves [66]. Therefore, there is a possibility that these TTS-producing mechanisms may not involve synthesis of excessive ROS. On the other hand, following noise exposure, lipid peroxidation (ROS breakdown of lipid molecules) has been seen in OHCs [5], and a lipid peroxidation inhibitor was found to attenuate TTS [67]. This finding may indicate that ROS is nevertheless involved in TTS. However, evaluation of the ability of a treatment strategy to reduce TTS exclusively is a complicated task because it should be based on a noise exposure paradigm that produces a hearing loss immediately after the exposure, without a residual loss after about 2 weeks (which would then be a PTS). (It seems that this experimental approach was not taken in the latter [67] study.) Therefore, the possible role of ROS and endogenous antioxidative mechanisms in the recovery following a noise exposure that results in a TTS only is less clear, as most research has focused on the more debilitating long-term noise-induced ROS damage in PTS.

If the noise exposure continues, and is intense, Ca^{2+} can enter the cells, triggering a metabolic cascade with excessive ROS formation, leading to cell death; this may represent an ROS-induced component of PTS [7].

11 Therapeutic Strategies

11.1 Prevention

Based on the premise that the hearing loss that appears after exposure to noise and following administration of ototoxic drugs is caused by synthesis of excessive levels of ROS, several therapeutic strategies have been suggested and evaluated. The first of these involves maneuvers designed to protect the ear before and during the noise exposure (i.e., prevention; Fig. 3); for example, avoidance of exposure to high noise levels and use of ear protectors. The protection may also involve (as already mentioned) injection of drugs that depress active vibrations in the cochlea before the noise exposure [54].

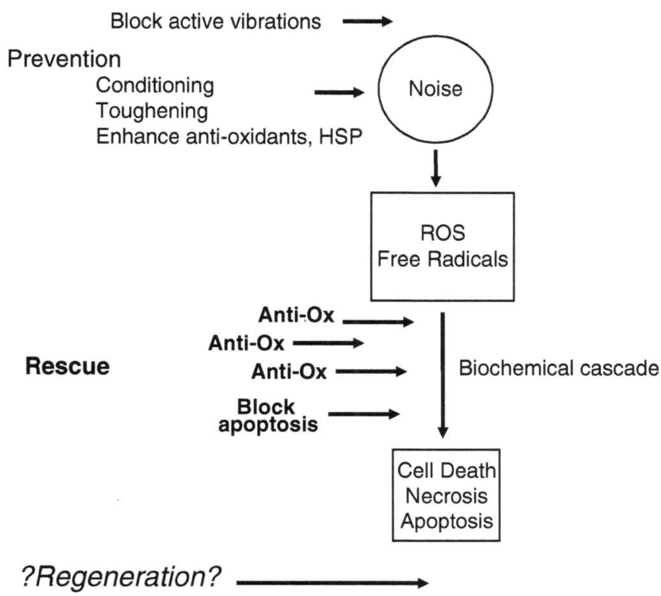

Fig. 3 Therapeutic maneuvers that may reduce the hearing loss caused by noise exposure: "prevention" before and during the exposure, and "rescue" after the exposure

11.2 Enhancing Endogenous Antioxidants

An additional protective strategy involves procedures designed to induce elevated levels of endogenous antioxidant agents in the cochlear tissues before the exposure to noise or ototoxic drug administration to reduce subsequent damage. Such procedures have been called conditioning or toughening. Toughening refers to the daily exposure to the same noise level that produces only a TTS. Following recovery from that TTS-inducing exposure on each day, the subject is again exposed to the same noise level. On each successive day, the degree of TTS is smaller [68, 69]. In conditioning, exposure to low levels of nondamaging noise for an extended period provides long-term protection from subsequent detrimental noise; that is, exposure to a much higher level of noise produces a smaller PTS than that in control animals in which there were no preceding "conditioning" noise exposures [69, 70]. The conditioning paradigm has been shown to lead to upregulation of several key antioxidant enzymes in the organ of Corti and the stria vascularis; this may be the mechanism for the attenuation of the NIHL in the conditioning procedure [69]. It has also been suggested that conditioning may involve upregulation of heat shock proteins, which can then stabilize cellular structures [70, 71].

Similarly, administration of low levels of an ototoxic drug (amikacin) over several days before the administration of the much higher therapeutic level led to structural protection from the hearing loss compared to that induced by the much higher therapeutic dose of the drug in animals given the higher therapeutic dose without the previous lower levels [72].

11.3 Cross-Tolerance

The method just described has lead to a highly interesting strategy based on the ability of low levels of various types of stress to induce elevated levels of endogenous antioxidant molecules in various tissues. These tissues are then able to withstand otherwise damaging levels of other types of stress, which has been called cross-tolerance. For example, with respect to the inner ear, acute, severe heat stress [71], or a period of more mild heat stress (heat acclimation) [73], can provide protection and reduce the hearing loss from a subsequent noise exposure. In addition, a previous period of mild physical restraint significantly reduced NIHL in mice [74]. An additional example of this strategy is represented by the preliminary study showing that exposure of chinchillas to low levels of noise (toughening), followed by administration of the antineoplastic ototoxic drug carboplatin, led to smaller auditory threshold elevations and less IHC loss [68]. It has been suggested that the mechanism for this protection may involve induction of elevated levels of heat shock proteins in the tissues, preventing or reducing structural damage. Furthermore, it is possible that this perhaps serves as an underlying mechanism for the phenomenon of conditioning, whereby a low level of stress can protect against a subsequent otherwise traumatic level of the same stress [71].

11.4 Exogenous Antioxidants

Attempts to protect the ear from noise damage have also involved inducing enhancement of levels of antioxidant agents in tissues before the exposure by injecting exogenous antioxidant agents. For example, pretreatment with free radical scavengers and with blockers of ROS protect against NIHL [75]. Multiple injections of several antioxidant drugs together (salicylic acid with *N*-acetyl cysteine, NAC [76] and salicylic acid with trolox) [77], beginning 2–3 days before a noise exposure and continuing for several days afterward, were effective in reducing threshold elevations and hair cell loss. However, if these injections began 5 days after the noise exposure, they were not effective in providing protection [77].

It cannot always be assumed that a reduction in NIHL following administration of a drug with antioxidant properties can be taken as evidence that the effect is a result of its antioxidative actions on ROS. For example, salicylic acid, although known to have antioxidant effects, was successful in reducing a NIHL only when it was administered in a single injection just before the noise exposure (so that it would depress the active mechanical vibrations of the cochlear amplifier by its effect on the OHC motor protein prestin). It did not lead to a smaller PTS when injected 1 h after cessation of the noise exposure [54], when elevated levels of ROS are already apparent in the cochlea [34].

12 Rescue from NIHL

An additional therapeutic strategy can be called rescuing, as the treatment begins essentially after the exposure (Fig. 3). It is based on the continued production of ROS following cessation of the noise [4, 34] and on the continuation of the biochemical cascade leading to structural lesions. The strategy involves administration of several antioxidant drugs, often in combination and over a period of several days after the exposure, to rescue the cochlea from the NIHL that would otherwise be induced [78]. There are several types of antioxidant agents, with different modes of action, which have been shown to be effective in reducing NIHL at different time periods after the noise exposure; this may be the rationale for experiments in which several drugs are administered together and over an extended time period. For example, injection of vitamins A, C, and E (each of these has antioxidative properties, although differing in mechanism), when coupled with magnesium (may reduce noise-induced vasoconstriction), led to smaller ABR threshold elevations and reduced hair cell loss than that in untreated animals [79]. An antioxidant drug that seems more promising for rescuing the ear from NIHL is *N*-acetyl cysteine (NAC). The efficacy of the drug, together with the antioxidant drug salicylic acid, has been assessed in animals [76]. Acetyl-L-carnitine, a drug that maintains mitochondrial bioenergetics, has also been studied [80]. When the drugs were administered 1 h after the noise and on the following 2 days, the PTS and hair cell loss were significantly reduced. Clinical trials with NAC are in progress [81]. As mentioned earlier, drugs such as leupeptin that block apoptosis by inhibiting the relevant proteolytic enzymes (calpain) may also be effective in "rescuing" from noise-induced structural damage [52]. The time window for effective rescuing by such a drug may be later than that by antioxidant drugs, as the antiapoptosis drugs apparently act on later stages of the cascade leading to cell death.

Which of these "rescue" maneuvers is the most cost-effective? An answer to this question is not yet available because the relevant research has not been conducted. Such an experiment would involve assessment of each rescue therapy (either alone or in combination) in the same species, using the same intensity and duration of noise exposure. The degree of protection provided by each therapy could then be statistically assessed.

A study has been conducted to assess whether "complete rest" from noise exposure following a possibly traumatic noise exposure contributes to recovery. Interestingly, the results were opposite to those expected: cochleas deprived of acoustic stimulation following acoustic trauma had significantly greater hair cell loss than the normally stimulated contralateral cochlea [82].

13 Impulse Noise

All these considerations presented here concerning the relationship between ROS and noise exposure have chiefly involved exposures to continuous noise. However, severe noise trauma is also caused by explosive sounds, such as terrorist bombings and impulse noise of firearms. This type of noise exposure has been studied less, so that it not yet clear if, in addition to direct mechanical injury to the ear, there is also ROS-initiated injury [68, 83, 84].

14 Final Remarks

In conclusion, because hearing loss, no matter what the cause, leads to a decrease in quality of life, attempts therefore should be made to prevent or treat it. Knowledge of the mechanisms leading to such loss is a prerequisite for designing prevention and treatment strategies, and this has been the main thrust of the present chapter. However, there are situations in which the hearing loss is a side effect of a treatment required for prolongation of life, as in antineoplastic drug therapy. It is then important to try and develop appropriate drugs that are less ototoxic or deliver them with drugs which will reduce the ototoxicity without decreasing therapeutic effects. Also, it is probably difficult to reduce or prevent the loss of hearing associated with aging, because it is likely that presbyacusis also has a genetic component [85]. In both these conditions (anticancer drugs and presbyacusis), efforts should be made to maintain quality of life. Finally, continued research into the mechanisms of ROS and free radical-induced hearing loss is required to lead to improved preventive and therapeutic strategies.

References

1. Morata TC. Young people: their noise and music exposures and the risk of hearing loss. Int J Audiol. 2007;46:111–2.
2. Saunders JC, Dear SP, Schneider ME. The anatomical consequences of acoustic injury: a review and tutorial. J Acoust Soc Am. 1985;78:833–60.
3. Ahmad M, Bohne BA, Harding GW. An in vivo tracer study of noise-induced damage to the reticular lamina. Hear Res. 2003;175:82–100.
4. Yamashita D, Jiang HY, Schacht J, Miller JM. Delayed production of free radicals following noise exposure. Brain Res. 2004;1019:201–9.
5. Ohinata Y, Miller JM, Altschuler RA, Schacht J. Intense noise induces formation of vasoactive lipid peroxidation products in the cochlea. Brain Res. 2000;878:163–73.
6. Henderson D, Bielefeld EC, Harris KC, Hu BH. The role of oxidative stress in noise-induced hearing loss. Ear Hear. 2006;27:1–19.

7. Kopke R, Allen KA, Henderson D, Hoffer M, Frenz D, Van de WT. A radical demise. Toxins and trauma share common pathways in hair cell death. Ann N Y Acad Sci. 1999;884:171–91.
8. Perez R, Freeman S, Sohmer H, Sichel JY. Vestibular and cochlear ototoxicity of topical antiseptics assessed by evoked potentials. Laryngoscope. 2000;110:1522–7.
9. Sohmer H, Elidan J, Plotnik M, Freeman S, Sockalingam R, Berkowitz Z et al. Effect of noise on the vestibular system: vestibular evoked potential studies in rats. Noise Health. 1999;2:41–52.
10. Biron A, Freeman S, Sichel JY, Sohmer H. The effect of noise exposure in the presence of canal fenestration on the amplitude of short-latency vestibular evoked potentials. Arch Otolaryngol Head Neck Surg. 2002;128:544–8.
11. Watson SR, Halmagyi GM, Colebatch JG. Vestibular hypersensitivity to sound (Tullio phenomenon): structural and functional assessment. Neurology. 2000;54:722–8.
12. Hawkins JE Jr. Comparative otopathology: aging, noise, and ototoxic drugs. Adv Otorhinolaryngol. 1973;20:125–41.
13. Izumikawa M, Minoda R, Kawamoto K, Abrashkin K, A, Swiderski DL, Dolan DF et al. Auditory hair cell replacement and hearing improvement by Atoh1 gene therapy in deaf mammals. Nat Med. 2005;11:271–6.
14. von Bekesy G. Experiments in hearing. New York: McGraw-Hill; 1960.
15. Kringlebotn M. The equality of volume displacements in the inner ear windows. J Acoust Soc Am. 1995;98:192–6.
16. Stenfelt S, Hato N, Goode RL. Fluid volume displacement at the oval and round windows with air and bone conduction stimulation. J Acoust Soc Am. 2004;115:797–812.
17. Sohmer H, Freeman S, Perez R. Semicircular canal fenestration: improvement of bone- but not air-conducted auditory thresholds. Hear Res. 2004;187:105–10.
18. Sohmer H, Sichel JY, Freeman S. Cochlear activation at low sound intensities by a fluid pathway. J Basic Clin Physiol Pharmacol. 2004;15:1–14.
19. Gillespie PG, Dumont RA, Kachar B. Have we found the tip link, transduction channel, and gating spring of the hair cell? Curr Opin Neurobiol. 2005;15:389–96.
20. Ding JP, Salvi RJ, Sachs F. Stretch-activated ion channels in guinea pig outer hair cells. Hear Res. 1991;56:19–28.
21. Rybalchenko V, Santos-Sacchi J. Cl^- flux through a non-selective, stretch-sensitive conductance influences the outer hair cell motor of the guinea-pig. J Physiol. 2003;547:873–91.
22. Cody AR, Russell IJ. The response of hair cells in the basal turn of the guinea-pig cochlea to tones. J Physiol. 1987;383:551–69.
23. Zheng J, Shen W, He DZ, Long KB, Madison LD, Dallos P. Prestin is the motor protein of cochlear outer hair cells. Nature (Lond) 2000; 405:149–155.
24. Brownell WE, Bader CR, Bertrand D, de Ribaupierre RY. Evoked mechanical responses of isolated cochlear outer hair cells. Science. 1985;227:194–6.
25. Oghalai JS. The cochlear amplifier: augmentation of the traveling wave within the inner ear. Curr Opin Otolaryngol Head Neck Surg. 2004;12:431–8.
26. Ashmore J. Cochlear outer hair cell motility. Physiol Rev. 2008;88:173–210.
27. Robles L, Ruggero MA. Mechanics of the mammalian cochlea. Physiol Rev. 2001;81:1305–52.
28. Ruggero MA, Rich NC. Furosemide alters organ of Corti mechanics: evidence for feedback of outer hair cells upon the basilar membrane. J Neurosci. 1991;11:1057–67.
29. Makimoto K, Takeda T, Silverstein H. Chemical composition in various compartments of inner ear fluid. Arch Otorhinolaryngol. 1978;220:259–64.
30. Kusakari J, Kambayashi J, Ise I, Kawamoto K. Reduction of the endocochlear potential by the new "loop" diuretic, bumetanide. Acta Otolaryngol. 1978;86:336–41.
31. Bosher SK. The nature of the negative endocochlear potentials produced by anoxia and ethacrynic acid in the rat and guinea-pig. J Physiol. 1979;293:329–45.
32. Spoendlin H. Anatomy of cochlear innervation. Am J Otolaryngol. 1985;6:453–67.

33. Eybalin M. Neurotransmitters and neuromodulators of the mammalian cochlea. Physiol Rev. 1993;73:309–73.
34. Ohlemiller KK, Wright JS, Dugan LL. Early elevation of cochlear reactive oxygen species following noise exposure. Audiol Neurootol. 1999;4:229–36.
35. McFadden SL, Ding D, Salvemini D, Salvi RJ. M40403, a superoxide dismutase mimetic, protects cochlear hair cells from gentamicin, but not cisplatin toxicity. Toxicol Appl Pharmacol. 2003;186:46–54.
36. Seidman MD, Khan MJ, Dolan DF, Quirk WS. Age-related differences in cochlear microcirculation and auditory brain stem response. Arch Otolaryngol Head Neck Surg. 1996; 122:1221–6.
37. McFadden SL, Ding D, Burkard RF, Jiang H, Reaume AG, Flood DG et al. Cu/Zn SOD deficiency potentiates hearing loss and cochlear pathology in aged 129,CD-1 mice. J Comp Neurol. 1999;413:101–12.
38. Keithley EM, Canto C, Zheng QY, Wang X, Fischel-Ghodsian N, Johnson KR. Cu/Zn superoxide dismutase and age-related hearing loss. Hear Res. 2005;209:76–85.
39. Jiang H, Talaska AE, Schacht J, Sha SH. Oxidative imbalance in the aging inner ear. Neurobiol Aging. 2007;28:1605–12.
40. Coling DE, Yu KC, Somand D, Satar B, Bai U, Huang TT et al. Effect of SOD1 overexpression on age- and noise-related hearing loss. Free Radic Biol Med. 2003;34:873–80.
41. Hirose K, Hockenbery DM, Rubel EW. Reactive oxygen species in chick hair cells after gentamicin exposure in vitro. Hear Res. 1997;104:1–14.
42. Schacht J. Antioxidant therapy attenuates aminoglycoside-induced hearing loss. Ann N Y Acad Sci. 1999;884:125–30.
43. Sha SH, Schacht J. Salicylate attenuates gentamicin-induced ototoxicity. Lab Invest. 1999;79:807–13.
44. Chen Y, Huang WG, Zha DJ, Qiu JH, Wang JL, Sha SH et al. Aspirin attenuates gentamicin ototoxicity: from the laboratory to the clinic. Hear Res. 2007;226:178–82.
45. Rybak LP, Somani S. Ototoxicity. Amelioration by protective agents. Ann N Y Acad Sci. 1999;884:143–51.
46. Minami SB, Sha SH, Schacht J. Antioxidant protection in a new animal model of cisplatin-induced ototoxicity. Hear Res. 2004;198:137–43.
47. Takeno S, Harrison RV, Ibrahim D, Wake M, Mount RJ. Cochlear function after selective inner hair cell degeneration induced by carboplatin. Hear Res. 1994;75:93–102.
48. Wake M, Takeno S, Ibrahim D, Harrison R. Selective inner hair cell ototoxicity induced by carboplatin. Laryngoscope. 1994;104:488–93.
49. Wake M, Anderson J, Takeno S, Mount RJ, Harrison RV. Otoacoustic emission amplification after inner hair cell damage. Acta Otolaryngol. 1996;116:374–81.
50. Ding D, Stracher A, Salvi RJ. Leupeptin protects cochlear and vestibular hair cells from gentamicin ototoxicity. Hear Res. 2002;164:115–26.
51. Momiyama J, Hashimoto T, Matsubara A, Futai K, Namba A, Shinkawa H. Leupeptin, a calpain inhibitor, protects inner ear hair cells from aminoglycoside ototoxicity. Tohoku J Exp Med. 2006;209:89–97.
52. Wang J, Ding D, Shulman A, Stracher A, Salvi RJ. Leupeptin protects sensory hair cells from acoustic trauma. Neuroreport. 1999;10:811–6.
53. Hamernik RP, Turrentine G, Roberto M, Salvi R, Henderson D. Anatomical correlates of impulse noise-induced mechanical damage in the cochlea. Hear Res. 1984;13:229–47.
54. Adelman C, Freeman S, Paz Z, Sohmer H. Salicylic acid injection before noise exposure reduces permanent threshold shift. Audiol Neurootol. 2008;13:266–72.
55. Nazarian Y, Perez R, Adelman C, Sichel JY, Sohmer H. Does temporary hearing threshold elevation protect the ear from noise induced damage? (Abstract). J Basic Clin Physiol Pharmacol. 2006;17:200.

56. Maison SF, Luebke AE, Liberman MC, Zuo J. Efferent protection from acoustic injury is mediated via α9 nicotinic acetylcholine receptors on outer hair cells. J Neurosci. 2002; 22:10838–46.
57. Darrow KN, Maison SF, Liberman MC. Selective removal of lateral olivocochlear efferents increases vulnerability to acute acoustic injury. J Neurophysiol. 2007;97:1775–85.
58. Hu BH, Henderson D, Nicotera TM. Involvement of apoptosis in progression of cochlear lesion following exposure to intense noise. Hear Res. 2002;166:62–71.
59. Prazma J, Vance SG, Bolster DE, Pillsbury HC, Postma DS. Cochlear blood flow. The effect of noise at 60 minutes' exposure. Arch Otolaryngol Head Neck Surg. 1987;113:36–9.
60. Lamm K, Arnold W. The effect of blood flow promoting drugs on cochlear blood flow, perilymphatic pO_2 and auditory function in the normal and noise-damaged hypoxic and ischemic guinea pig inner ear. Hear Res. 2000;141:199–219.
61. Ehrenberger K, Felix D. Receptor pharmacological models for inner ear therapies with emphasis on glutamate receptors: a survey. Acta Otolaryngol. 1995;115:236–40.
62. Puel JL, Pujol R, Tribillac F, Ladrech S, Eybalin M. Excitatory amino acid antagonists protect cochlear auditory neurons from excitotoxicity. J Comp Neurol. 1994;341:241–56.
63. Puel JL, Ruel J, Gervais d'Aldin C, Pujol R. Excitotoxicity and repair of cochlear synapses after noise-trauma induced hearing loss. Neuroreport. 1998;9:2109–14.
64. Khan MJ, Seidman MD, Quirk WS, Shivapuja BG. Effects of kynurenic acid as a glutamate receptor antagonist in the guinea pig. Eur Arch Otorhinolaryngol. 2000;257:177–81.
65. Nordmann AS, Bohne BA, Harding GW. Histopathological differences between temporary and permanent threshold shift. Hear Res. 2000;139:13–30.
66. Patuzzi RB, Yates GK, Johnstone BM. Changes in cochlear microphonic and neural sensitivity produced by acoustic trauma. Hear Res. 1989;39:189–202.
67. Quirk WS, Shivapuja BG, Schwimmer CL, Seidman MD. Lipid peroxidation inhibitor attenuates noise-induced temporary threshold shifts. Hear Res. 1994;74:217–20.
68. Henderson D, McFadden SL, Liu CC, Hight N, Zheng XY. The role of antioxidants in protection from impulse noise. Ann N Y Acad Sci. 1999;884:368–80.
69. Jacono AA, Hu B, Kopke RD, Henderson D, Van De Water TR, Steinman HM. Changes in cochlear antioxidant enzyme activity after sound conditioning and noise exposure in the chinchilla. Hear Res. 1998;117:31–8.
70. Niu X, Canlon B. Protective mechanisms of sound conditioning. Adv Otorhinolaryngol. 2002;59:96–105.
71. Yoshida N, Kristiansen A, Liberman MC. Heat stress and protection from permanent acoustic injury in mice. J Neurosci. 1999;19:10116–24.
72. Oliveira JA, Canedo DM, Rossato M, Andrade MH. Self-protection against aminoglycoside ototoxicity in guinea pigs. Otolaryngol Head Neck Surg. 2004;131:271–9.
73. Paz Z, Freeman S, Horowitz M, Sohmer H. Prior heat acclimation confers protection against noise-induced hearing loss. Audiol Neurootol. 2004;9:363–9.
74. Wang Y, Liberman MC. Restraint stress and protection from acoustic injury in mice. Hear Res. 2002;165:96–102.
75. Seidman MD, Shivapuja BG, Quirk WS. The protective effects of allopurinol and superoxide dismutase on noise-induced cochlear damage. Otolaryngol Head Neck Surg. 1993;109: 1052–6.
76. Kopke RD, Weisskopf PA, Boone JL, Jackson RL, Wester DL, Hoffer ME et al. Reduction of noise-induced hearing loss using L-NAC and salicylate in the chinchilla. Hear Res. 2000;149:138–46.
77. Yamashita D, Jiang HY, Le Prell CG, Schacht J, Miller JM. Post-exposure treatment attenuates noise-induced hearing loss. Neuroscience. 2005;134:633–42.
78. Le Prell CG, Yamashita D, Minami SB, Yamasoba T, Miller JM. Mechanisms of noise-induced hearing loss indicate multiple methods of prevention. Hear Res. 2007;226:22–43.
79. Le Prell CG, Hughes LF, Miller JM. Free radical scavengers vitamins A, C, and E plus magnesium reduce noise trauma. Free Radic Biol Med. 2007;42:1454–63.

80. Coleman JK, Kopke RD, Liu J, Zheng J, Jackson R, Henderson D et al. Pharmacological rescue of noise induced hearing loss using *N*-acetylcysteine and acetyl-L-carnitine. Hear Res. 2007;226:104–13.
81. Kopke RD, Jackson RL, Coleman JK, Liu J, Bielefeld EC, Balough BJ. NAC for noise: from the bench top to the clinic. Hear Res. 2007;226:114–25.
82. Fukushima N, White P, Harrison RV. Influence of acoustic deprivation on recovery of hair cells after acoustic trauma. Hear Res. 1990;50:107–18.
83. Bielefeld EC, Kopke RD, Jackson RL, Coleman JK, Liu J, Henderson D. Noise protection with *N*-acetyl-L-cysteine (NAC) using a variety of noise exposures, NAC doses, and routes of administration. Acta Otolaryngol. 2007;127:914–9.
84. Ylikoski J. Acute acoustic trauma in Finnish conscripts. Etiological factors and characteristics of hearing impairment. Scand Audiol. 1989;18:161–5.
85. Van Eyken E., Van CG, Van LL. The complexity of age-related hearing impairment: contributing environmental and genetic factors. Audiol Neurootol. 2007;12:345–58.

Chapter 15
Reactive Oxygen Species in Mitochondrial Encephalomyopathy: Mechanisms and Effects

Sun Young Park and Ronald G. Haller

Abstract Mitochondria produce energy in the form of ATP by oxidative phosphorylation. In addition to supplying energy, mitochondria are the major source of cellular reactive oxygen species (ROS) because of leakage of electrons from the mitochondrial respiratory chain in the process of aerobic ATP production. The production of ROS is an integral function of O_2 consumption. ROS are important for many life-sustaining processes of cells and tissues, but they also can induce cell damage and death. If their production and level in cells are not effectively controlled, then the harmful effects of oxidative stress can accumulate within the cells. The accumulation of ROS is enhanced in many pathological conditions in which the respiratory chain is impaired. In addition, mitochondria are particularly vulnerable to oxidative stress. Thus, mitochondria are both the major source of ROS generation and a major target site of oxidative stress. Mitochondrial defects are commonly associated with prominent brain and skeletal muscle symptoms and pathology. The vulnerability of brain and skeletal muscle is generally attributed to the high requirement of these tissues for mitochondrial oxidative phosphorylation to meet cellular energy requirements. Skeletal muscle is unique among tissues is its range of oxidative metabolism. From rest to peak exercise, rates of oxidative phosphorylation in skeletal muscle may increase more than 50-fold. When muscle oxidative metabolism is restricted, exercise intolerance is typically a central feature. In brain, the range of oxidative metabolism is much more narrow, but cellular structures within the brain are highly vulnerable to restricted oxidative metabolism. The dominant consequence of mitochondrial disorders is considered to be a limited rate of oxidative phosphorylation. However, mitochondria have other metabolic functions that may contribute to

S.Y. Park (✉)
Department of Neurology, University of Texas Southwestern Medical Center, Dallas, TX, USA
and
Neuromuscular Center, Institute for Exercise and Environmental Medicine, Dallas, TX 75231, USA
e-mail: sun.park@utsouthwestern.edu

N. Gadoth and H.H Göbel (eds.), *Oxidative Stress and Free Radical Damage in Neurology*, Oxidative Stress in Applied Basic Research and Clinical Practice, DOI 10.1007/978-1-60327-514-9_15,

the pathophysiology of mitochondrial disorders, including producing reactive oxygen. In this chapter, we review the mechanism of production and metabolism in reactive oxygen as well as possible pathological consequences of oxidative stress in skeletal muscle and brain; and we also review evidence for a role of oxidative stress in the cellular pathology of mitochondrial myopathies and encephalomyopathies.

Keywords Mitochondria · Reactive oxygen species (ROS) · Antioxidant · Fenton reaction · Apoptosis · Calcium · Mitochondrial myopathy · Mitochondrial encepalomyopathy · Oxidative phosphorylation (OXPHOS)

1 Energy Metabolism and Generation of Reactive Oxygen Species in Mitochondria

1.1 Oxidative Phosphorylation

The main function of mitochondria is to generate energy in the form of adenosine triphosphate (ATP) through OXPHOS, in which mitochondrial ATP production is coupled to oxygen consumption. Mitochondrial ATP production occurs by the flow of electrons derived from the tricarboxylic acid (TCA) cycle and beta-oxidation capable of reducing of nicotinamide adenine dinucleotide (NAD) and flavin adenine dinucleotide (FAD). These electrons are passed along the molecular complex in the inner mitochondrial membrane known as the electron transport chain, which is composed of complexes I–IV and two small electron carriers, coenzyme Q (ubiquinones, CoQ_{10}) and cytochrome *c*. As a result of this electron transfer, protons are transferred across the inner mitochondrial membrane producing a large membrane potential. Complex I (reduced nicotinamide adenine dinucleotide (NADH)–ubiquinone reductase) accepts electrons from the citric acid cycle from NADH, with part of the energy of this electron flow used to pump a proton (i.e., acidity) across the inner membrane, after which the electron is passed to complex III via coenzyme Q. Complex II (succinate dehydrogenase (SDH) complex) accepts electrons from the reduced form of flavin adenine dinucleotide ($FADH_2$) and also passes them to complex III (ubiquinol–cytochrome *c* reductase) via coenzyme Q. Complex III uses another part of the energy of the electron to pump another proton across the inner mitochondrial membrane. The electron is then passed to complex IV (cytochrome *c* oxidase) via cytochrome *c*, where it uses most of the remaining energy to pump the third proton across the membrane. The reduced electron is then transferred to oxygen to generate water [1–3] (Fig. 1). The last complex, V (ATP synthase), couples proton flow from the intermembrane space back to the matrix to the conversion of adenosine diphosphate (ADP) to ATP. Matrix ATP is then exported out of the mitochondria, in exchange for cytosol ADP, by the adenine nucleotide translocators (ANT).

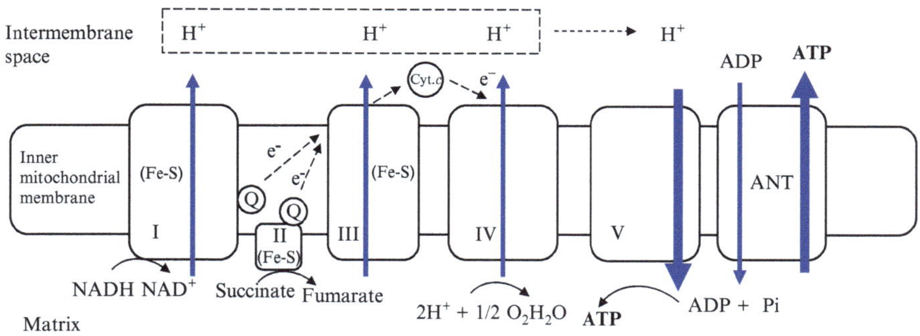

Fig. 1 Mitochondrial oxidative phosphorylation. The reduced nicotinamide adenine dinucleotide (NADH) and succinate generated in the citric acid cycle are oxidized, releasing energy to power the ATP synthase. Electrons (e^-) from carbon oxidations are transferred via NADH into oxidative phosphorylation (OXPHOS) complex I, which is in the inner mitochondrial membrane, then transported to coenzyme Q. Some electrons from organic acid oxidations are transferred, via other flavin-containing enzyme complexes, directly to coenzyme Q. Coenzyme Q delivers electrons via complex III and cytochrome *c* (Cyt *c*) to the final electron acceptor complex IV. Here, oxygen is reduced to water. The electrons lose free energy at each transfer step, and in complexes I, III, and IV, the energy is harnessed and coupled to the movement of H^+ (*blue* and *dashed lines*) from the mitochondrial matrix to the intermembrane space. The proton gradient thus generated is used for the production of ATP by complex V. Except for complex II, all complexes contain some proteins encoded by the mitochondrial genome and others encoded by the nuclear genome. The number is indicated for each complex. *Q*, coenzyme Q

1.2 Generation of Free Radicals

During OXPHOS, single electrons sometimes escape and result in a single electron reduction of molecular oxygen to form a superoxide anion (O_2^-). It is estimated that between 0.2 and 2% of the oxygen consumed is reduced to form reactive oxygen species (ROS) such as superoxide anions, with complex I and the ubiquinone of complex III being the two major sites of ROS.

In generation [4], in normal circumstances superoxide anions (O_2^-) are reduced to hydrogen peroxide (H_2O_2) by superoxide dismutase (SOD) [5, 6] (Fig. 2). Then, H_2O_2 is converted to water by glutathione peroxidase (GPX) or catalase (CAT) [7]. In addition to these enzymes, cytochrome *c*, ubiquinol, vitamin E, and cytochrome *c* oxidase (COX) play roles as antioxidants [8]. However, when these enzymes cannot convert ROS to H_2O fast enough, oxidative damage may occur and accumulate in the mitochondria [9]. Superoxide can also release ferrous iron from aconitase, an enzyme in the TCA cycle [10]. This step exposes iron, which can react with H_2O_2 to produce hydroxyl radicals ($OH^\bullet$) by way of the Fenton reaction [10–12].

Additionally, nitric oxide ($NO^\bullet$) is produced within mitochondria by mitochondrial nitric oxide synthase (mtNOS) [13] and also freely diffuses into mitochondria from the cytosol [7]. NO reacts with O_2^- to produce another radical, peroxynitrite ($ONOO^-$) [7, 12]. The oxidants derived from $NO^\bullet$ are termed reactive nitrogen species (RNS). Together, these two radicals can damage mitochondria and other

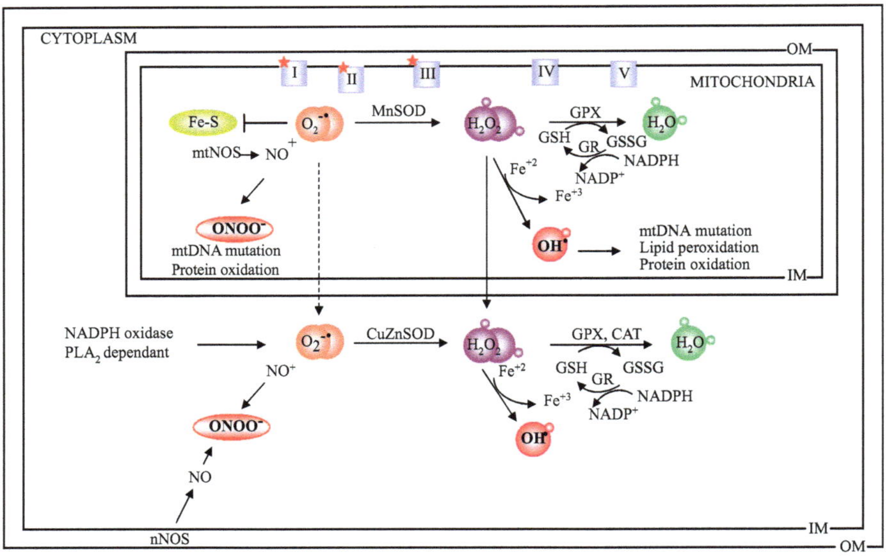

Fig. 2 Mechanism of reactive oxygen species (ROS) and reactive nitrogen species (RNS) generation and detoxification. Mitochondrial damage is caused by reactive oxygen and nitrogen species

cellular components. H_2O_2 and $NO^\bullet$ are membrane permeable and diffuse from the mitochondria into the cytosol. O_2^- cannot cross the inner mitochondrial membrane except in the protonated form, which is only a very small fraction of superoxide formed in physiological conditions. However, some O_2^- radicals generated during mitochondrial respiration can be released into the intermembrane space [14].

Mitochondria are also a major site for accumulation of low molecular weight Fe^{2+} complexes, which promote oxidative damage of membrane lipid. Lipid peroxide is another important mitochondrial ROS. The lipid peroxidation reaction is initiated at two sites in the respiratory chain: complexes I and II (succinate cytochrome *c* oxidoreductase) [15–17]. Free Fe^{2+} abstracts hydrogen from unsaturated phospholipids and initiates lipid peroxidation. The prerequisite reduction of free Fe^{3+} to Fe^{2+} is performed directly by the respiratory chain, not by the O_2^- produced by the respiratory chain. These ROS and their reaction products, such as the hydroxyl radical and lipid peroxide, are very harmful to cells, as they oxidize proteins and cause mutations in DNA. Acute ROS exposure can inactivate the iron–sulfur (Fe–S) centers of electron transfer chain complexes I, II, and III and the TCA cycle enzyme, aconitase, resulting in shutdown of mitochondrial energy production [18]. Complex I is especially susceptible to nitric oxide damage, and animals administered natural and synthetic complex I antagonists develop death of neurons [19, 20]. Complex I dysfunction has been associated with a variety of mitochondrial diseases such as Leber's hereditary optic neuropathy (LHON), and Parkinson's and other neurodegenerative diseases [20–23].

ROS generation and oxidative stress participate in cell signaling processes [24] and apoptosis, and oxidative stress has also been linked to aging and many pathologies such as atherosclerosis, hypertension, inflammation, cancer, diabetes, and neurodegenerative diseases, as well as mitochondrial myopathies and encephalomyopathies [25–35].

2 The Oxidant-Sensitive Site of the Krebs Cycle: Mitochondrial Aconitase

2.1 *Structure and Function*

Mitochondrial aconitase (*m*-aconitase) catalyzes the stereo-specific dehydration–rehydration of citrate to isocitrate in the Krebs cycle. The active site of aconitase contains a cubane diamagnetic $[4Fe–4S]^{2+}$ clusters [36–42]. The $[4Fe–4S]^{2+}$ cluster of aconitase is highly susceptible to oxidation by superoxide, generating the inactive $[3Fe–4S]^{+}$ aconitase. In this reaction, the likely products are iron (Fe^{2+}) and hydrogen peroxide. Consequently, the inactivation of *m*-aconitase by superoxide may increase the formation of hydroxyl radical ($OH^{\bullet}$) through the Fenton reaction in mitochondria. Two aconitase isozymes are present in mammalian cells: the mitochondrial enzyme (*m*-aconitase), which functions in the citric acid cycle, and the bifunctional cytosolic enzyme (c-aconitase/IRP1), which also plays a role in the regulation of iron metabolism. In the cytosol, Fe–S cluster assembly and disassembly controls the function of one of the iron regulatory proteins, IRP1. In the mitochondria, disruption of Fe–S cluster biogenesis may result in mitochondrial iron overload, cytosolic iron depletion, activation of cellular iron uptake, and increased oxidative stress [43–46]. In patients with Friedreich's ataxia (FRDA), decreased frataxin levels result in reduced heme and Fe–S cluster biogenesis, defective mitochondrial respiration, mitochondrial iron overload, and increased oxidative damage [47–49].

2.2 *Oxidative Stress with Aconitase*

Mitochondrial aconitase is inactivated by conditions that increase the production of O_2^- and is protected from inactivation by MnSOD. An aconitase inactivation–reactivation cycle involves opposing roles of O_2^- and iron in modulating aconitase activity. Active mitochondrial aconitase, as well as cytoplasmic aconitase, contains a $[4Fe–4S]^{2+}$ cluster that is attacked and oxidized by endogenous O_2^- to form inactive aconitase containing stable oxidized [3Fe–4S] clusters. Release of the solvent-exposed iron atom in the ferrous state, cluster oxidation, and formation of H_2O_2 are thought to occur during the inactivation process [50, 51]. Regeneration of the active aconitase $[4Fe–4S]^{2+}$ cluster is efficiently achieved in vivo by the

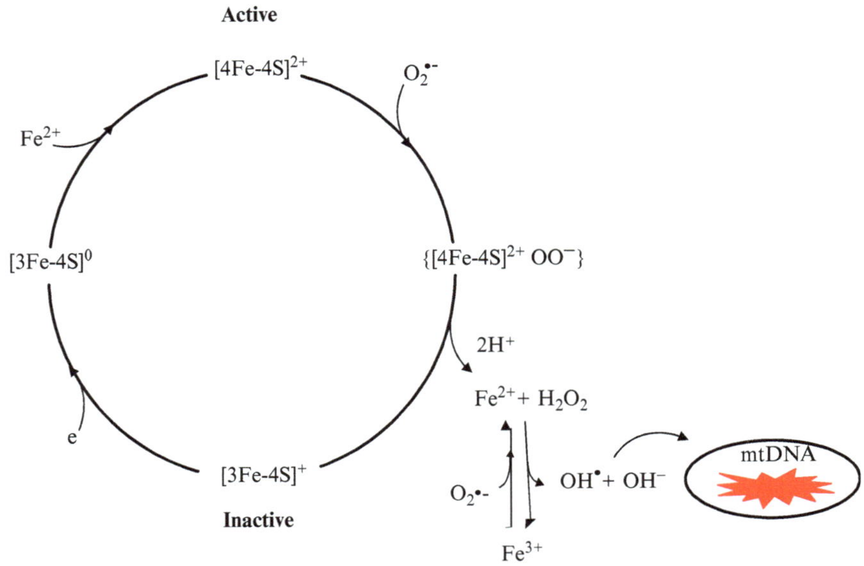

Fig. 3 Mechanism of generation of hydroxyl radical from aconitase

univalent reduction of the oxidized $[3Fe–4S]^+$ cluster to form a $[3Fe–4S]^0$ cluster [18] and subsequent insertion of a ferrous ion from the iron pool (Fig. 3). The cyclical inactivation–reactivation of aconitase may modulate aconitase function under stress and normal conditions. The citric acid cycle flux capacity and homeostatic regulation of iron metabolism may be affected during oxidative stress [52, 53], nutritional or immune-induced iron deficiencies, or SOD deficiencies. In cytoplasmic aconitase, the control of the aconitase iron–sulfur state by intracellular $[O_2^-]$ and $[Fe^{2+}]$ also modulates iron homeostasis through the RNA binding of IRP1 [54, 55]. Interestingly, the opposing effects of iron and O_2^- on aconitase, and other [4Fe–4S]-containing (de)hydratases, can explain the previously puzzling and paradoxical benefit of MnSOD upregulation during iron limitation in *Escherichia coli* [56]; it may also provide a rationale for the increase in mitochondrial MnSOD synthesis [57] that accompanies increased ferritin synthesis [58] and hypoferremia during the inflammatory cytokine-mediated immune response in mammals.

3 Antioxidant Defense Mechanisms in Mitochondria

Mammalian mitochondria possess a multilevel ROS defense network of enzymes and nonenzymatic antioxidants. Primary antioxidant enzymes include SOD, GPX, and catalase (CAT); these enzymes are responsible for removing superoxide radicals, H_2O_2 and organic hydroperoxides, and H_2O_2, respectively. Important nonenzymic antioxidants include vitamins E and C, β-carotene, GSH, and ubiquinones. Vitamin E, β-carotene, and ubiquinone are located in lipid regions of the cell, whereas glutathione (GSH) and vitamin C are in aqueous compartments of the cell. This chapter describes only selected, primarily enzymatic subsystems that are thought to represent mainstream mitochondrial ROS-detoxifying pathways.

3.1 Enzymatic Antioxidant Defenses

3.1.1 Superoxide Dismutase

SODs are a family of metalloenzymes that catalyze the conversion of two superoxides into hydrogen peroxide and oxygen [59]. The benefit here is that hydrogen peroxide is substantially less toxic than superoxide. Although the cytosol of cells contains a copper–zinc SOD (CuZn-SOD) [60], the mitochondrial matrix has a specific form of SOD, with manganese in the active site (MnSOD), which eliminates the O_2^- formed in the matrix or on the inner side of the inner mitochondrial membrane[59].

Because O_2^- may either reduce transition metals, which in turn can react with H_2O_2 producing the highly toxic OH^- or spontaneously react with $NO^\bullet$ to produce peroxynitrite, it is important to maintain the steady-state concentration of O_2^- at the lowest possible level. Thus, although the dismutation of O_2^- to H_2O_2 and O_2 can also occur spontaneously, the role of SODs is to increase the rate of the reaction to that of a diffusion-controlled process.

The critical function of the mitochondrial form of SOD, MnSOD, in protecting mitochondria from oxidative stress has been illuminated using a MnSOD-knockout mouse model that is neonatal lethal and associated with a dilated cardiomyopathy and neuronal and skeletal muscle pathology. Biochemically there is marked reduction of the activity of the iron–sulfur cluster containing enzymes, particularly aconitase (89% reduction) and SDH, in skeletal muscle, heart, and brain. Mitochondria from heterozygous (MnSOD±) animals also have been shown to have increased oxidative stress as marked by reduced levels of mitochondrial aconitase (35% reduction) and complex I and increased levels of protein carbonyls [61]. In contrast, knockouts of the cytosolic Cu/ZnSOD are viable and the animals grow normally, although lifespan is reduced.

3.1.2 Glutathione Peroxidase

GPX is located in cytosolic and mitochondrial compartments. It is a group of enzymes, the most abundant of which contain selenium. Hydrogen peroxide, the product of O_2^- dismutation and the main precursor of $OH^\bullet$ in the presence of reduced transition metals, is mostly decomposed by the glutathione peroxidase (GPX1) in the cytosol [62]. Secondary GPX associated with the mitochondrial membrane, known as phospholipid-hydroperoxide glutathione peroxidase (GPX4, PHGPx), is specifically involved in reducing lipid peroxides associated with the membrane [63, 64].

3.1.3 Catalase (CAT)

Catalase, a major H_2O_2-detoxifying enzyme found in peroxisomes, is also present in heart and brain mitochondria [65, 66]. However, this enzyme has not been found in mitochondria from other tissues, including skeletal muscle [67]. Catalase is a heme protein that decomposes H_2O_2 to oxygen and water.

3.2 Nonenzymatic Antioxidants

3.2.1 Coenzyme Q (Ubiquinol-10)

Ubiquinone is localized in the inner mitochondrial membrane, serving as an electron and proton carrier in the mitochondrial respiratory chain, and ubisemiquinone can promote O_2^-production at complex III [68]. In addition to cytochrome *c*, other electron carriers appear to have a detoxifying role against ROS. Ubiquinol (QH_2) has been shown to act as a reducing agent in the elimination of various peroxides in the presence of succinate [17, 69]. Thus, coenzyme Q is a source of O_2^- when partially reduced (semiquinone form) and an antioxidant when fully reduced [70]. The inner mitochondrial membrane also contains vitamin E, a powerful antioxidant that interferes with the propagation of free radical-mediated chain reactions [71]. Finally, cytochrome *c* oxidase (complex IV) may also act as a peroxidase although, given the high K_m for H_2O_2 (0.18 mM), the relevance of this reaction may be negligible [72].

3.2.2 Vitamin C and Vitamin E

Vitamin C or ascorbate is a water-soluble antioxidant that can reduce radicals from a variety of sources. It also appears to participate in recycling vitamin E radicals. Interestingly, vitamin C also function as a pro-oxidant under certain conditions. Vitamin E is the major lipid-soluble antioxidant and plays a vital role in protecting membranes from oxidative damage. Its primary activity is to trap peroxyl radical in cellular membranes.

3.2.3 Glutathione (GSH)

Glutathione is maintained as a thiol by the action of glutathione reductase, an enzyme that reduces oxidized glutathione (GSSG) to GSH. Glutathione plays an important antioxidant role in the mitochondrial matrix. In addition to participation as a substrate in the reaction of GPX and GPHx, GSH can directly scavenge free radicals and other oxidants including OH, O_2^-, H_2O_2, and $ONOO^-$ [73–75]. It is also a major target of peroxynitrite [73, 76].

In addition to the antioxidant defenses already mentioned, the mitochondrion has a variety of DNA-repairing enzymes to correct errors resulting from oxidative damage. This mechanism is very important because, although 95% of the mitochondrial proteins are encoded by nuclear DNA (nDNA), the mitochondrial genome (mtDNA) contains genes for several important proteins including subunits of NADH dehydrogenase, cytochrome *c* oxidase, and ATP synthase and for cytochrome *b*.

4 Role of ROS in Apoptosis

Mitochondria also represent a major site for initiation of apoptosis. Apoptosis, or programmed cell death, is a cellular suicide program in which individual cells are destroyed while the integrity and architecture of surrounding tissue are preserved. This targeted cell destruction is critical in many physiological contexts including embryogenesis, immune cell maturation and response, and tissue homeostasis, and in the cellular response to injury. High levels of ROS, especially in the presence of calcium, can induce the mitochondrial permeability transition pore (mtPTP). Induction of the mtPTP involves the calcium-mediated opening of the pore in the inner mitochondrial membrane, which allows the diffusion of small molecules; this contributes to cytotoxicity via apoptosis. Apoptosis can be initiated by various external and internal signals and executed through several related signaling pathways controlled by the regulated expression of apoptosis-associated genes and proteins. The mitochondria pathway, which is regulated by the B-cell-leukemia/lymphoma-2 (BCL-2) family of proteins, is particularly important. The Bcl-2 family includes proteins that inhibit apoptosis (e.g., Bcl-2 and Bcl-X_L) and proteins which promote apoptosis (e.g., Bcl-2-associated X protein (BAX) and Bcl-2 homologous antagonist/killer (BAK)) [77, 78]. The mitochondrial inner membrane space contains a number of cell death promotion factors, including cytochrome *c*, apoptosis-induction factor (AIF), and the latent form of specialized proteases called caspases. Opening of the mtPTP causes collapse of the mitochondrial membrane potential ($\Delta\Psi$m), swelling of the mitochondria inner membrane, and release of these death-promoting factors. AIF translocates to the nucleus. Cytochrome *c* released into the cytosol binds to apoptotic protease activity factor-1 (Apaf-1) and pro-caspase-9 to form the apoptosome and catalyzes the activation of caspase-9. Cleaved caspase-9 activates caspase-3. Activated caspase-3 in turn activates other caspases and leads to the characteristic features of apoptosis, including chromatin condensation and nuclear fragmentation [79, 80] (Fig. 4). When cytochrome *c* is released from the mitochondria, another protein, *second mitochondrial derived activator of caspase* (Smac), known also as Diabolo, is also released. This protein neutralizes a set of caspase inhibitors known as inhibitor of apoptosis (IAPs), thereby releasing the caspases.

Early studies examining the relationship between mitochondrial function and apoptosis focused on mtDNA-less (ρ^o) cells, with sometimes conflicting results. The total lack of OXPHOS in human ρ^o fibroblasts did not affect their sensitivity to the protein kinase inhibitor staurosporine, with apoptotic cell death maintained [81]. Conversely, some studies with SK-Hepl hepatoma ρ^o and osteosarcoma ρ^o cells showed resistance to ROS- and staurosporine-induced cell death with normal release of cytochrome *c* but decreased caspase-3 activation [82, 83].

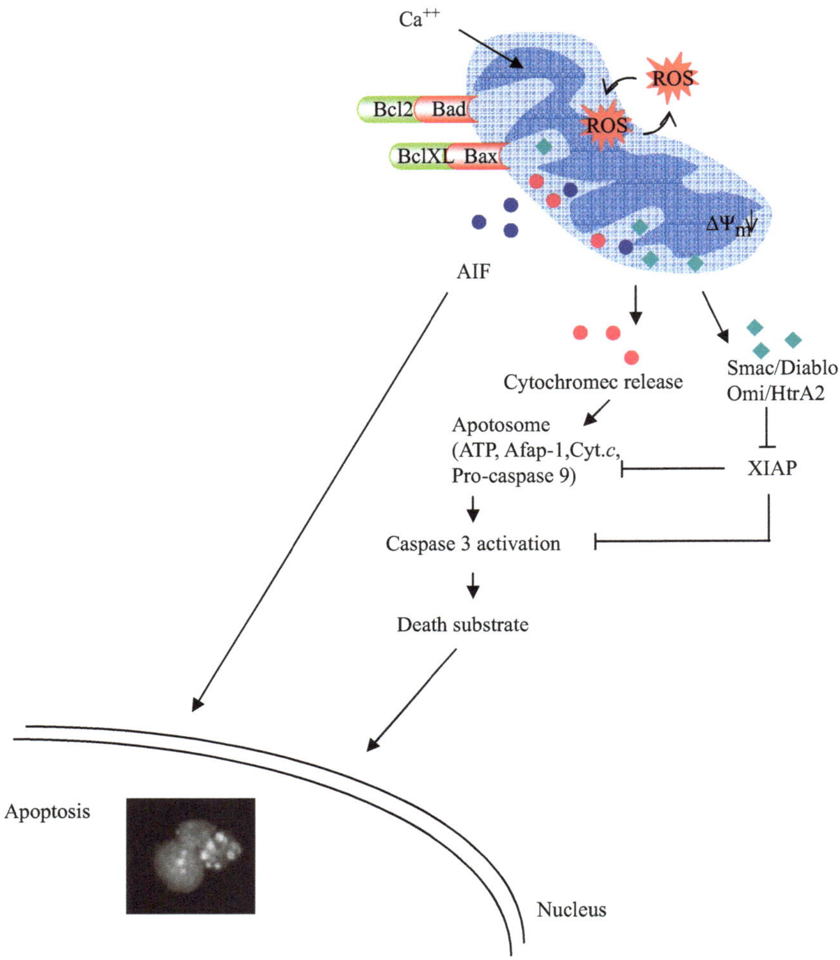

Fig. 4 Apoptosis pathway. ROS, Ca^{2+}, and other stimuli can initiate apoptosis through the mitochondrial pathway. Proapoptotic. BCL2 family proteins, for example, BAX and BAD (*red boxes*), are important mediators of these signals. Activation of mitochondria leads to the release of cytochrome *c* (Cyt *c*) into the cytosol, where it binds apoptotic protease activating factor-1 (Apaf-1) to form the apoptosome. At the apoptosome, the initiator caspase-9 is activated. Apoptosis through mitochondria can be inhibited on different levels by antiapoptotic proteins, including the anti-apoptotic Bcl2 family members Bcl2 and Bcl-XL (*green boxes*) and inhibitors of apoptosis proteins (IAPs), which are regulated by Smac/Diablo (second mitochondria-derived activator of caspase/direct IAP binding protein). AIF is also released from mitochondria and it can activate apoptosis via an unknown, caspase-independent pathway. Activated caspases lead to the fragmentation and condensation of chromosome in the nucleus through caspase-3

The importance of mitochondrial OXPHOS function in cell death signaling has also been examined in vivo. Germline or tissue-specific disruption of the nuclear-encoded mitochondrial transcriptional factor A (T*fam*) gene in mice results in a gradual reduction of mtDNA levels, leading to complete loss of OXPHOS, analogous to ρ^o cells [84]. This loss results in massive apoptosis in the mouse embryo with homozygous disruption of T*fam* [85] or delayed apoptosis of neurons or cardiomyocytes in mice with neuronal or heart-specific disruption of T*fam* [86–89]. These results suggest that reduced or absent mtDNA, which results in abnormal mitochondrial morphology and severely reduced/absent OXPHOS enzyme activities, results in increased cell death in vivo.

The accumulation of oxidative damage may lead to apoptosis [90, 91]. During apoptosis, there is an increase in intracellular ROS production, probably including the hydroxyl radical [92–94]. There is also an increase in lipid peroxidation during apoptosis [95]. Overexpression of GPX inhibits apoptosis and overexpression of Cu/ZnSOD inhibits apoptosis in both the neural cell line and primary neurons in culture [96, 97]. Expression of the anti-*bcl-2* gene decreases the net cellular generation of ROS [98–100]. Moreover, expression of *bcl-2* rescues the growth of yeast mutants that are null for Cu-Zn SOD [98].

Calcium is an activator of the mtPTP [101]. The mtPTP is a multiprotein complex spanning the inner and outer mitochondrial membranes; its activation by calcium results in dissipation of the mitochondrial inner membrane potential ($\Delta\Psi_m$) and eventual disruption of the outer mitochondrial membrane. Moreover, mitochondrial calcium levels have been observed to increase before the dissipation of $\Delta\Psi_m$ following staurosporine treatment of cultured neural cells [102]. Calcium may indirectly contribute to mitochondrial cytochrome *c* release through interplay with Bcl-2 family members. Calcineurin, a Ca^{2+}-activated protein phosphatase, was demonstrated to dephosphorylate *Bcl-2-associated death promoter* (Bad), a proapoptotic Bcl-2 family member [87], allowing Bad to translocate to mitochondria. Bcl-2 is known to modulate mitochondrial and endoplasmic reticulum (ER) calcium homeostasis [103, 104] and can regulate calcium fluxes into mitochondria in response to apoptotic stimuli [105, 106]. The calcium-activated protease calpain has been shown to cleave the proapoptotic Bcl-2 family member *Bcl-2-associated X protein* (Bax), although the physiological significance of this event is unclear [107, 108].

5 ROS and Calcium

One of the functions of mitochondria is to store free calcium. Mitochondria are essential for cytosolic Ca^{2+} homeostasis, and required indirectly for Ca^{2+} signaling pathways including muscle contraction, hormonal and neuronal signaling, and apoptosis. Release of this stored calcium back into the interior of the cell can initiate calcium spikes or waves. These events coordinate various processes in different types of cells, for example, neurotransmitter release in nerve cells and release of hormones in endocrine cells. Disturbed Ca^{2+} homeostasis may be the consequence of deficient respiratory chain activity as well as ATP deficiency. Ca^{2+} import into the mitochondria depends on an intact mitochondrial membrane potential, which requires a normally functioning respiratory chain, and ATP is needed for pumping Ca^{2+} out of the cell or into intracellular stores. The results of many investigations have shown that calcium is essential for the production of ROS [109, 110]. Elevation of intracellular calcium levels results in the formation of free radicals by the mitochondrial respiratory chain. On the other hand, an increase in intracellular calcium concentration may be stimulated by ROS [111]. Excess calcium ions stored in mitochondria can inhibit OXPHOS. In neural cells this can cause an irreversible reduction in the energy status of nerve terminals, which can initiate pathological processes in those cells.

Mitochondrial Ca^{2+} signaling is also essential for mitochondrial ATP production. The production of ROS and altered Ca^{2+} metabolism have been linked to nerve cell death associated with various neurodegenerative diseases such as Alzheimer's disease [112], Parkinson's disease, and acute central nervous system (CNS) trauma and stroke. Considerable effort has been expended on trying to determine the relationship between increases in intracellular Ca^{2+} and oxidative stress. However, there is still no general agreement about whether increases in ROS precede increases in intracellular Ca^{2+} or vice versa. A number of studies of the relationship between increases in intracellular Ca^{2+} and oxidative stress in nerve cells using various models of cell death have generally found a close relationship between the two events. Furthermore, many studies point to a pivotal role for mitochondria both in generating ROS and in mediating the toxic effects of excess Ca^{2+} influx.

6 Biomarkers of Oxidative Stress in Mitochondrial Diseases

Oxidative injury to macromolecules occurs in a wide range of pathological conditions. These ROS can lead directly to the peroxidation of lipid and the production of highly reactive aldehyde species, such as 4-hydroxy-2, 3, trans-nonenal (4HNE), 8-hydroxy-deoxyguanosine (8-OHDG), a marker for oxidative DNA damage, and protein carbonyls, a marker for oxidative protein damage [113–118] (Table 1).

6.1 DNA Oxidation: 8-OHDG

Cellular DNA damage can be caused by ROS, especially by hydroxyl radicals. Intracellular ROS can cause strand breaks in DNA and base modification, including the oxidation of guanine residues to 8-hydroxyl-deoxyguanosine (8-OHDG) [119–121]. 8-OHDG is biologically significant because it induces mutations and has a specific repair system. 8-OHDG levels have been used to evaluate oxidative stress. 8-OHDG might serve as a sensitive biomarker of oxidative modification of DNA in vivo and in vitro [121, 122], and increased 8-OHDG levels have been found in patients with mitochondrial diseases [123, 124]. Several techniques have been developed to measure 8-OHDG, including high pressure liquid chromatography (HPLC), gas chromatography-mass spectrometry (GC-MS), and liquid chromatography-mass spectrometry (LC-MS) [119, 120, 122, 125]. Antibody-based

Table 1 Mitochondrial DNA diseases that were found to be associated with increased oxidative stress on the basis of (potential) biomarkers of oxidative damage

	Biomarker	Reactive compound
DNA	8-Hydroxy 2-deoxyguanosine	Oxidation of guanine residues (hydroxy-deoxyguanosine)
Protein	Carbonylated proteins	Oxidation of amino acid (Lys, Arg, Pro, and Thr)
	Nitrotyrosine	Nitration of tyrosine (peroxynitrite)
Lipid	4-Hydroxy-2-nonenal	α,β-unsaturated aldehydes
	Malondialdehyde	Reactive aldehydic fragments
	Lipid peroxidation	Lipid radical ($L^{\bullet}$, $LO^{\bullet}$, $LOO^{\bullet}$)
Oxidant/antioxidant	GSSG:GSH	

GSSG oxidized glutathione; *GSH* glutathione; *Lys* lysine; *Arg* arginine; *Pro* proline; *Thr* threonine; $L^{\bullet}$ lipid radical; $LO^{\bullet}$ alkoyl radical; $LOO^{\bullet}$ peroxy radical

methods have also been developed to detect 8-OHDG and are useful for visualization of oxidative damage in histological preparations, but they are likely to be only semiquantitative [119].

6.2 *Protein Oxidation: Protein Carbonyl Compounds*

Oxidative damage to proteins is important because it can contribute to secondary damage to other biomolecules. Protein carbonyls are generated by the oxidation of several amino acid side chains (e.g., in Lys, Arg, Pro, His, and Thr) [11, 126, 127]. The formation of carbonyl compounds is the most general and widely used marker of significant protein oxidation both in vitro and in vivo, with several assays developed for the quantification of these species [126, 128, 129]. Carbonyls can be produced as a result of protein glycosylation by sugars, by the binding of aldehydes (including many of those formed during lipid peroxidation) to proteins, and by the direct oxidation of amino acid side chains by ROS to generate such products as glutamate and aminoadipic semialdehydes [126, 130]. Carbonyls can be readily measured spectrophotometrically, with enzyme-linked immunosorbent assay (ELISA) techniques, and by Western blotting using DNPH, a compound that reacts with carbonyl group in proteins. Tissue or plasma protein carbonyl levels have been shown to be elevated in many human diseases [131]. As a marker of oxidative damage to proteins, carbonyls have been shown to accumulate with aging and in mitochondrial diseases [118, 125, 126].

6.3 *Lipid Peroxidation*

Oxidative stress-induced peroxidation of membrane lipids can be very damaging because it leads to alterations in the biological properties of the membrane, such as the degree of fluidity, integrity, permeability, and metabolic functions. Lipid peroxidation can also lead to inactivation of membrane-bound receptors or enzymes, which in turn may impair normal cellular function and increase tissue permeability. Moreover, lipid peroxidation may contribute to and amplify cellular damage resulting from generation of oxidized products, some of which are chemically reactive and covalently modify critical macromolecules. Lipid peroxidation results in the formation of highly reactive and unstable hydroperoxides of both saturated and unsaturated fatty acids. Lipid peroxidation generates a variety of relatively stable decomposition end products, mainly α- and β-unsaturated reactive aldehydes, such as malondialdehyde (MDA), 4-hydroxy-2-nonenal (HNE), and 2-propenal (acrolein) [132–134], and isoprostanes [135], which can then be measured in patients with mitochondrial diseases as an indirect index of oxidative stress.

4-Hydroxynonenal (HNE) is a major aldehydic product formed by membrane peroxidation of omega-6 unsaturated fatty acids. It is a key mediator of neuronal apoptosis induced by oxidative stress and is regarded as a specific marker of lipid peroxidation. MDA is a highly reactive three-carbon dialdehyde produced as a byproduct of polyunsaturated fatty acid peroxidation and arachidonic acid metabolism. MDA reacts with functional groups of nucleic acid bases, proteins, and lipoproteins. It reacts with DNA to form adducts to deoxyguanosine and deoxyadenosine, and it reacts with lysine residues of proteins [132, 136, 137].

6.4 Nitrotyrosine

Tyrosine residues in proteins are a prominent target for reactive oxygen. As a substrate for tyrosine kinases, oxidation of tyrosine residues has profound effects on the biological function of proteins in cellular signaling. In addition to direct oxidative damage, cells experience nitrative damage from RNS, which occurs as a consequence of the reaction of superoxide with NO, a cellular messenger found in many tissues, to generate the highly reactive peroxynitrite anion. Peroxynitrite modification of proteins generates nitrotyrosine, which contributes significantly to protein alterations as an overall part of oxidative damage. Analysis of 3-nitrotyrosine (NO_2-Tyr), a stable marker for $NO^{\bullet}$-derived oxidants, products such as 3-chlorotyrosine (Cl-Tyr), or 3-bromotyrosine has been performed in various mitochondrial diseases, and diverse methods have been developed for such measurements [136, 138].

6.5 Glutathione

Glutathione (GSH) acts as an antioxidant to protect the cell from ROS and RNS [139, 140]. Within the cell, mitochondria are a major intracellular source of ROS and are highly enriched with GSH [141]. To maintain mitochondrial GSH status, a high-affinity transport system imports GSH from the cytosol into the mitochondria [142]. Glutathione exists in reduced (GSH) and oxidized (GSSG) states. In the reduced state, the thiol group of cysteine is able to donate a reducing equivalent (H^{++} e^{-}) to other unstable molecules, such as ROS. In donating an electron, glutathione itself becomes reactive, but readily reacts with another reactive glutathione to form glutathione disulfide (GSSG). Such a reaction is possible because of the relatively high concentration of glutathione in cells (up to 5 mM in the liver). GSH can be regenerated from GSSG by the enzyme glutathione reductase. In healthy cells and tissue, more than 90% of the total glutathione pool is in the reduced form (GSH) and less than 10% exists in the disulfide form (GSSG). An increased GSSG:GSH ratio is considered to indicate oxidative stress. Glutathione works as an antioxidant, defending cells against the potentially destructive process of oxidative stress caused by both normal metabolism and environmental contaminants, and glutathione levels or the ratio of reduced to oxidized glutathione are biomarkers of oxidative stress. Prolonged depletion of reduced glutathione (GSH) in the brain is associated with oxidative neuronal death and may play a role in many of the neurodegenerative diseases associated with aging. Blood glutathione concentrations may reflect glutathione status in accessible tissues. Measurement of both reduced glutathione (GSH) and glutathione disulfide (GSSG) in blood has been considered an index of whole-body GSH status and may be a useful indicator of oxidative stress status in human mitochondrial disease [143, 144].

7 Oxidative Stress in Mitochondrial Diseases

Mitochondria serve several important cellular functions, including essential pathways of intermediary metabolism, amino acid biosynthesis, fatty acid oxidation, steroid metabolism, and apoptosis, but the most important function of mitochondria is the production of ATP by OXPHOS. Defects of OXPHOS affect the function of numerous tissues and predict that mitochondrial OXPHOS defects will affect many organ systems [145]. However, tissue dependence upon OXPHOS for normal function and survival varies widely. Brain, muscle, and heart are highly dependent, bone and fibroblasts less so. Defective OXPHOS has been associated with a large variety of diseases, most of which involve muscle and CNS symptoms and are collectively referred to as mitochondrial encephalomopathies [146, 147]. Moreover, mitochondrial OXPHOS defects also promote the generation of ROS that can represent an additional mechanism of cellular dysfunction. For example, mitochondrial DNA (mtDNA) has a much higher mutation rate than nDNA, because it lacks protective DNA-binding proteins such as histones, its mtDNA repair system is limited, and it is also exposed to the major cellular source of ROS, the mitochondrial electron transport chain.

There are three general genetic categories of mitochondrial diseases: (1) mutations in mtDNA, (2) nuclear gene defects that affect the maintenance of mtDNA (nuclear–mitochondrial signaling defects), and (3) defects in nuclear genes coding for structural subunits of the respiratory chain complexes or factors involved in the function, maintenance, or assembly of respiratory chain complexes and related mitochondrial enzymes. Blocks of the electron transfer chain resulting from these genetic defects restrict the rate of OXPHOS but also increase ROS production and oxidative stress.

8 Mitochondrial Biology and Genetics

8.1 *Structure*

Mitochondria are the only organelle in the cell, aside from the nucleus, that has their own DNA. The human mitochondrial genome is a 16,569-bp circular double-stranded DNA that contains 37 genes [148]; it encodes 22 transfer RNAs (tRNAs), 2 ribosomal RNAs (12S and 16S), and 13 polypeptides. The respiratory chain is composed of five enzyme complexes made up of approximately 90 polypeptide subunits, 13 of which are coded in mtDNA. Reduced nicotinamide adenine dinucleotide (NADH)–ubiquinone oxidoreductase (complex I) is composed of 45 subunits that are mostly encoded by nDNA [149], but 7 of the subunits (ND1–ND6) are mtDNA encoded. Succinate–ubiquinone oxidoreductase (complex II) is unique among respiratory chain complexes in that all four of its subunits are encoded by nDNA, and thus it is not affected by mtDNA mutations. Its function is to oxidize $FADH_2$ derived from the TCA cycle. Ubiquinone–cytochrome *c* oxidoreductase (complex III) is composed of 10 nDNA-encoded subunits and cytochrome *b* (Cyt b), which is encoded by mtDNA. Cytochrome *c* oxidase (COX, complex IV) contains 13 subunits, 3 of which (COX1, COXII, and COXIII) are mtDNA encoded. ATP synthase (complex V) is composed of 16 subunits, 2 of which (ATP6 and ATP8) are mtDNA encoded [2, 3]. All the subunits of complex II, the remaining subunits of the other mitochondrial respiratory chain complex, as well as the factors involved in mtDNA replication, transcription, and translation, and the remainder of the mitochondrial proteins are encoded by nDNA (Fig. 5).

mtDNA replication of the heavy (outer) and light (inner) strands occurs from separate sites (O_H and O_L) and is under nuclear control. mtDNA is totally dependent upon nuclear-encoded proteins for its maintenance and transcription. Most of the genetic information is encoded in the heavy (purine-rich) strand (2 rRNAs, 14 tRNAs, and 12 polypeptides). The light (pyrimidine-rich) strand contains genetic information for only one polypeptide and 8 tRNAs. Mitochondrial genes have no introns, and intergenic sequences are absent or limited to a few bases. Some genes overlap and, in some instances, termination codons are not encoded but are generated posttranscriptionally by polyadenylation.

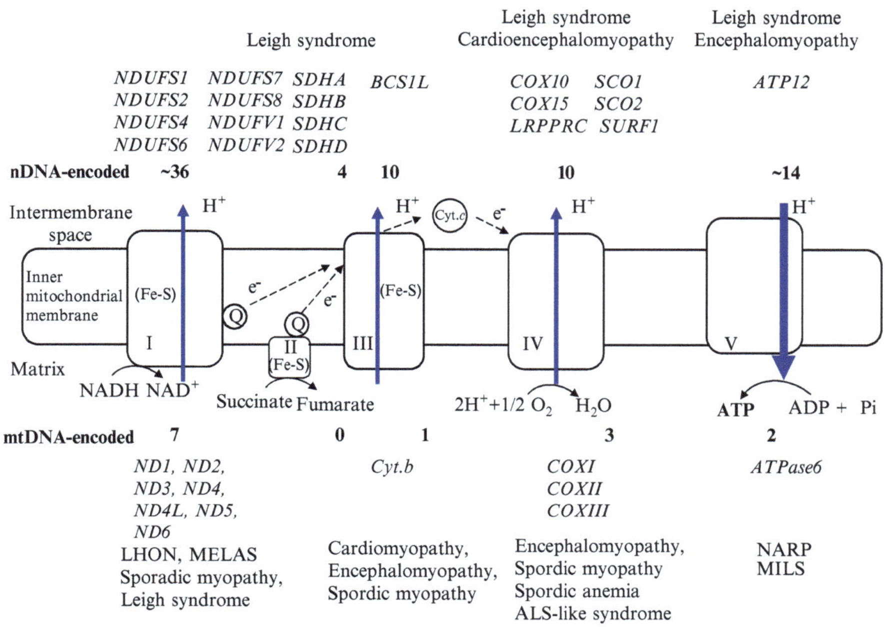

Fig. 5 Respiratory chain and human oxidative phosphorylation disorders. Diseases associated with defects of the respiratory chain complexes can be caused by mutations in the genes of the structural subunits of the five complexes, or to the inactivation of complex specific assembly factors. The ADP/ATP carrier mediates the transfer of nucleotides across the inner mitochondrial membrane and is therefore tightly connected to energy production through oxidative phosphorylation (OXPHOS)

8.2 Genetics

Mitochondrial genetics differ from Mendelian genetics in some fundamental aspects. First, mtDNA is inherited from the mother, and diseases caused by pathogenic point mutations in mtDNA are most commonly maternally inherited with mitochondria derived from the oocyte [150]. An exception to this general rule has been reported [151], but the occurrence of paternal inheritance in humans seems to be an extremely rare event [152]. Second, mtDNA molecules are present in multiple copies (polyplasmy). There are thousands of mtDNA molecules per cell and many millions per individual. Third, pathogenic mutations usually affect only a proportion of mtDNA (heteroplasmy), and a minimum critical percentage of mutated DNA has to be present to impair OXPHOS (threshold effect). In normal individuals only identical wild-type mtDNA molecules exist (homoplasmy). Fourth, organelle division and mtDNA replication operate independently of the cell cycle, both in dividing cells (such as glia) and in postmitotic nondividing cells (such as neurons). Fifth, during cellular division, mtDNA is randomly distributed into daughter cells (replicative segregation). The percentage of mutant relative to wild-type mtDNA in daughter cells determines the expression of mitochondrial disease. Some somatic mtDNA mutations that arise and accumulate with aging could have a role in the senescence of tissues. The most common mechanism of somatic mutation in mtDNA is the effect of free radicals produced by the respiratory chain itself.

Mitochondrial diseases caused by mtDNA mutations may be maternally inherited or sporadic, arising from de novo mtDNA mutations; they are characterized by the coexistence of wild-type and mutant mtDNA in the same cell. Respiratory chain function is not impaired if normal mtDNA is sufficient to compensate for the presence of mutant DNA to maintain a functional respiratory chain. When the ratio of mutant to normal mtDNA exceeds a certain critical threshold, respiratory chain assembly and function will be impaired. This threshold is different in different tissues. Skeletal muscle, brain, heart, and liver have a high energy demand, and their threshold is lower than in tissues with a low energy demand. mtDNA are in continuous replication, even in nondividing cells, with a potential of changing levels of heteroplasmy and with the possibility that the phenotypic effects of a mitochondrial disease could change over the lifespan of an affected individual. Mitochondria are not equally distributed to daughter cells during mitotic cell division, and an individual could have different amounts of normal and mutant mtDNA in different tissues or different cells in the same tissue. mtDNA accumulates mutations more than ten times faster than the nuclear genome owing to the abundance of ROS with mitochondria and the relative inefficiency of mtDNA repair mechanisms.

9 Diseases of Oxidative Phosphorylation

The OXPHOS diseases can be divided into three main groups: (i) pathogenic mtDNA mutations including point mutations in polypeptides, Trna, or rRNA-encoding regions, which are often maternally inherited, and single large-scale rearrangements, duplications, or deletions, which are usually sporadic; (ii) intergenomic communication defects, resulting in multiple mtDNA deletions or mtDNA depletion; and (iii) nuclear defects affecting the synthesis, assembly, or maintenance of function of respiratory chain complexes and related mitochondrial enzymes (Table 2). The pathogenesis of each class of mitochondrial disease relates to impaired production of ATP via OXPHOS and is associated with increased anaerobic metabolism as marked by lactic acidosis. Most mtDNA or nDNA mutations that impair OXPHOS also induce the overproduction of ROS [146, 147]. Clinical and experimental studies support the conclusion that oxidative stress, resulting from the imbalance between ROS formation and antioxidant defenses, plays a major role in the pathogenesis of mitochondrial diseases related to mitochondrial dysfunction [145].

Table 2 Diseases of oxidative phosphorylation defects

		Disorder	Mutation	Gene	Defective complex
mtDNA mutation	*Point mutation*	LHON	G3460A	ND1	Complex I
	Protein-coding Genes mutation		T4160C	ND1	
			G11778A	ND4	
			T14484C	ND6	
			G15257A	Cytochrome *b*	Complex III
		NARP	T8993G/C	ATPase-6	Complex V
		MILS	T8993G/C	ATPase-6	Complex V
		MELAS	G3380A, G13513A	ND1 and DN5	Complex I
	tRNA	MELAS	A3243G	tRNALEU (UUR)	Complex IV
		MERRF	A8344G	tRNALys	
	Rearrangements				
		CPEO	A single large-scale deletion		
		KSS	A single large-scale deletion		
nDNA mutation		adPEO	4q34–35 (multiple deletion)	ANT1	Adenine nucleotide translocator
		adPEO/ataxia	10q24 (multiple deletion)	Twinkle	mtDMA helicase
		adPEO/Alpers/ataxia	15q25 (mutiple deletion)	POLG	mtDNA polymerase-gamma
		MNGIE	22q13.32	TP	Thymidine phosphorylase
		mtDNA depletion syndrome (MDS)	2p13	DGUOK	Deoxyguanosine kinase
			16q22	TK	Thymidine kinase
				PRM2B	p53-induced ribonucleotide reductase B subunit
				SUCLA1	Succinyl-coenzyme synthase B
				SUCLA2	Succinyl-coenzyme synthase A
				MPV17	Mitochondrial inner membrane protein
		LS		SURF1	Complex IV
				SCO1	complex IV
				BCS1L	Complex III
				COX15	Complex IV
		Friedreich's ataxia	GAA repeat	Frataxin	Complex I, II, and III

9.1 Mitochondrial DNA Defects

9.1.1 Point Mutations

More than 100 pathogenic point mutations or microdeletions affecting mtDNA-encoded subunits of complexes I, III, IV, and V and tRNAs have been described. The clinical syndromes associated with these mutations are vast: included is LHON, which in most cases is attributed to pathogenic mutation at nucleotides 3,460 and 4,160 in ND1, 11,778 in ND4, and 14,484 in ND6 in mtDNA-encoded subunits of complex I. A G15257A mutation affecting complex III has also been linked to LHON [25, 153]

Neurogenic muscle weakness, ataxia, and retinitis pigmentosa (NARP) and maternally inherited Leigh syndrome (MILS) are progressive neurodegenerative disorders of infancy and early childhood; both are caused by mutations at the 8,993-bp point in mtDNA [154–156]. Two point mutations (T8993G and T8993C) affecting the ATPase-6 subunit of the F_1F_0-ATPase (ATP synthase; complex V) have been associated with both syndromes and are always found as heteroplasmic mutations [157, 158].

Mitochondrial encephalomyopathy, lactic acidosis, and stroke-like episodes (MELAS) syndrome is associated with at least five different mutations in $tRNA^{Leu(UUR)}$, A3243G, A3252G, C3256T, T3271C, and T3291C, the most frequent being A3243G [159, 160]. MELAS syndrome has also been described in association with mutations in other tRNA genes and in mtDNA-encoded subunits of complex I, G3380A in ND1, and G13513A in ND5 [161, 162]. A point mutation in subunit III of cytochrome *c* oxidase has been accompanied by a syndrome resembling MELAS [163–165].

Myoclonus epilepsy with ragged red fibers (MERRF) syndrome is associated with various mutations in $tRNA^{Lys}$, most frequently with the A8344G mutation. A MELAS/MERRF overlap syndrome has been described in association with the $tRNA^{Leu(UUR)}$ A3243G and other tRNA mutations [166–168]. An immunohistochemical analysis of the expression of respiratory chain proteins has shown widespread deficiency of subunit II of COX.

9.1.2 Rearrangements

Among the most common forms of mtDNA mutations are single large-scale mtDNA deletions or duplications. The mutations are typically sporadic and commonly predominate in skeletal muscle. The typical phenotype involves chronic progressive external ophthalmoplegia (CPEO), usually associated with symptoms of more widespread myopathy. Single large-scale deletions that are more abundant and more widely distributed are usually responsible for Kearns–Sayre syndrome (KSS) in which CPEO is associated with retinitis pigmentosa, heart block, and CNS manifestations. KSS has occasionally been associated with tRNA point mutations. The most common deletion is defined by accumulation of truncated mtDNA molecules lacking a specific 4,977-bp fragment.

9.2 Nuclear DNA and Mitochondrial Diseases

Mitochondrial encephalomyopathies also are caused by nDNA mutations, affecting respiratory chain proteins, assembly factors, or proteins involved in the crosstalk between the two genomes that regulate mtDNA integrity and copy number (Table 2).

9.2.1 Intergenomic Communication Defects

In 1989, Zeviani and colleagues described autosomal dominant progressive external ophthalmoplegia (adPEO) with multiple mtDNA deletions [169]. Autosomal dominant PEO has proven to be genetically heterogeneous, most commonly attributable to mutations in mtDNA polymerase gamma (POLG1), the muscle-heart-specific adenine nucleotide translocase (ANT1), or to mutations in the mtDNA helicase, Twinkle [170–174].

Mitochondrial neurogastrointestinal encephalomyopathy (MNGIE) disease is an autosomal recessive human disease associated with both depletion and multiple deletions of skeletal muscle mtDNA. The disease locus was mapped to chromosome 22q13.32, which encodes thymidine phosphorylase (TP) [175–177]. TP contributes to the maintenance of plasma thymidine homeostasis by phosphorylytic catabolism of thymidine to the thymine and 2-deoxy D-ribose 1-phosphate. In MNGIE, TP mutations are associated with loss of enzyme function and the systemic accumulation of thymidine and deoxyuridine, leading to deoxynucleotide pool imbalance resulting in mtDNA depletion, multiple deletions, and point mutations [175].

A number of other genetic mechanisms have been identified as causes of mtDNA depletion syndrome (MDS). These disorders are autosomal recessive and cause depletion of mtDNA, which impairs synthesis of respiratory chain complexes containing mitochondrial encoded subunits [178, 179].

Mutations in thymidine kinase 2 (TK2) and in RRM2B, a subunit of the p53-inducible ribonucleotide reductase, have been identified as causing MDS associated with early-onset myopathy. Mutations in SUCLA1 and SUCLA2, isoforms of the Krebs cycle enzyme succinyl coenzyme synthase, have been associated with MDS and infantile encephalopathy, whereas mutations in deoxyguanosine kinase (DGUOK) and in MPV17, a mitochondrial inner membrane protein, cause a hepatocerebral form of MDS.

9.2.2 Nuclear DNA Mutation

Mutations affecting nuclear-encoded subunits in complexes I and II have been associated with Leigh syndrome (LS), an often fatal, necrotizing encephalopathy of infancy and early childhood that affects the basal ganglia, thalamus, and brainstem [180, 181–185]. Other nuclear defects causing LS include mutations in the COX assembly gene, SURF1, and mutations in an assembly factor for complex III, BCS1L. Mutations in another COX assembly gene, SCO1, and in COX 15 have been shown to cause infantile encephalopathy and cardiomyopathy [186, 187].

A novel class of mitochondrial disorders attributable to nuclear mutations is represented by FRDA. FRDA is the most common of the hereditary ataxias. It is an autosomal recessive disorder associated with progressive ataxia, sensory neuropathy, absent tendon reflexes, and pyramidal signs. Cardiomyopathy, typical skeletal abnormalities, and diabetes are also features of this disease. FRDA is caused by a loss-of-function mutation (typically a trinucleotide (GAA) repeat expansion) in the gene that encodes frataxin. Frataxin is a mitochondrially targeted protein involved in the synthesis of mitochondrial iron-containing proteins [140, 188], and frataxin mutations result in deficiency of iron–sulfur (Fe–S) cluster-containing proteins, including aconitase and SDH (complex II), as well as Fe–S-containing subunits of complex I and III. Frataxin is particularly abundant in the cerebellum and spinal cord [189, 190]. FRDA causes mitochondrial iron accumulation, which, via the Fenton reaction, promotes an increase in ROS.

10 Site of the Mitochondrial Defect: Implications for the Production of ROS

10.1 Selective Respiratory Chain Defects

Complex I: Complex I is a major site of generation of superoxide, and experimental inhibition of complex I increases O^- production. Selective defects in complex I are the most commonly identified biochemical defect in OXPHOS disorders, and there is strong evidence that these defects promote oxidative stress. Although available evidence suggests that most if not all complex I defects promote increased levels of mitochondrial ROS production, antioxidant defense mechanisms are critical in determining the levels and patterns of ROS accumulation. In mitochondria isolated from fibroblasts of patients with selective complex I defects, steady-state levels of superoxide were inversely related to levels of MnSOD. In patients with levels of MnSOD comparable to control subjects, steady-state levels of superoxide were substantially elevated compared to controls whereas patients with normal or low levels of superoxide exhibited significant elevations in MnSOD levels, in some cases two- to threefold higher than in control subjects [191–194]. Subsequent studies in cultured fibroblasts showed that cells with isolated complex I deficiency associated with substantial MnSOD induction had increased levels of toxic hydroxyl radicals and lipid peroxidation [191]. These results were interpreted as an indication that increased levels of MnSOD promoted increased formation of peroxide, which, in the presence of transition metals, led to hydroxyl radical formation. Variables responsible for regulating MnSOD and related oxidant defense mechanisms in such mitochondrial defects are poorly understood. In a transmitochondrial cell line homoplasmic for a mutation in the mtDNA-encoded ND6 subunit of complex I, increased peroxide production and increased levels of lipid peroxidation and protein carbonyls were documented without changes in MnSOD or GPX [195]. A more recent study in fibroblasts from children with selective complex I defects attributable to mutations in the NDUFS1, NDUFS2, NDUFS7, NDUFS8, or NDUFV1 gene displayed increased ROS production that was markedly reduced by treatment with a vitamin E analogue [196]. Remarkably, the antioxidant treatment increased levels and activities of complex I, suggesting that one effect of increased ROS is to magnify the degree of complex I deficiency. A recent review of selective complex I deficiency from the same authors confirms a consistent increase in ROS in cell lines from affected patients [197].

LHON is caused by three main mutations in genes for complex I subunits. Skin fibroblasts of LHON patients show production of hydroxyl radical and lipid peroxidation and an increased level of SOD1 activity [191]. Even if the catalytic activity of complex I is maintained except in cells carrying the 3460/ND1 mutation, in all cases there is a change in sensitivity to complex I inhibitors and an impairment of mitochondrial respiration, eliciting the possibility of generation of ROS by the complex.

8-OHDG DNA content is increased in leukocytes, and the activity of SOD2 and lipid peroxidation is increased in skin fibroblasts, of LHON patients [123, 191] (Table 3). In addition, apoptosis is increased in patient lymphocytes.

Complex II: Isolated complex II (SDH) defects are rare and complex II is not generally recognized as a site of free radical formation. However, experimental inhibition of complex II has been shown to impair cellular function in part by increasing oxidative stress. Malonate, an inhibitor of mitochondrial complex II), induced apoptosis via p38 MAPK in human SH-SY5Y neuroblastoma cells. In addition, malonate reduced the mitochondrial membrane potential and increased the levels of MDA, a marker of lipid peroxidation [198]. Treatment of cell lines with vitamin E decreased ROS production and reduced malonate-induced apoptosis. During exposure to malonate, 2′,7′-dichlorofluorescin (DCF) fluorescence, a ROS indicator, increased in cultured mesencephalic cells. Ascorbate treatment decreased malonate-induced DCF fluorescence, prevented the efflux of GSH, and attenuated the efflux of GSSG in these cells [199].

Coenzyme Q_{10}: Deficiency of coenzyme Q_{10} (CoQ) is a rare human genetic condition that has been associated with variety of clinical phenotypes, including myopathy, encephalopathy/LS, and ataxia. CoQ is a small lipophilic molecule mediating electron transport from complex I and II to complex III in the mitochondrial respiratory chain. It also plays a role in antioxidant defense. In one study, fibroblasts from CoQ-deficient patients showed reduced CoQ-dependent respiratory enzyme activities in association with a reduced mitochondrial membrane potential and an increase in ROS generation [200]. In another study of cultured fibroblasts from patients with CoQ_{10} deficiency caused by mutations in two separate genes involved in CoQ_{10} synthesis (COQ2 and PDSS2), variable effects upon cellular ATP synthesis and ROS production were found. Fibroblasts harboring the PDSS2 mutation had 12% of normal CoQ_{10} levels and a markedly reduced rate of ATP synthesis but did not show increased ROS production or signs of oxidative stress. In contrast, fibroblasts harboring the COQ2 mutation had a more moderate reduction in CoQ_{10} (30% of control) and ATP synthesis and demonstrated increased ROS production in association with increased levels of protein carbonyls and

Table 3 Evidence of reactive oxygen species (ROS) in mitochondrial myopathies and encephalomyopathies

Patients

Subjects	Tissue	Biochemical assay	Histochemical assay	References
Large-scale deletion	Skeletal muscle	SOD1•, SOD2↑		[218]
		SOD1↑, SOD2↑, 8-OHDG↑	SOD1↑, SOD2↑ in COX-negative RRF	[219]
			SOD1•, SOD2↑, GSH↑ in COX-negative RRF	[224]
		SOD1↑, SOD2↑-protein, 8-OHDG↑	SOD2↑, 8-OHDG↑, 4HNE↑	[220]
		SOD1↑, SOD2↑, CAT↑, GPX↑	SOD1↑, SOD2↑, CAT↑, GPX↑, ATPase↓ in COX-negative RRF	[222]
	Myoblast	SOD1•, SOD2↑, CAT•, GPX•-activity, protein, 8-OHDG↑		[221]
	Skin fibroblast	SOD1•, SOD2↑, CAT•, GPX•-activity, RNA, protein		[223]
	Blood	GSH↓ Lipid peroxidation↑		[225]
LHON	Skin fibroblast	SOD2↑-activity Hydroxyl radical and lipid peroxidation↑ 4HNE, MDA↑		[191]
	Blood	8-OHDG↑		[123]
MELAS (A3243G)	Muscle	SOD1•, SOD2↑		[218]
	Skeletal muscle		SOD1, SOD2↑, GSH↑ in COX-negative RRF	[224]
	Myoblast, blood	SOD↑, CAT↑-activity		[226]
	Skin fibroblast	ATP/ADP and GSH/GSSG ratio↓, malondiadehyde content↑		[227]
MERRF (A8344G)	Skin fibroblast	SOD2↑-protein	RRF	[228]
NARP (T8993G)	Lymphocytes	SOD1↑, SOD2↑, CAT•		[157]
		ATPase synthase↑-activity		[25]
Friedreich ataxia	Blood	SOD↑, GST↑		[231]
		Total GSH•, free GSH↓		[232]
	Urine	8-OHDG↑		[239]
	Plasma	Malondialdehyde↑		[233]

Animal model

Subject	Genotype	Biochemical assay	Tissue	References
Mouse	Ant1tm2Mgr−/−	SOD2↑, GPx1↑-protein, mRNA	Skeletal muscle, heart	[214]
	Fratxin −/−	SOD2↓-Western Cytosolic aconitase↓-activity	Heart, cerebellum	[47]
Yeast	Fratxin −/−	Protein carbonylation↑ ATPase syntase↓		[235]
		Protein carbonylation		[236]
Drosophila	*fh*-RNAi	Aconitase activity↓		[234]

↑, Increased; ↓, decreased; •, not changed

increased lipid peroxidation [201]. These results suggested the possibility that more severe CoQ_{10} deficiency may prevent ROS production by limiting electron flux necessary to promote electron leaks. Further evidence for the role of ROS in the pathophysiology of CoQ_{10} deficiency in vivo is suggested by the finding of increased apoptosis and upregulation of antioxidant defenses in skeletal muscles of affected patients [202–204].

Complex III: Complex III, similar to complex I, is a recognized site of superoxide formation. Isolated complex III deficiency most commonly occurs as a result of pathogenic mtDNA mutations affecting cytochrome *b* and is associated with myopathy or encephalomyopathy. In transmitochondrial cybrids containing mtDNA harboring high levels of a cytochrome *b* mutation known to be associated with features of parkinsonism and MELAS, decreased complex III activity was associated with increased production of hydrogen peroxide [205].

Complex IV: Few studies have evaluated ROS in selective complex IV (cytochrome *c* oxidase) defects (from mtDNA mutations affecting mitochondrially encoded cytochrome *c* oxidase subunits I, II, or III or assembly defects). Di Giovanni and coworkers evaluated skeletal muscle biopsies from 28 patients with mitochondrial encephalomyopathy, 17 with biochemical and histochemical cytochrome *c* oxidase (COX) deficiency (3 with isolated COX deficiency from SURF 1 mutations, 2 with mtRNA mutations, 1 with a single large-scale deletion, and the rest not characterized genetically) and 11 with biochemical evidence of complex I, II, or III deficiency (none genetically characterized) without COX deficiency [206]. They found that COX-deficient muscle fibers commonly showed an increase in immunostaining for apoptosis-related proteins and the presence of TUNEL staining of nuclei (an indicator of apoptosis). These fibers also often displayed increased immunostaining for SOD1 and SOD2, and these muscle samples had increased GPX enzymatic activity. In contrast, muscle from the 11 patients without COX-deficient fibers showed neither TUNEL-positive myonuclei or increase in apoptosis-related proteins nor increase in antioxidant enzymes [206].

Complex V: Defects of complex V (ATP synthase or mitochondrial ATPase) caused by mtDNA and nuclear mutations have been described [1, 207]. When ATP synthase is defective, the low ability of mitochondria to utilize the respiration-generated proton gradient for ATP synthesis causes the mitochondrial membrane to be hyperpolarized, which promotes the generation of ROS by the respiratory chain [208]. It has been demonstrated that fibroblasts and transmitochondrial cybrids with mitochondrially encoded *ATP6* mutations as well as fibroblasts with nuclear ATPase synthase defects maintain high $\Delta\Psi_m$ [147, 209]. In both types of ATP synthase disorders, high $\Delta\Psi_m$ values were shown to cause increased ROS production. mtDNA T8993G and T8993C mutations and nuclear ATP synthase defects caused an increase in ROS production, as assessed with the fluorescent probe DCFDA, and resulted in an increase in antioxidant defense mechanisms as marked primarily by an increase in MnSOD (SOD2) [63, 147, 157, 210]. Importantly, increased ROS production in both mitochondrial and nDNA defects was reduced by the uncoupler FCCP, indicating that the high membrane potential was an important mechanism for upregulation of ROS production. Interestingly, there is evidence that the T8993G and T8993C mutations differ in pathophysiological effects. The T8993C causes a lesser restriction in ATP synthesis but a greater increase in ROS production whereas the T8993G mutation causes a more severe restriction in OXPHOS and lesser increase in ROS [157].

ANT1 defects: Mutations in the muscle-heart isoform of the adenine nucleotide translocator (ANT1) most commonly are associated with autosomal dominant multiple mtDNA deletions. Autosomal recessive ANT1 deficiency associated with exercise intolerance and cardiomyopathy has also been described. ANT is a bifunctional protein that transports ADP and ATP across the mitochondrial inner membrane and regulates the mtPTP, which initiates apoptosis. Treatment with acrolein, an ANT inhibitor, induced ROS generation and decreased glutathione content and aconitase activity in isolated brain mitochondria [211]. These findings suggest that deficits in ANT reduce matrix ADP and limit matrix ADP-dependent proton translocation through F_1-F_0-ATPase. The reduction in proton transport to the mitochondrial matrix promotes hyperpolarization of the mitochondrial membrane, which further limits electron transfer through the respiratory chain. Electrons would then accumulate and be available for production of O_2^- from O_2 and render the mitochondria more vulnerable to ROS production. A mouse model of mitochondrial myopathy/cardiomyopathy has been constructed by 'knocking out' the heart/muscle form of the adenine nucleotide translocator [212, 213]. This strategy resulted in viable mice ($Ant1^{tm2Mgr}-/-$) that developed classical mitochondrial myopathy and hypertrophic cardiomyopathy [212, 214, 215]. Skeletal muscle from $Ant1^{tm2Mgr}-/-$ mice exhibit ragged red fibers (RRF) with increased SDH and COX staining. A six- to eightfold increase in H_2O_2 generation is seen in skeletal muscle and heart mitochondria, with a sixfold increase in MnSOD expression and a threefold increase in Gpx1 expression [214] whereas in the heart little or no increase in MnSOD levels was found. Accordingly, the increase in ROS generation in the heart was associated with increases in mtDNA damage, with the levels of mtDNA rearrangements in $Ant1^{tm2Mgr}-/-$ heart higher than in skeletal muscle and much higher than in age-matched controls [214]. Thus, it appears that the inhibition of OXPHOS increases mitochondrial ROS production in this mouse model, and if the antioxidant mechanisms are overloaded, an increase in the mtDNA mutation rate results. The oxidative stress is one of the contributing factors to the pathogenesis observed in the $Ant1^{tm2Mgr}-/-$ *knockout* mouse.

10.2 Mitochondrial Disorders Affecting Multiple Sites of the Respiratory Chain

Large-scale mtDNA deletions: Most CPEO and KSS patients have a heteroplasmic single large-scale mtDNA deletion, classically 4.9–9 kb in size, which can be readily identified by Southern blot analysis or a specifically designed polymerase chain reaction (PCR)-based assay [216]. Rarely, this phenotype is caused by a mitochondrial tRNA point mutation, whereas others may harbor a nDNA mutation [217].

Large-scale mtDNA mutations impair the function of multiple mtDNA-encoded tRNAs as well as well as multiple mtDNA-encoded protein subunits, commonly including subunits of complex I, IV, and V. Thus, although fiber segments that contain high percentages of these large-scale deletions are commonly identified histochemically simply as COX-negative/SDH-positive fibers, the biochemical defect in these muscle fibers is complex, involving a respiratory chain that is defective at multiple sites. Several studies indicate that single large-scale deletion mutations increase levels of antioxidant defense enzymes including MnSOD (SOD2), CAT, and GPX in skeletal muscle, and cultured fibroblasts [218–223] (Table 3). Additionally, 8-OHDG and 4HNE, as oxidative biomarkers, are increased, particularly in COX-negative, RRF that contain high percentages of the mtDNA mutation [219, 224]. Increases in oxidative stress have been shown in transmitochondrial cybrids harboring high concentrations of the 4,977-bp mtDNA deletion [221, 225]. Cybrids that contained greater than 65% of the deletion showed a significant increase in 8-hydroxy 2-deoxyguanosine and lipid peroxides compared to cybrids containing undetectable mutant mtDNA [221]. In an immunohistochemical study in KSS, a reduced expression of mtDNA-encoded subunits of COX and ND1 were found in the dentate nucleus, similar to findings in RRF of skeletal muscle.

Mitochondrial tRNA mutations: Pathogenic mutations affecting mtDNA-encoded tRNAs also cause complex biochemical defects affecting multiple respiratory chain complexes that commonly are associated with evidence of increased oxidative stress as well as impaired OXPHOS. The A3243G mutation associated with MELAS has been shown to increase total SOD and catalase activities in skeletal muscle and cultured myoblasts [218, 226]. In skeletal muscle, fibers containing high percentages of the pathogenic mutation are highlighted as RRF that usually are COX negative. Immunocytochemisty reveals that SOD2 and GSH increase specifically in COX-negative RRFs, consistent with increased oxidative stress in these fiber segments [224] (Table 3). Cybrids harboring more than 90% of mtDNA with the A3243G mutation are found to have significantly lower oxygen consumption rates and ATP: ADP ratios. The defective respiratory function in these cells is associated with increased oxidative stress as indicated by a decreased GSH:GSSG ratio and enhanced oxidative damage to lipids [227]. There are few reports on brain pathology in patients with MELAS syndrome and an established mtDNA mutation. Cultured fibroblasts from patients with MELAS have been showed to have increased cytosolic calcium and a decreased mitochondrial membrane potential [227].

MERRF syndrome is one of the major maternally inherited mitochondrial encephalomyopathies, which is most commonly attributed to the A8344G mtDNA mutation affecting the *tRNA lys*. Cybrids containing high percentages of this mutation have decreased respiratory chain activity, increased ROS production, and induction of antioxidant enzyme activities [228–230] (see Table 3).

FRDA: FRDA is an autosomal recessively inherited disorder attributable to loss-of-function mutations in the gene for frataxin. In vitro and in vivo data indicate that frataxin is necessary for iron incorporation in Fe–S clusters (ISC) and for heme biosynthesis. As a consequence, frataxin deficiency causes deficiency of iron–sulfur cluster enzymes in respiratory chain complexes I, II, and III as well as deficiency of the mitochondrial enzyme aconitase. In addition, frataxin deficiency results in the mitochondrial accumulation of iron. An increase in oxidative stress is an important component in the pathophysiology of FRDA. Contributing factors include impaired electron transport that promotes electron leakage as well as an increase in ROS promoted by an increase in free iron via the Fenton reaction. Furthermore, there is evidence that the normal induction of MnSOD in response to increased production of superoxide is impaired. Evidence of increased oxidative stress includes the finding of increased plasma MAD (a marker of lipid peroxidation), increased urinary 8-OHDG (a marker of oxidative DNA damage), and reduced plasma glutathione, which is also increased in urine of FRDA patients [231–233]. Additional features of oxidative stress have been identified in models of frataxin deficiency. In study of frataxin deficiency induced by RNA interference in flies, hyperoxia caused selective severe deficiency of mitochondrial aconitase, implying that frataxin acts to specifically protect aconitase from inactivation by oxidative stress [234]. The frataxin-deficient (Δyfh1) yeast model show increases of protein carbonylation of specific proteins, including ATP synthase [235, 236] (see Table 3). It was found that magnesium-binding enzymes were selectively vulnerable to oxidative modification. The authors postulated that free iron displaces magnesium and promotes oxidative modification at the metal-binding site of affected enzymes.

Evidence of an important role of oxidative stress in FRDA has emerged from trials of antioxidant treatment. Oral treatment of the coenzyme Q analogue idebenone has been shown to decrease concentrations of 8-OHDG [237–239]. Cardiac and muscle bioenergetics as monitored by ^{31}P magnetic resonance spectroscopy have shown sustained improvement with treatment with vitamin E and coenzyme Q_{10}, and cardiac function has shown improvement with treatment with idebenone [240–242].

11 Conclusions

The evidence that mitochondrial disease causes increased production of ROS is overwhelming, and the conclusion that increased oxidative stress is an important contributor to the pathophysiology of mitochondrial encephalomyopathies seems inescapable. However, there remain many limitations to our understanding. The most comprehensive studies have been conducted in cultured cells or cybrids whereas our knowledge of the contribution of oxidative stress to human mitochondrial disease in vivo, particularly as it relates to the nervous system, is, at best, indirect. Although increased oxidative stress has been identified in virtually all varieties of human mitochondrial disorders, the variables in respiratory chain function that are most critical in determining increased ROS production are incompletely understood. Data from studies of coenzyme Q_{10} defects and the T8993G vs. T899C NARP/MILS mutations suggest dissociation between the severity of impaired OXPHOS and the level of ROS production; but the relationship between the severity of impaired OXPHOS and levels of oxidative stress across the spectrum of mitochondrial diseases remains unknown. A critical variable in determining the net accumulation of ROS in mitochondrial diseases is the level and pattern of antioxidant defenses, particularly of MnSOD. There is strong evidence that MnSOD is normally induced in response to increased production of superoxide, but the mechanism is poorly understood, and the factors responsible for limiting increases in oxidative defense mechanisms are unknown. Furthermore, although a variety of biomarkers of oxidative stress have been identified, clear guidelines for the quantitative assessment of the presence and magnitude of oxidative stress and the implications for therapeutic intervention and determination of therapeutic efficacy remain to be established. Although these challenges remain, there is reason for optimism that continued research in this field along with the development of more effective antioxidants will provide new directions for effective therapy of mitochondrial disease.

Acknowledgments This work was supported by grants to R.G.H. from the National Institutes of Health (NAIMS, RO1-AR050597), a VA Merit Review, and NORD.

References

1. Dimauro S, Davidzon G. Mitochondrial DNA and disease. Ann Med. 2005;37:222–32.
2. DiMauro S, Schon EA. Mitochondrial respiratory-chain diseases. N Engl J Med. 2003; 348:2656–68.
3. Schmiedel J, Jackson S, Schafer J, Reichmann H. Mitochondrial cytopathies. J Neurol. 2003;250:267–77.
4. Jezek P, Hlavata L. Mitochondria in homeostasis of reactive oxygen species in cell, tissues, and organism. Int J Biochem Cell Biol. 2005;37:2478–503.
5. Cadenas E, Davies KJA. Mitochondrial free radical generation, oxidative stress, and aging. Free Radic Biol Med. 2000;29:222–30.
6. McKenzie M, Liolitsa D, Hanna MG. Mitochondrial disease: mutations and mechanisms. Neurochem Res. 2004;29:589–600.
7. Green K, Brand MD, Murphy MP. Prevention of mitochondrial oxidative damage as a therapeutic strategy in diabetes. Diabetes. 2004;53 Suppl 1:S110–8.
8. Turrens JF. Mitochondrial formation of reactive oxygen species. J Physiol. 2003; 552:335–44.
9. Sies H. Strategies of antioxidant defense. Eur J Biochem. 1993;215:213–9.
10. Vasquez-Vivar J, Kalyanaraman B, Kennedy MC. Mitochondrial aconitase is a source of hydroxyl radical. An electron spin resonance investigation. J Biol Chem. 2000;275:14064–9.
11. Dean RT, Fu S, Stocker R, Davies MJ. Biochemistry and pathology of radical-mediated protein oxidation. Biochem J. 1997;324(Pt 1):1–18.
12. Lipton SA, Nicotera P. Calcium, free radicals and excitotoxins in neuronal apoptosis. Cell Calcium. 1998;23:165–71.
13. Carreras MC, Franco MC, Peralta JG, Poderoso JJ. Nitric oxide, complex I, and the modulation of mitochondrial reactive species in biology and disease. Mol Aspects Med. 2004;25:125–39.
14. Fariss MW, Chan CB, Patel M, Van Houten B, Orrenius S. Role of mitochondria in toxic oxidative stress. Mol Interv. 2005;5:94–111.
15. Takayanagi R, Takeshige K, Minakami S. NADH- and NADPH-dependent lipid peroxidation in bovine heart submitochondrial particles. Dependence on the rate of electron flow in the respiratory chain and an antioxidant role of ubiquinol. Biochem J. 1980;192:853–60.

16. Takeshige K, Takayanagi R, Minakami S. Lipid peroxidation and the reduction of ADP-Fe3$^+$ chelate by NADH-ubiquinone reductase preparation from bovine heart mitochondria. Biochem J. 1980;192:861–6.
17. Eto Y, Kang D, Hasegawa E, Takeshige K, Minakami S. Succinate-dependent lipid peroxidation and its prevention by reduced ubiquinone in beef heart submitochondrial particles. Arch Biochem Biophys. 1992;295:101–6.
18. Gardner PR, Raineri I, Epstein LB, White CW. Superoxide radical and iron modulate aconitase activity in mammalian cells. J Biol Chem. 1995;270:13399–405.
19. Betarbet R, Sherer TB, MacKenzie G, Garcia-Osuna M, Panov AV, Greenamyre JT. Chronic systemic pesticide exposure reproduces features of Parkinson's disease. Nat Neurosci. 2000;3:1301–6.
20. Qi X, Lewin AS, Hauswirth WW, Guy J. Suppression of complex I gene expression induces optic neuropathy. Ann Neurol. 2003;53:198–205.
21. Schon EA, Manfredi G. Neuronal degeneration and mitochondrial dysfunction. J Clin Invest. 2003;111:303–12.
22. Perier C, Tieu K, Guegan C, et al. Complex I deficiency primes Bax-dependent neuronal apoptosis through mitochondrial oxidative damage. Proc Natl Acad Sci USA. 2005;102:19126–31.
23. Chan TS, Teng S, Wilson JX, Galati G, Khan S, O'Brien PJ. Coenzyme Q cytoprotective mechanisms for mitochondrial complex I cytopathies involves NAD (P) H: quinone oxidoreductase 1(NQO1). Free Radic Res. 2002;36:421–7.
24. Brookes PS, Yoon Y, Robotham JL, Anders MW, Sheu SS. Calcium, ATP, and ROS: a mitochondrial love-hate triangle. Am J Physiol Cell Physiol. 2004;287:C817–33.
25. Lenaz G, Baracca A, Carelli V, D'Aurelio M, Sgarbi G, Solaini G. Bioenergetics of mitochondrial diseases associated with mtDNA mutations. Biochim Biophys Acta. 2004;1658:89–94.
26. Mroczek-Tonska K, Kisiel B, Piechota J, Bartnik E. Leber hereditary optic neuropathy: a disease with a known molecular basis but a mysterious mechanism of pathology. J Appl Genet. 2003;44:529–38.
27. Qi X, Lewin AS, Hauswirth WW, Guy J. Optic neuropathy induced by reductions in mitochondrial superoxide dismutase. Invest Ophthalmol Vis Sci. 2003;44:1088–96.
28. Amaral S, Oliveira PJ, Ramalho-Santos J. Diabetes and the impairment of reproductive function: possible role of mitochondria and reactive oxygen species. Curr Diabetes Rev. 2008;4:46–54.
29. Bai RK, Leal SM, Covarrubias D, Liu A, Wong LJ. Mitochondrial genetic background modifies breast cancer risk. Cancer Res. 2007;67:4687–94.
30. Coughlan MT, Cooper ME, Forbes JM. Renal microvascular injury in diabetes: RAGE and redox signaling. Antioxid Redox Signal. 2007;9:331–42.
31. Davidson SM, Duchen MR. Effects of NO on mitochondrial function in cardiomyocytes: pathophysiological relevance. Cardiovasc Res. 2006;71:10–21.
32. Droge W. Free radicals in the physiological control of cell function. Physiol Rev. 2002; 82:47–95.
33. Emre Y, Hurtaud C, Karaca M, Nubel T, Zavala F, Ricquier D. Role of uncoupling protein UCP2 in cell-mediated immunity: how macrophage-mediated insulitis is accelerated in a model of autoimmune diabetes. Proc Natl Acad Sci USA. 2007;104:19085–90.
34. Kang D, Hamasaki N. Alterations of mitochondrial DNA in common diseases and disease states: aging, neurodegeneration, heart failure, diabetes, and cancer. Curr Med Chem. 2005; 12:429–41.
35. Brand MD, Affourtit C, Esteves TC, et al. Mitochondrial superoxide: production, biological effects, and activation of uncoupling proteins. Free Radic Biol Med. 2004;37:755–67.
36. Robbins AH, Stout CD. Structure of activated aconitase: formation of the [4Fe-4S] cluster in the crystal. Proc Natl Acad Sci USA. 1989;86:3639–43.

37. Beinert H, Emptage MH, Dreyer JL, et al. Iron-sulfur stoichiometry and structure of iron-sulfur clusters in three-iron proteins: evidence for [3Fe-4S] clusters. Proc Natl Acad Sci USA. 1983;80:393–6.
38. Kennedy MC, Emptage MH, Dreyer JL, Beinert H. The role of iron in the activation-inactivation of aconitase. J Biol Chem. 1983;258:11098–105.
39. Robbins AH, Stout CD. Iron-sulfur cluster in aconitase. Crystallographic evidence for a three-iron center. J Biol Chem. 1985;260:2328–33.
40. Tong J, Feinberg BA. Direct square-wave voltammetry of superoxidized $[4Fe\text{-}4S]3^{+}$ aconitase and associated 3Fe/4Fe cluster interconversions. J Biol Chem. 1994;269:24920–7.
41. Beinert H, Holm RH, Munck E. Iron-sulfur clusters: nature's modular, multipurpose structures. Science. 1997;277:653–9.
42. Johnson DC, Dean DR, Smith AD, Johnson MK. Structure, function, and formation of biological iron-sulfur clusters. Annu Rev Biochem. 2005;74:247–81.
43. Babcock M, de Silva D, Oaks R, et al. Regulation of mitochondrial iron accumulation by Yfh1p, a putative homolog of frataxin. Science. 1997;276:1709–12.
44. Knight SA, Sepuri NB, Pain D, Dancis A. Mt-Hsp70 homolog, Ssc2p, required for maturation of yeast frataxin and mitochondrial iron homeostasis. J Biol Chem. 1998;273:18389–93.
45. Wolfgang MJ, Lane MD. Control of energy homeostasis: role of enzymes and intermediates of fatty acid metabolism in the central nervous system. Annu Rev Nutr. 2006;26:23–44.
46. Foury F, Talibi D. Mitochondrial control of iron homeostasis. A genome wide analysis of gene expression in a yeast frataxin-deficient strain. J Biol Chem. 2001;276:7762–8.
47. Seznec H, Simon D, Monassier L, et al. Idebenone delays the onset of cardiac functional alteration without correction of Fe-S enzymes deficit in a mouse model for Friedreich ataxia. Hum Mol Genet. 2004;13:1017–24.
48. Napoli E, Taroni F, Cortopassi GA. Frataxin, iron-sulfur clusters, heme, ROS, and aging. Antioxid Redox Signal. 2006;8:506–16.
49. Pandolfo M. Friedreich ataxia. Semin Pediatr Neurol. 2003;10:163–72.
50. Flint DH, Tuminello JF, Emptage MH. The inactivation of Fe-S cluster containing hydro-lyases by superoxide. J Biol Chem. 1993;268:22369–76.
51. Hausladen A, Fridovich I. Superoxide and peroxynitrite inactivate aconitases, but nitric oxide does not. J Biol Chem. 1994;269:29405–8.
52. Sato H, Yamaguchi M, Bannai S. Regulation of ferritin synthesis in macrophages by oxygen and a sulfhydryl-reactive agent. Biochem Biophys Res Commun. 1994;201:38–44.
53. Gardner PR, Nguyen DD, White CW. Aconitase is a sensitive and critical target of oxygen poisoning in cultured mammalian cells and in rat lungs. Proc Natl Acad Sci USA. 1994;91:12248–52.
54. Klausner RD, Rouault TA, Harford JB. Regulating the fate of mRNA: the control of cellular iron metabolism. Cell. 1993;72:19–28.
55. Haile DJ, Rouault TA, Tang CK, Chin J, Harford JB, Klausner RD. Reciprocal control of RNA-binding and aconitase activity in the regulation of the iron-responsive element binding protein: role of the iron-sulfur cluster. Proc Natl Acad Sci USA. 1992;89:7536–40.
56. Gardner PR, Fridovich I. Inactivation-reactivation of aconitase in *Escherichia coli*. A sensitive measure of superoxide radical. J Biol Chem. 1992;267:8757–63.
57. Wong GH, Elwell JH, Oberley LW, Goeddel DV. Manganous superoxide dismutase is essential for cellular resistance to cytotoxicity of tumor necrosis factor. Cell. 1989;58:923–31.
58. Tsuji Y, Miller LL, Miller SC, Torti SV, Torti FM. Tumor necrosis factor-alpha and interleukin 1-alpha regulate transferrin receptor in human diploid fibroblasts. Relationship to the induction of ferritin heavy chain. J Biol Chem. 1991;266:7257–61.
59. Fridovich I. Superoxide radical and superoxide dismutases. Annu Rev Biochem. 1995; 64:97–112.
60. Okado-Matsumoto A, Fridovich I. Subcellular distribution of superoxide dismutases (SOD) in rat liver: Cu, Zn-SOD in mitochondria. J Biol Chem. 2001;276:38388–93.

61. Williams MD, Van Remmen H, Conrad CC, Huang TT, Epstein CJ, Richardson A. Increased oxidative damage is correlated to altered mitochondrial function in heterozygous manganese superoxide dismutase knockout mice. J Boil Chem. 1998;273:28510–5.
62. Chance B, Sies H, Boveris A. Hydroperoxide metabolism in mammalian organs. Physiol Rev. 1979;59:527–605.
63. Nomura K, Imai H, Koumura T, Kobayashi T, Nakagawa Y. Mitochondrial phospholipid hydroperoxide glutathione peroxidase inhibits the release of cytochrome c from mitochondria by suppressing the peroxidation of cardiolipin in hypoglycaemia-induced apoptosis. Biochem J. 2000;351:183–93.
64. Ursini F, Heim S, Kiess M, et al. Dual function of the selenoprotein PHGPx during sperm maturation. Science. 1999;285:1393–6.
65. Radi R, Turrens JF, Chang LY, Bush KM, Crapo JD, Freeman BA. Detection of catalase in rat heart mitochondria. J Biol Chem. 1991;266:22028–34.
66. Del Maestro R, McDonald W. Subcellular localization of superoxide dismutases, glutathione peroxidase and catalase in developing rat cerebral cortex. Mech Ageing Dev. 1989;48:15–31.
67. Phung CD, Ezieme JA, Turrens JF. Hydrogen peroxide metabolism in skeletal muscle mitochondria. Arch Biochem Biophys. 1994;315:479–82.
68. Turrens JF, Alexandre A, Lehninger AL. Ubisemiquinone is the electron donor for superoxide formation by complex III of heart mitochondria. Arch Biochem Biophys. 1985;237:408–14.
69. Bindoli A, Cavallini L, Jocelyn P. Mitochondrial lipid peroxidation by cumene hydroperoxide and its prevention by succinate. Biochim Biophys Acta. 1982;681:496–503.
70. Beyer RE. The participation of coenzyme Q in free radical production and antioxidation. Free Radic Biol Med. 1990;8:545–65.
71. Ham AJ, Liebler DC. Vitamin E oxidation in rat liver mitochondria. Biochemistry. 1995;34:5754–61.
72. Orii Y. The cytochrome c peroxidase activity of cytochrome oxidase. J Biol Chem. 1982;257:9246–8.
73. Radi R, Beckman JS, Bush KM, Freeman BA. Peroxynitrite oxidation of sulfhydryls. The cytotoxic potential of superoxide and nitric oxide. J Biol Chem. 1991;266:4244–50.
74. Dikalov S, Khramtsov V, Zimmer G. Determination of rate constants of the reactions of thiols with superoxide radical by electron paramagnetic resonance: critical remarks on spectrophotometric approaches. Arch Biochem Biophys. 1996;326:207–18.
75. Winterbourn CC, Metodiewa D. The reaction of superoxide with reduced glutathione. Arch Biochem Biophys. 1994;314:284–90.
76. Koppenol WH, Moreno JJ, Pryor WA, Ischiropoulos H, Beckman JS. Peroxynitrite, a cloaked oxidant formed by nitric oxide and superoxide. Chem Res Toxicol. 1992;5:834–42.
77. Green DR, Kroemer G. The pathophysiology of mitochondrial cell death. Science. 2004;305:626–9.
78. Spierings D, McStay G, Saleh M, et al. Connected to death: the (unexpurgated) mitochondrial pathway of apoptosis. Science. 2005;310:66–7.
79. Earnshaw WC. Apoptosis. A cellular poison cupboard. Nature. 1999;397:387–9.
80. Liu X, Kim CN, Yang J, Jemmerson R, Wang X. Induction of apoptotic program in cell-free extracts: requirement for dATP and cytochrome *c*. Cell. 1996;86:147–57.
81. Jacobson MD, Burne JF, King MP, Miyashita T, Reed JC, Raff MC. Bcl-2 blocks apoptosis in cells lacking mitochondrial DNA. Nature. 1993;361:365–9.
82. Park SY, Chang I, Kim JY, et al. Resistance of mitochondrial DNA-depleted cells against cell death: role of mitochondrial superoxide dismutase. J Biol Chem. 2004;279:7512–20.
83. Dey R, Moraes CT. Lack of oxidative phosphorylation and low mitochondrial membrane potential decrease susceptibility to apoptosis and do not modulate the protective effect of Bcl-x (L) in osteosarcoma cells. J Biol Chem. 2000;275:7087–94.
84. Larsson NG, Wang J, Wilhelmsson H, et al. Mitochondrial transcription factor A is necessary for mtDNA maintenance and embryogenesis in mice. Nat Genet. 1998;18:231–6.

85. Wang Y, Michikawa Y, Mallidis C, et al. Muscle-specific mutations accumulate with aging in critical human mtDNA control sites for replication. Proc Natl Acad Sci USA. 2001;98:4022–7.
86. Sorensen L, Ekstrand M, Silva JP, et al. Late-onset corticohippocampal neurodepletion attributable to catastrophic failure of oxidative phosphorylation in MILON mice. J Neurosci. 2001;21:8082–90.
87. Wang HG, Pathan N, Ethell IM, et al. Ca^{2+}-induced apoptosis through calcineurin dephosphorylation of BAD. Science. 1999;284:339–43.
88. Wang J, Wilhelmsson H, Graff C, et al. Dilated cardiomyopathy and atrioventricular conduction blocks induced by heart-specific inactivation of mitochondrial DNA gene expression. Nat Genet. 1999;21:133–7.
89. Lin PC, Lin JK, Yang SH, Wang HS, Li AF, Chang SC. Expression of beta-F1-ATPase and mitochondrial transcription factor A and the change in mitochondrial DNA content in colorectal cancer: clinical data analysis and evidence from an in vitro study. Int J Colorectal Dis. 2008.
90. Gramzinski RA, Parchment RE, Pierce GB. Evidence linking programmed cell death in the blastocyst to polyamine oxidation. Differentiation. 1990;43:59–65.
91. Pierce GB, Parchment RE, Lewellyn AL. Hydrogen peroxide as a mediator of programmed cell death in the blastocyst. Differentiation. 1991;46:181–6.
92. Giardino I, Fard AK, Hatchell DL, Brownlee M. Aminoguanidine inhibits reactive oxygen species formation, lipid peroxidation, and oxidant-induced apoptosis. Diabetes. 1998; 47:1114–20.
93. Ravati A, Ahlemeyer B, Becker A, Krieglstein J. Preconditioning-induced neuroprotection is mediated by reactive oxygen species. Brain Res. 2000;866:23–32.
94. Cole KK, Perez-Polo JR. Poly (ADP-ribose) polymerase inhibition prevents both apoptotic-like delayed neuronal death and necrosis after H(2)O(2) injury. J Neurochem. 2002; 82:19–29.
95. Hockenbery DM, Oltvai ZN, Yin XM, Milliman CL, Korsmeyer SJ. Bcl-2 functions in an antioxidant pathway to prevent apoptosis. Cell. 1993;75:241–51.
96. Rabizadeh S, Gralla EB, Borchelt DR, et al. Mutations associated with amyotrophic lateral sclerosis convert superoxide dismutase from an antiapoptotic gene to a proapoptotic gene: studies in yeast and neural cells. Proc Natl Acad Sci USA. 1995;92:3024–8.
97. Greenlund LJ, Deckwerth TL, Johnson EM, Jr. Superoxide dismutase delays neuronal apoptosis: a role for reactive oxygen species in programmed neuronal death. Neuron. 1995;14:303–15.
98. Kane DJ, Sarafian TA, Anton R, et al. Bcl-2 inhibition of neural death: decreased generation of reactive oxygen species. Science. 1993;262:1274–7.
99. Wiedau-Pazos M, Goto JJ, Rabizadeh S, et al. Altered reactivity of superoxide dismutase in familial amyotrophic lateral sclerosis. Science. 1996;271:515–8.
100. Wiedau-Pazos M, Trudell JR, Altenbach C, Kane DJ, Hubbell WL, Bredesen DE. Expression of bcl-2 inhibits cellular radical generation. Free Radic Res. 1996;24:205–12.
101. Crompton M. The mitochondrial permeability transition pore and its role in cell death. Biochem J. 1999;341(Pt 2):233–49.
102. Kruman I, Guo Q, Mattson MP. Calcium and reactive oxygen species mediate staurosporine-induced mitochondrial dysfunction and apoptosis in PC12 cells. J Neurosci Res. 1998; 51:293–308.
103. He H, Lam M, McCormick TS, Distelhorst CW. Maintenance of calcium homeostasis in the endoplasmic reticulum by Bcl-2. J Cell Biol. 1997;138:1219–28.
104. Zhu L, Ling S, Yu XD, et al. Modulation of mitochondrial Ca(2+) homeostasis by Bcl-2. J Biol Chem. 1999;274:33267–73.
105. Distelhorst CW, McCormick TS. Bcl-2 acts subsequent to and independent of Ca^{2+} fluxes to inhibit apoptosis in thapsigargin- and glucocorticoid-treated mouse lymphoma cells. Cell Calcium. 1996;19:473–83.

106. Murphy AN, Bredesen DE, Cortopassi G, Wang E, Fiskum G. Bcl-2 potentiates the maximal calcium uptake capacity of neural cell mitochondria. Proc Natl Acad Sci USA. 1996; 93:9893–8.
107. Wood DE, Newcomb EW. Cleavage of Bax enhances its cell death function. Exp Cell Res. 2000;256:375–82.
108. Wood DE, Thomas A, Devi LA, et al. Bax cleavage is mediated by calpain during drug-induced apoptosis. Oncogene. 1998;17:1069–78.
109. Camello-Almaraz C, Gomez-Pinilla PJ, Pozo MJ, Camello PJ. Mitochondrial reactive oxygen species and Ca^{2+} signaling. Am J Physiol Cell Physiol. 2006;291:C1082–8.
110. Leski ML, Hassinger LC, Valentine SL, Baer JD, Coyle JT. L-type calcium channels reduce ROS generation in cerebellar granule cells following kainate exposure. Synapse. 2002;43:30–41.
111. Gordeeva AV, Zvyagilskaya RA, Labas YA. Cross-talk between reactive oxygen species and calcium in living cells. Biochemistry (Mosc). 2003;68:1077–80.
112. Sheehan JP, Swerdlow RH, Miller SW, et al. Calcium homeostasis and reactive oxygen species production in cells transformed by mitochondria from individuals with sporadic Alzheimer's disease. J Neurosci. 1997;17:4612–22.
113. Hayakawa M, Torii K, Sugiyama S, Tanaka M, Ozawa T. Age-associated accumulation of 8-hydroxydeoxyguanosine in mitochondrial DNA of human diaphragm. Biochem Biophys Res Commun. 1991;179:1023–9.
114. Shigenaga MK, Ames BN. Assays for 8-hydroxy-2′-deoxyguanosine: a biomarker of in vivo oxidative DNA damage. Free Radic Biol Med. 1991;10:211–6.
115. Poon HF, Shepherd HM, Reed TT, et al. Proteomics analysis provides insight into caloric restriction mediated oxidation and expression of brain proteins associated with age-related impaired cellular processes: mitochondrial dysfunction, glutamate dysregulation and impaired protein synthesis. Neurobiol Aging. 2006;27:1020–34.
116. Sompol P, Ittarat W, Tangpong J, et al. A neuronal model of Alzheimer's disease: an insight into the mechanisms of oxidative stress-mediated mitochondrial injury. Neuroscience. 2008;153:120–30.
117. Chen H, Tappel AL Protection by vitamin E selenium, trolox C, ascorbic acid palmitate, acetylcysteine, coenzyme Q, beta-carotene, canthaxanthin, and (+)-catechin against oxidative damage to liver slices measured by oxidized heme proteins. Free Radic Biol Med. 1994;16:437–44.
118. Blumberg J. Use of biomarkers of oxidative stress in research studies. J Nutr. 2004; 134:3188S–9S.
119. Halliwell B, Whiteman M. Measuring reactive species and oxidative damage in vivo and in cell culture: how should you do it and what do the results mean? Br J Pharmacol. 2004;142:231–55.
120. Dizdaroglu M, Jaruga P, Birincioglu M, Rodriguez H. Free radical-induced damage to DNA: mechanisms and measurement. Free Radic Biol Med. 2002;32:1102–15.
121. Halliwell B. Effect of diet on cancer development: is oxidative DNA damage a biomarker? Free Radic Biol Med. 2002;32:968–74.
122. Collins AR, Cadet J, Moller L, Poulsen HE, Vina J. Are we sure we know how to measure 8-oxo-7, 8-dihydroguanine in DNA from human cells? Arch Biochem Biophys. 2004; 423:57–65.
123. Yen MY, Kao SH, Wang AG, Wei YH. Increased 8-hydroxy-2′-deoxyguanosine in leukocyte DNA in Leber's hereditary optic neuropathy. Invest Ophthalmol Vis Sci. 2004; 45:1688–91.
124. Umaki Y, Mitsui T, Endo I, Akaike M, Matsumoto T. Apoptosis-related changes in skeletal muscles of patients with mitochondrial diseases. Acta Neuropathol. 2002;103:163–70.
125. Griffiths HR, Moller L, Bartosz G, et al. Biomarkers. Mol Aspects Med. 2002;23:101–208.
126. Dalle-Donne I, Giustarini D, Colombo R, Rossi R, Milzani A. Protein carbonylation in human diseases. Trends Mol Med. 2003;9:169–76.

127. Davies MJ. The oxidative environment and protein damage. Biochim Biophys Acta. 2005;1703:93–109.
128. Winterbourn CC, Buss IH. Protein carbonyl measurement by enzyme-linked immunosorbent assay. Methods Enzymol. 1999;300:106–11.
129. Yan LJ, Levine RL, Sohal RS. Effects of aging and hyperoxia on oxidative damage to cytochrome c in the housefly, *Musca domestica*. Free Radic Biol Med. 2000;29:90–7.
130. Reznick AZ, Cross CE, Hu ML, et al. Modification of plasma proteins by cigarette smoke as measured by protein carbonyl formation. Biochem J. 1992;286(Pt 2):607–11.
131. Frank J, Pompella A, Biesalski HK. Histochemical visualization of oxidant stress. Free Radic Biol Med. 2000;29:1096–105.
132. Uchida K. 4-Hydroxy-2-nonenal: a product and mediator of oxidative stress. Prog Lipid Res. 2003;42:318–43.
133. Hyun DH, Lee MH, Halliwell B, Jenner P. Proteasomal dysfunction induced by 4-hydroxy-2,3-trans-nonenal, an end-product of lipid peroxidation: a mechanism contributing to neurodegeneration? J Neurochem. 2002;83:360–70.
134. Carini M, Aldini G, Facino RM. Mass spectrometry for detection of 4-hydroxy-trans-2-nonenal (HNE) adducts with peptides and proteins. Mass Spectrom Rev. 2004;23:281–305.
135. Cracowski JL, Durand T, Bessard G. Isoprostanes as a biomarker of lipid peroxidation in humans: physiology, pharmacology and clinical implications. Trends Pharmacol Sci. 2002;23:360–6.
136. Butterfield DA, Lauderback CM. Lipid peroxidation and protein oxidation in Alzheimer's disease brain: potential causes and consequences involving amyloid beta-peptide-associated free radical oxidative stress. Free Radic Biol Med. 2002;32:1050–60.
137. Barnham KJ, Masters CL, Bush AI. Neurodegenerative diseases and oxidative stress. Nat Rev Drug Discov. 2004;3:205–14.
138. Brennan ML, Wu W, Fu X, et al. A tale of two controversies: defining both the role of peroxidases in nitrotyrosine formation in vivo using eosinophil peroxidase and myeloperoxidase-deficient mice, and the nature of peroxidase-generated reactive nitrogen species. J Biol Chem. 2002;277:17415–27.
139. Azbill RD, Mu X, Bruce-Keller AJ, Mattson MP, Springer JE. Impaired mitochondrial function, oxidative stress and altered antioxidant enzyme activities following traumatic spinal cord injury. Brain Res. 1997;765:283–90.
140. Calabrese V, Lodi R, Tonon C, et al. Oxidative stress, mitochondrial dysfunction and cellular stress response in Friedreich's ataxia. J Neurol Sci. 2005;233:145–62.
141. Melov S, Schneider JA, Day BJ, et al. A novel neurological phenotype in mice lacking mitochondrial manganese superoxide dismutase. Nat Genet. 1998;18:159–63.
142. Martensson J, Goodwin CW, Blake R. Mitochondrial glutathione in hypermetabolic rats following burn injury and thyroid hormone administration: evidence of a selective effect on brain glutathione by burn injury. Metabolism. 1992;41:273–7.
143. Reed MC, Thomas RL, Pavisic J, James SJ, Ulrich CM, Nijhout HF. A mathematical model of glutathione metabolism. Theor Biol Med Model. 2008;5:8.
144. Inoue M, Sato EF, Nishikawa M, et al. Mitochondrial generation of reactive oxygen species and its role in aerobic life. Curr Med Chem. 2003;10:2495–505.
145. Wallace DC. A mitochondrial paradigm of metabolic and degenerative diseases, aging, and cancer: a dawn for evolutionary medicine. Annu Rev Genet. 2005;39:359–407.
146. Lenaz G, Baracca A, Fato R, Genova ML, Solaini G. New insights into structure and function of mitochondria and their role in aging and disease. Antioxid Redox Signal. 2006;8:417–37.
147. Mracek T, Pecina P, Vojtiskova A, Kalous M, Sebesta O, Houstek J. Two components in pathogenic mechanism of mitochondrial ATPase deficiency: energy deprivation and ROS production. Exp Gerontol. 2006;41:683–7.
148. Anderson S, Bankier AT, Barrell BG, et al. Sequence and organization of the human mitochondrial genome. Nature. 1981;290:457–65.

149. Carroll J, Fearnley IM, Skehel JM, Shannon RJ, Hirst J, Walker JE Bovine complex I. is a complex of 45 different subunits. J Biol Chem. 2006;281:32724–7.
150. DiMauro S, Bonilla E, De Vivo DC. Does the patient have a mitochondrial encephalomyopathy? J Child Neurol. 1999;14 Suppl 1:S23–35.
151. Schwartz M, Vissing J. Paternal inheritance of mitochondrial DNA. N Engl J Med. 2002;347:576–80.
152. Filosto M, Mancuso M, Vives-Bauza C, et al. Lack of paternal inheritance of muscle mitochondrial DNA in sporadic mitochondrial myopathies. Ann Neurol. 2003;54:524–6.
153. Melberg A, Moslemi AR, Palm O, Raininko R, Stalberg E, Oldfors A. A patient with two mitochondrial DNA mutations causing PEO and LHON. Eur J Med Genet. 2009;52:47–8.
154. Holt IJ, Harding AE, Petty RK, Morgan-Hughes JA. A new mitochondrial disease associated with mitochondrial DNA heteroplasmy. Am J Hum Genet. 1990;46:428–33.
155. Lodi R, Montagna P, Iotti S, et al. Brain and muscle energy metabolism studied in vivo by ^{31}P-magnetic resonance spectroscopy in NARP syndrome. J Neurol Neurosurg Psychiatry. 1994;57:1492–6.
156. Poulton J, Marchington DR. Segregation of mitochondrial DNA (mtDNA) in human oocytes and in animal models of mtDNA disease: clinical implications. Reproduction. 2002;123:751–5.
157. Baracca A, Sgarbi G, Mattiazzi M, et al. Biochemical phenotypes associated with the mitochondrial ATP6 gene mutations at nt8993. Biochim Biophys Acta. 2007;1767:913–9.
158. Rojo A, Campos Y, Sanchez JM, et al. NARP-MILS syndrome caused by 8993 T>G mitochondrial DNA mutation: a clinical, genetic and neuropathological study. Acta Neuropathol. 2006;111:610–6.
159. Hao R, Yao YN, Zheng YG, Xu MG, Wang ED. Reduction of mitochondrial tRNALeu (UUR) aminoacylation by some MELAS-associated mutations. FEBS Lett. 2004;578:135–9.
160. Koga Y, Akita Y, Takane N, Sato Y, Kato H. Heterogeneous presentation in A3243G mutation in the mitochondrial tRNA(Leu(UUR)) gene. Arch Dis Child. 2000;82:407–11.
161. Horvath R, Reilmann R, Holinski-Feder E, Ringelstein EB, Klopstock T. The role of complex I genes in MELAS: a novel heteroplasmic mutation 3380G>A in ND1 of mtDNA. Neuromusculoskel Disord. 2008;18:553–6.
162. Santorelli FM, Tanji K, Kulikova R, et al. Identification of a novel mutation in the mtDNA ND5 gene associated with MELAS. Biochem Biophys Res Commun. 1997;238:326–8.
163. Mariotti C, Savarese N, Suomalainen A, et al. Genotype to phenotype correlations in mitochondrial encephalomyopathies associated with the A3243G mutation of mitochondrial DNA. J Neurol. 1995;242:304–12.
164. Morgan-Hughes JA, Sweeney MG, Cooper JM, et al. Mitochondrial DNA (mtDNA) diseases: correlation of genotype to phenotype. Biochim Biophys Acta. 1995;1271:135–40.
165. Vielhaber S, Varlamov DA, Kudina TA, et al. Expression pattern of mitochondrial respiratory chain enzymes in skeletal muscle of patients harboring the A3243G point mutation or large-scale deletions of mitochondrial DNA. J Neuropathol Exp Neurol. 2002;61:885–95.
166. Horvath R, Kley RA, Lochmuller H, Vorgerd M. Parkinson syndrome, neuropathy, and myopathy caused by the mutation A8344G (MERRF) in tRNALys. Neurology. 2007; 68:56–8.
167. Mancuso M, Petrozzi L, Filosto M, et al. MERRF syndrome without ragged-red fibers: the need for molecular diagnosis. Biochem Biophys Res Commun. 2007;354:1058–60.
168. Antonicka H, Floryk D, Klement P, et al. Defective kinetics of cytochrome c oxidase and alteration of mitochondrial membrane potential in fibroblasts and cytoplasmic hybrid cells with the mutation for myoclonus epilepsy with ragged-red fibers ('MERRF') at position 8344 nt. Biochem J. 1999;342(Pt 3):537–44.
169. Zeviani M, Servidei S, Gellera C, Bertini E, DiMauro S, DiDonato S. An autosomal dominant disorder with multiple deletions of mitochondrial DNA starting at the D-loop region. Nature. 1989;339:309–11.
170. Rowland LP, Blake DM, Hirano M, et al. Clinical syndromes associated with ragged red fibers. Rev Neurol (Paris). 1991;147:467–73.

171. Li FY, Tariq M, Croxen R, et al. Mapping of autosomal dominant progressive external ophthalmoplegia to a 7-cM critical region on 10q24. Neurology. 1999;53:1265–71.
172. Li YP, Atkins CM, Sweatt JD, Reid MB. Mitochondria mediate tumor necrosis factor-alpha/NF-kappaB signaling in skeletal muscle myotubes. Antioxid Redox Signal. 1999; 1:97–104.
173. Di Fonzo A, Bordoni A, Crimi M, et al. POLG mutations in sporadic mitochondrial disorders with multiple mtDNA deletions. Hum Mutat. 2003;22:498–9.
174. Spinazzola A, Zeviani M. Disorders from perturbations of nuclear-mitochondrial intergenomic cross-talk. J Intern Med. 2009;265:174–92.
175. Nishino I, Spinazzola A, Hirano M. Thymidine phosphorylase gene mutations in MNGIE, a human mitochondrial disorder. Science. 1999;283:689–92.
176. DiMauro S, Hirano M. Mitochondria and heart disease. Curr Opin Cardiol. 1998;13:190–7.
177. Carrozzo R, Hirano M, Fromenty B, et al. Multiple mtDNA deletions features in autosomal dominant and recessive diseases suggest distinct pathogeneses. Neurology. 1998;50:99–106.
178. Villarroya J, de Bolos C, Meseguer A, Hirano M, Vila MR. Altered gene transcription profiles in fibroblasts harboring either TK2 or DGUOK mutations indicate compensatory mechanisms. Exp Cell Res. 2009;315:1429–38.
179. Brahimi N, Jambou M, Sarzi E, et al. The first founder DGUOK mutation associated with hepatocerebral mitochondrial DNA depletion syndrome. Mol Genet Metab. 2009.
180. Martin MA, Blazquez A, Gutierrez-Solana LG, et al. Leigh syndrome associated with mitochondrial complex I deficiency due to a novel mutation in the NDUFS1 gene. Arch Neurol. 2005;62:659–61.
181. Moslemi AR, Tulinius M, Darin N, Aman P, Holme E, Oldfors A. SURF1 gene mutations in three cases with Leigh syndrome and cytochrome c oxidase deficiency. Neurology. 2003;61:991–3.
182. Pecina P, Capkova M, Chowdhury SK, et al. Functional alteration of cytochrome c oxidase by SURF1 mutations in Leigh syndrome. Biochim Biophys Acta. 2003;1639:53–63.
183. Pecina P, Gnaiger E, Zeman J, Pronicka E, Houstek J. Decreased affinity for oxygen of cytochrome-c oxidase in Leigh syndrome caused by SURF1 mutations. Am J Physiol Cell Physiol. 2004;287:C1384–8.
184. van Riesen AK, Antonicka H, Ohlenbusch A, Shoubridge EA, Wilichowski EK. Maternal segmental disomy in Leigh syndrome with cytochrome c oxidase deficiency caused by homozygous SURF1 mutation. Neuropediatrics. 2006;37:88–94.
185. Williams SL, Valnot I, Rustin P, Taanman JW. Cytochrome c oxidase subassemblies in fibroblast cultures from patients carrying mutations in COX10, SCO1, or SURF1. J Biol Chem. 2004;279:7462–9.
186. Poyau A, Buchet K, Bouzidi MF, et al. Missense mutations in SURF1 associated with deficient cytochrome c oxidase assembly in Leigh syndrome patients. Hum Genet. 2000;106:194–205.
187. Petruzzella V, Tiranti V, Fernandez P, Ianna P, Carrozzo R, Zeviani M. Identification and characterization of human cDNAs specific to BCS1, PET112, SCO1, COX15, and COX11, five genes involved in the formation and function of the mitochondrial respiratory chain. Genomics. 1998;54:494–504.
188. Isaya G, O'Neill HA, Gakh O, Park S, Mantcheva R, Mooney SM. Functional studies of frataxin. Acta Paediatr Suppl. 2004;93:68–71; discussion 2–3.
189. Pandolfo M. Iron and Friedreich ataxia. J Neural Transm Suppl. 2006;143–6.
190. Calabrese V, Guagliano E, Sapienza M, Mancuso C, Butterfield DA, Stella AM. Redox regulation of cellular stress response in neurodegenerative disorders. Ital J Biochem. 2006;55:263–82.
191. Luo X, Pitkanen S, Kassovska-Bratinova S, Robinson BH, Lehotay DC. Excessive formation of hydroxyl radicals and aldehydic lipid peroxidation products in cultured skin fibroblasts from patients with complex I deficiency. J Clin Invest. 1997;99:2877–82.
192. Robinson BH. Human complex I deficiency: clinical spectrum and involvement of oxygen free radicals in the pathogenicity of the defect. Biochim Biophys Acta. 1998;1364:271–86.

193. Pitkanen S, Merante F, McLeod DR, Applegarth D, Tong T, Robinson BH. Familial cardiomyopathy with cataracts and lactic acidosis: a defect in complex I (NADH-dehydrogenase) of the mitochondria respiratory chain. Pediatr Res. 1996;39:513–21.
194. Robinson BH, Luo XP, Pitkanen S, et al. Diagnosis of mitochondrial energy metabolism defects in tissue culture. Induction of MnSOD and bcl-2 in mitochondria from patients with complex I (NADH-CoQ reductase) deficiency. Biofactors. 1998;7:229–30.
195. Gonzalo R, Garcia-Arumi E, Llige D, et al. Free radicals-mediated damage in transmitochondrial cells harboring the T14487C mutation in the ND6 gene of mtDNA. FEBS Lett. 2005;579:6909–13.
196. Koopman WJ, Verkaart S, van Emst-de Vries SE, et al. Mitigation of NADH: ubiquinone oxidoreductase deficiency by chronic Trolox treatment. Biochim Biophys Acta. 2008; 1777:853–9.
197. Distelmaier F, Koopman WJ, van den Heuvel LP, et al. Mitochondrial complex I deficiency: from organelle dysfunction to clinical disease. Brain. 2009;132:833–42.
198. Gomez-Lazaro M, Galindo MF. Melero-Fernandez de Mera RM, et al. Reactive oxygen species and p38 mitogen-activated protein kinase activate Bax to induce mitochondrial cytochrome c release and apoptosis in response to malonate. Mol Pharmacol. 2007; 71:736–43.
199. Ehrhart J, Zeevalk GD. Cooperative interaction between ascorbate and glutathione during mitochondrial impairment in mesencephalic cultures. J Neurochem. 2003;86:1487–97.
200. Rodriguez-Hernandez A, Cordero MD, Salviati L, et al. Coenzyme Q deficiency triggers mitochondria degradation by mitophagy. Autophagy. 2009;5:19–32.
201. Quinzii CM, Lopez LC, Von-Moltke J, et al. Respiratory chain dysfunction and oxidative stress correlate with severity of primary CoQ10 deficiency. FASEB J. 2008;22:1874–85.
202. Salviati L, Sacconi S, Murer L, et al. Infantile encephalomyopathy and nephropathy with CoQ10 deficiency: a CoQ10-responsive condition. Neurology. 2005;65:606–8.
203. Sobreira C, Hirano M, Shanske S, et al. Mitochondrial encephalomyopathy with coenzyme Q10 deficiency. Neurology. 1997;48:1238–43.
204. Di Giovanni S, Mirabella M, Spinazzola A, et al. Coenzyme Q10 reverses pathological phenotype and reduces apoptosis in familial CoQ10 deficiency. Neurology. 2001;57:515–8.
205. Rana M, de Coo I, Diaz F, Smeets H, Moraes CT. An out-of-frame cytochrome b gene deletion from a patient with parkinsonism is associated with impaired complex III assembly and an increase in free radical production. Ann Neurol. 2000;48:774–81.
206. Di Giovanni S, Mirabella M, Papacci M, Odoardi F, Silvestri G, Servidei S. Apoptosis and ROS detoxification enzymes correlate with cytochrome c oxidase deficiency in mitochondrial encephalomyopathies. Mol Cell Neurosci. 2001;17:696–705.
207. DiMauro S, Moraes CT. Mitochondrial encephalomyopathies. Arch Neurol. 1993; 50:1197–208.
208. Korshunov SS, Skulachev VP, Starkov AA. High protonic potential actuates a mechanism of production of reactive oxygen species in mitochondria. FEBS Lett. 1997;416:15–8.
209. Houstek J, Klement P, Floryk D, et al. A novel deficiency of mitochondrial ATPase of nuclear origin. Hum Mol Genet. 1999;8:1967–74.
210. Houstek J, Mracek T, Vojtiskova A, Zeman J. Mitochondrial diseases and ATPase defects of nuclear origin. Biochim Biophys Acta. 2004;1658:115–21.
211. Luo J, Shi R. Acrolein induces oxidative stress in brain mitochondria. Neurochem Int. 2005;46:243–52.
212. Graham BH, Waymire KG, Cottrell B, Trounce IA, MacGregor GR, Wallace DC. A mouse model for mitochondrial myopathy and cardiomyopathy resulting from a deficiency in the heart/muscle isoform of the adenine nucleotide translocator. Nat Genet. 1997; 16:226–34.
213. Flierl A, Chen Y, Coskun PE, Samulski RJ, Wallace DC. Adeno-associated virus-mediated gene transfer of the heart/muscle adenine nucleotide translocator (ANT) in mouse. Gene Ther. 2005;12:570–8.

214. Esposito LA, Melov S, Panov A, Cottrell BA, Wallace DC. Mitochondrial disease in mouse results in increased oxidative stress. Proc Natl Acad Sci USA. 1999;96:4820–5.
215. Guo Q, Fu W, Sopher BL, et al. Increased vulnerability of hippocampal neurons to excitotoxic necrosis in presenilin-1 mutant knock-in mice. Nat Med. 1999;5:101–6.
216. Manfredi G, Vu T, Bonilla E, et al. Association of myopathy with large-scale mitochondrial DNA duplications and deletions: which is pathogenic? Ann Neurol. 1997;42:180–8.
217. Shoffner JM. An introduction: oxidative phosphorylation diseases. Semin Neurol. 2001;21:237–50.
218. Ohkoshi N, Mizusawa H, Shiraiwa N, Shoji S, Harada K, Yoshizawa K. Superoxide dismutases of muscle in mitochondrial encephalomyopathies. Muscle Nerve. 1995; 18:1265–71.
219. Mitsui T, Kawai H, Nagasawa M, et al. Oxidative damage to skeletal muscle DNA from patients with mitochondrial encephalomyopathies. J Neurol Sci. 1996;139:111–6.
220. Mitsui T, Azuma H, Nagasawa M, et al. Chronic corticosteroid administration causes mitochondrial dysfunction in skeletal muscle. J Neurol. 2002;249:1004–9.
221. Lu CY, Wang EK, Lee HC, Tsay HJ, Wei YH. Increased expression of manganese-superoxide dismutase in fibroblasts of patients with CPEO syndrome. Mol Genet Metab. 2003;80:321–9.
222. Kunishige M, Mitsui T, Akaike M, et al. Overexpressions of myoglobin and antioxidant enzymes in ragged-red fibers of skeletal muscle from patients with mitochondrial encephalomyopathy. Muscle Nerve. 2003;28:484–92.
223. Wei YH, Lu CY, Wei CY, Ma YS, Lee HC. Oxidative stress in human aging and mitochondrial disease-consequences of defective mitochondrial respiration and impaired antioxidant enzyme system. Chin J Physiol. 2001;44:1–11.
224. Filosto M, Tonin P, Vattemi G, Spagnolo M, Rizzuto N, Tomelleri G. Antioxidant agents have a different expression pattern in muscle fibers of patients with mitochondrial diseases. Acta Neuropathol. 2002;103:215–20.
225. Piccolo G, Banfi P, Azan G, et al. Biological markers of oxidative stress in mitochondrial myopathies with progressive external ophthalmoplegia. J Neurol Sci. 1991;105:57–60.
226. Rusanen H, Majamaa K, Hassinen IE. Increased activities of antioxidant enzymes and decreased ATP concentration in cultured myoblasts with the 3243A→G mutation in mitochondrial DNA. Biochim Biophys Acta. 2000;1500:10–6.
227. Pang CY, Lee HC, Wei YH. Enhanced oxidative damage in human cells harboring A3243G mutation of mitochondrial DNA: implication of oxidative stress in the pathogenesis of mitochondrial diabetes. Diabetes Res Clin Pract. 2001;54 Suppl 2:S45–56.
228. Byrne E, Trounce I, Marzuki S, et al. Functional respiratory chain studies in mitochondrial cytopathies. Support for mitochondrial DNA heteroplasmy in myoclonus epilepsy and ragged red fibers (MERRF) syndrome. Acta Neuropathol. 1991;81:318–23.
229. Masucci JP, Davidson M, Koga Y, Schon EA, King MP. In vitro analysis of mutations causing myoclonus epilepsy with ragged-red fibers in the mitochondrial tRNA(Lys)gene: two genotypes produce similar phenotypes. Mol Cell Biol. 1995;15:2872–81.
230. Villani G, Attardi G. In vivo control of respiration by cytochrome c oxidase in wild-type and mitochondrial DNA mutation-carrying human cells. Proc Natl Acad Sci USA. 1997; 94:1166–71.
231. Tozzi G, Nuccetelli M, Lo Bello M, et al. Antioxidant enzymes in blood of patients with Friedreich's ataxia. Arch Dis Child. 2002;86:376–9.
232. Piemonte F, Pastore A, Tozzi G, et al. Glutathione in blood of patients with Friedreich's ataxia. Eur J Clin Invest. 2001;31:1007–11.
233. Bradley JL, Homayoun S, Hart PE, Schapira AH, Cooper JM. Role of oxidative damage in Friedreich's ataxia. Neurochem Res. 2004;29:561–7.
234. Llorens JV, Navarro JA, Martinez-Sebastian MJ, et al. Causative role of oxidative stress in a *Drosophila* model of Friedreich ataxia. FASEB J. 2007;21:333–44.

235. Irazusta V, Moreno-Cermeno A, Cabiscol E, Ros J, Tamarit J. Major targets of iron-induced protein oxidative damage in frataxin-deficient yeasts are magnesium-binding proteins. Free Radic Biol Med. 2008;44:1712–23.
236. Gakh O, Park S, Liu G, et al. Mitochondrial iron detoxification is a primary function of frataxin that limits oxidative damage and preserves cell longevity. Hum Mol Genet. 2006;15:467–79.
237. Myers L, Farmer JM, Wilson RB, et al. Antioxidant use in Friedreich ataxia. J Neurol Sci. 2008;267:174–6.
238. Rustin P, von Kleist-Retzow JC, Chantrel-Groussard K, Sidi D, Munnich A, Rotig A. Effect of idebenone on cardiomyopathy in Friedreich's ataxia: a preliminary study. Lancet. 1999;354:477–9.
239. Schulz JB, Dehmer T, Schols L, et al. Oxidative stress in patients with Friedreich ataxia. Neurology. 2000;55:1719–21.
240. Lodi R, Rajagopalan B, Blamire AM, et al. Cardiac energetics are abnormal in Friedreich ataxia patients in the absence of cardiac dysfunction and hypertrophy: an in vivo ^{31}P magnetic resonance spectroscopy study. Cardiovasc Res. 2001;52:111–9.
241. Hart PE, Lodi R, Rajagopalan B, et al. Antioxidant treatment of patients with Friedreich ataxia: four-year follow-up. Arch Neurol. 2005;62:621–6.
242. Schulz JB, Di Prospero NA, Fischbeck K. Clinical experience with high-dose idebenone in Friedreich ataxia. J Neurol. 2009;256 Suppl 1:42–5.

Chapter 16
Potential Role of Oxidative Damage in Neurological Manifestations of Acute Intermittent Porphyria

Elena Pischik and Raili Kauppinen

Abstract Accumulation of 5-aminolevulinic acid (ALA) is the main defect in acute porphyria and the most likely potential candidate to cause acute neurological manifestations during an acute porphyric attack via multiple direct and indirect mechanisms. ALA is a potential endogenous source of reactive oxygen species (ROS). After administration of ALA or inducers of ALA-synthase in vitro conditions or in animal models, the main pro-oxidants detected have been superoxide, hydrogen peroxide, hydroxyl radicals, and 4,5-dioxovaleric acid (DOVA), which has produced oxidative damage (OD) to lipids, proteins, and DNA. At the organelle level, ALA-induced OD affects the permeability of the biological membranes, probably as a result of protein polymerization and lipid peroxidation. In vitro exposure to ALA has caused OD to Schwann cells and inhibited myelin formation. Magnetic resonance imaging (MRI) of patients with acute intermittent porphyria (AIP) who suffer from severe reversible posterior encephalopathy syndrome (PRES) during the acute attack shows features of impaired permeability of the blood–brain barrier (BBB); this could be the result of oxidative stress (OS) allowing neurotoxins such as ALA to damage neurons. Peripheral demyelination found in heterozygote or homozygote patients with AIP could be caused by direct OD caused by ALA, which produces pro-oxidants that may affect Schwann cells and myelin. Because ALA is not the most potent pro-oxidant, the OD is only a minor contributor to the neurological manifestations of AIP in general. It could, however, explain the fact that encephalopathy and peripheral demyelination are present only during severe attacks of AIP, in which the high level of serum ALA results in significant auto-oxidation.

E. Pischik (✉)
Department of Medicine, Division of Endocrinology, Porphyria Research Unit, University Central Hospital of Helsinki, Helsinki, Finland
and
Department of Neurology, Neuromuscular Unit, City Hospital #2, Pavlov State Medical University, Saint Petersburg, Russia
e-mail: pischike@yahoo.com

N. Gadoth and H.H. Göbel (eds.), *Oxidative Stress and Free Radical Damage in Neurology*, Oxidative Stress in Applied Basic Research and Clinical Practice,
DOI 10.1007/978-1-60327-514-9_16,

Keywords Porphyria · Acute intermittent porphyria · Neuropathy · Encephalopathy

Abbreviations

AIP	Acute intermittent porphyria
ALA	5-Aminolevulinic acid
ALAS	ALA synthase
BBB	Blood–brain barrier
CNS	Central nervous system
CSF	Cerebrospinal fluid
DHPY	3,6-Dihydropyrazine-2,5-dipropanoic acid
DOVA	4,5-Dioxovaleric acid
GABA	Gamma amino butyric acid
HCP	Hereditary coproporphyria
HMBS	Hydroxymethylbilane synthase
MRI	Magnetic resonance imaging
OD	Oxidative damage
OS	Oxidative stress
PBG	Porphobilinogen
PNP	Peripheral neuropathy, polyneuropathy
ROS	Reactive oxygen species
SIADH	Syndrome of inadequate secretion of antidiuretic hormone
VP	Variegate porphyria

1 Introduction

Acute intermittent porphyria (AIP) is an inherited metabolic disease caused by a deficiency of the hydroxymethylbilane synthase (HMBS), the third enzyme in heme biosynthesis (Fig. 1). AIP manifests with occasional neurovisceral crises associated with overproduction of porphyrin precursors such as aminolevulinic acid (ALA) and porphobilinogen (PBG), which are released from the liver to the circulation [1]. The majority of acute attacks manifest as a combination of abdominal pain, mild mental symptoms, and autonomic dysfunction. In addition, both acute peripheral neuropathy and severe encephalopathy may develop if an acute attack proceeds. AIP is usually caused by administration of porphyrinogenic drugs used when the diagnosis of acute porphyria was not considered [2].

Currently, an excess of ALA is the most potential candidate to cause neurological manifestations in acute porphyria via multiple direct and indirect mechanisms [3]. Porphyrins, especially uroporphyrin I and III, are also accumulated during an

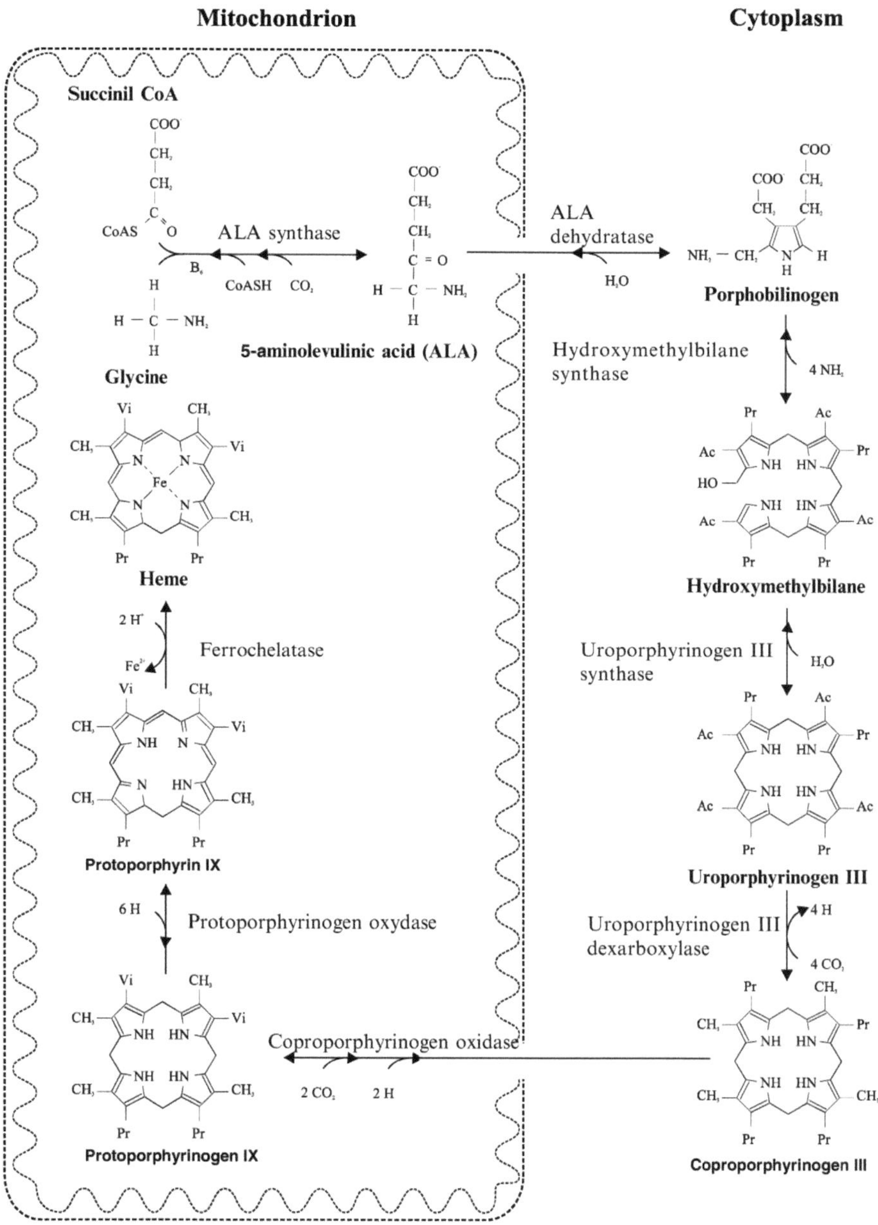

Fig. 1 Heme biosynthesis in the cell. *Ac*–CH_2COOH; *Pr*–CH_2CH_2COOH; *Vi*–$CH{=}CH_2$

acute attack [4], but it is unlikely that they are neurotoxic because patients with porphyria cutanea tarda, who have massive increase in plasma uroporphyrins associated with skin symptoms and liver disease but *do not* accumulate ALA, are neurologically unaffected. However, massive accumulation of porphyrins in the skin induces a photoreaction in which transfer of a hydrogen atom or electron from the singlet excited state porphyrin to molecular oxygen produces free radicals and cellular damage [5]. Interestingly, in acute porphyrias other than AIP, such as variegate porphyria (VP) and hereditary coproporphyria (HCP), which may also manifest skin symptoms, acute attacks and skin symptoms occur independently from each other [6]. This situation is in contrast to acute neuropathy or encephalopathy, which, if these occur, are always signs of an acute attack in all types of acute porphyrias (AIP, VP, and HCP).

ALA is a pro-oxidizing alpha-aminoketone that undergoes enolization and subsequent aerobic oxidation both in vitro and in vivo [7]. This reaction yields oxyradicals and several other pro-oxidants. Thus, oxidative stress (OS) is one of the potential mechanisms that could cause neurological manifestations during an acute porphyric attack. This review is focused on reported evidence of massive formation of pro-oxidants from ALA and its potential implications in developing acute neuropathy or encephalopathy in acute porphyria.

2 Accumulation of ALA During an Acute Attack

AIP is an autosomal dominant disorder caused by several different mutations in the *HMBS* gene (http://www.hgmd.cf.ac.uk). Heterozygous patients have approximately 50% of the total HMBS activity measured in their erythrocytes, lymphocytes, fibroblasts, and hepatocytes [8–11]. This amount is sufficient to maintain the normal demand for heme, but the half-normal enzyme activity causes accumulation of heme precursors when the synthesis of heme is activated.

Accumulation of porphyrins and their precursors is initialized via activation of ALA synthase (ALAS1) in the liver; this is the first and rate-limiting enzyme in heme biosynthesis (Fig. 1). The induction of ALAS1 results in symptoms only if an additional enzyme causing acute porphyria such as HMBS is deficient (see Fig. 1) [1]. ALAS1 can be induced directly at the transcriptional and translational level by many drugs, chemicals, and alcohol [12] or indirectly by low glucose concentration via *peroxisome-proliferator-activated receptor* γ *co activator 1*α (PCG-1α) [13] and stress, by activating heme oxygenase [14]. These factors may provoke accumulation of porphyrin precursors. On the other hand, ALAS1 can be inhibited via negative feedback mechanism using heme, the end product of the biosynthesis [12]. Therefore, both heme preparations and glucose infusions have been used to treat acute attacks because they both lower the level of ALA [15, 16].

Neurological manifestations of acute porphyria could be precipitated either directly by excess of ALA, or by deficiency of neural heme-containing enzymes, or both [3]. However, currently there is no evidence for abnormal activity of the heme-containing

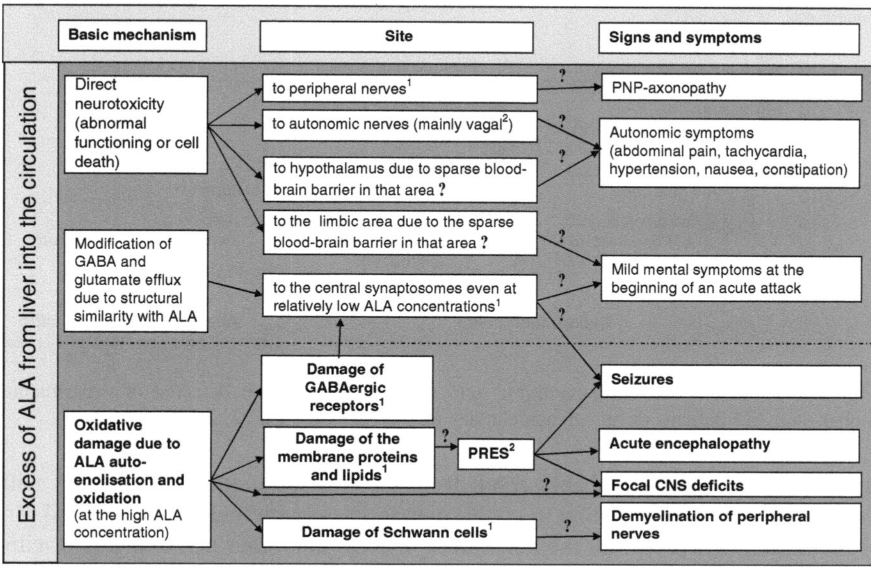

Fig. 2 The place of oxidative damage in the current understanding of the pathogenesis of neurological manifestations of acute intermittent porphyria (AIP) (main hypotheses). *1* evidence in vitro; *2* evidence in vivo; *?* hypothesis

enzymes in neural tissues as measured in brain homogenates of compound heterozygous HMBS (–/–) mouse model [17] and patients with AIP. The postulated main mechanisms that could explain why excess of ALA is responsible for acute autonomic and peripheral neuropathy and encephalopathy include direct neurotoxicity of ALA, ALA-induced modification of release of the inhibitory neurotransmitter, gamma-amino butyric acid (GABA), or the excitatory neurotransmitter glutamate, and, finally, oxidative damage (OD) to neuronal and glial membrane structures associated with ALA-induced reactive oxygen species (ROS) [3] (Fig. 2).

The assumption of direct neurotoxicity of ALA is based on studies in vitro [3, 18–20] at ALA concentrations compatible with its increased levels measured in patients with AIP during an acute attack [21] (Fig. 3). The increased plasma concentration of ALA as a sole mechanism causing neurological impairment is unlikely because injection of ALA caused no clinical manifestations in a healthy volunteer, or in patients with neuroblastoma, who were given ALA during photodynamic therapy [22, 23]. Moreover, the majority of patients with AIP have constantly increased plasma levels and urinary excretion of ALA even during the asymptomatic phase, but the level of ALA still increases substantially during an acute attack [4, 21]. Intracellular ALA is most likely much higher in patients with AIP than in controls with no enzymatic deficiency in every cell despite the same plasma level.

ALA-induced modification of the release of neurotransmitters, GABA and glutamate, from central synaptosomes in a dose-dependent manner has been demonstrated in brains of rats treated with ALA or ALAS inducers in vitro and

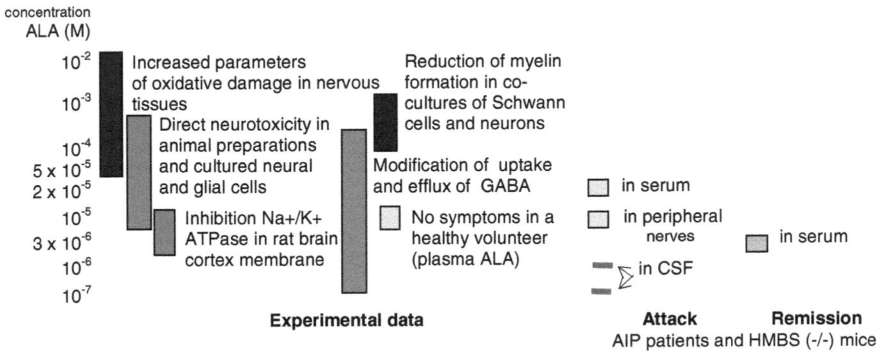

Fig. 3 Comparison of 5-aminolevulinic acid (ALA) concentration detected in experimental models and AIP patients during an acute attack and in remission [3, 22, 24–27]

in vivo [24, 25]; this probably result from the structural similarity of ALA with GABA and glutamate leading to concurrent agonism with their receptors [25].

Because ALA promotes the generation of ROS and other pro-oxidants in vitro and in animal models in vivo, OD to membrane structures could contribute to transient dysfunction of the central nervous system (CNS) and probably of the peripheral nervous system [25, 26].

3 Pro-Oxidizing Nature of ALA

Several studies have shown that ALA can cause OD to lipids [27], proteins [28], and DNA [29] in solution and in vitro and in vivo conditions, as well as to myelin formation in vitro [30]. The potentially damaging effects of ALA are gained by metal-catalyzed oxidation of ALA via molecular oxygen, which yields very reactive intermediates and products such as peroxides, radicals, triplet species, and α-oxoaldehydes [7].

3.1 Pathways of Oxyradical Production from ALA

Three main pathways of oxyradical production from ALA have been postulated (Fig. 4):

(a) Direct production of superoxide and oxaloaldehydes such as the keto-aldehyde, 4,5-dioxovaleric acid (DOVA), via ALA enolization and subsequent aerobic oxidation.
(b) Iron-catalyzed production of highly reactive ROS (hydrogen peroxide and hydroxyl radicals) from superoxide arising from ALA-enoyl oxidation; iron release from ferritin is also promoted by superoxide.

(c) Production of pyrazine and oxyradicals from dehydropyrazine (3,6-dihydropyrazine-2,5-dipropanoic acid [DHPY]), which is formed by ALA dimerization.

ALA (see Fig. 1) is an aminoketone, and as such it is highly reactive and unstable in solution forming ALA-enoyl at physiological pH [7, 31]. After subsequent oxidation by dioxygen, the ALA-enoyl radical is produced while oxygen is reduced to superoxide (O_2•) (see Fig. 4). This is the first and the main pathway of oxyradical production from ALA. Further oxidation of ALA-enoyl radical produces an additional molecule of superoxide and ALA-derived-imine, which after hydrolysis produces DOVA and ammonium ion in amounts comparable with consumed O_2 [7]. The end product of ALA oxidation, DOVA, is an efficient alkylating agent of the guanine moieties known to damage DNA [32]. The other side metabolites of ALA auto-oxidation are triplet carbonyls [7] (Fig. 4). Because they have high alkoxyl radical-like reactivity in addition to a long lifetime, they may promote deleterious reactions in the cell [33].

In the presence of the iron storage protein, ferritin, these oxidative reactions, when studied in vitro, have been accompanied by a release of free iron [7], which is responsible for the second pathway of oxyradical production from ALA. Free iron (Fe^{2+}) provokes conversion of superoxide to a much more aggressive ROS, hydrogen peroxide, and thereafter hydroxyl radicals [7]. This process is known as the Fenton reaction.

In contrast to superoxide, which is the least reactive form of ROS and can be directly deactivated by antioxidant enzymes such as superoxide dismutase (SOD) [27], hydrogen peroxide and hydroxyl radicals have very high oxygenation potential and cannot be eliminated by SOD; however, intracellular low molecular antioxidants such as glutathione may decrease their reactivity [34]. Increased amounts (4.5 fold) of hydrogen peroxide, hydroxyl radicals, and DOVA have been demonstrated both in vitro and in vivo in rats treated with ALA or ALAS inducers [7, 29, 35, 36].

The third considered mechanism of ROS excess in acute porphyria is the production of superoxide from DHPY, which is formed by a spontaneous ALA dimerization [31] (see Fig. 4). This mechanism is only hypothetical because spontaneous ALA dimerization was shown only in solution but never in vivo. The main conditions of spontaneous ALA dimerization include high ALA concentration (10^{-4}–10^{-2} M), a lack of ALA dehydrase probably in combination with dysfunction of other enzymes in heme biosynthesis, and neutral or slightly basic pH [31]. Dimerization of ALA does not require metal ion participation. The potential oxidative toxicity of DHPY depends on pyrazine formation in the presence of oxygen, resulting in generation of ROS as side metabolites. In the presence of cupric ions, DHPY can also form carbon-centered radicals and hydroxyl radicals. It can also break single-strand DNA and promote apoptosis in vitro [31]. This mechanism has not been studied in vivo, and a high concentration of ALA and low pH limit the potential likelihood of OD in patients with AIP caused by DHPY.

In summary, the main pro-oxidants produced after ALA or ALAS inducer administration in vitro and in vivo in "healthy" rodents are superoxide, hydrogen peroxide, hydroxyl radicals, and DOVA.

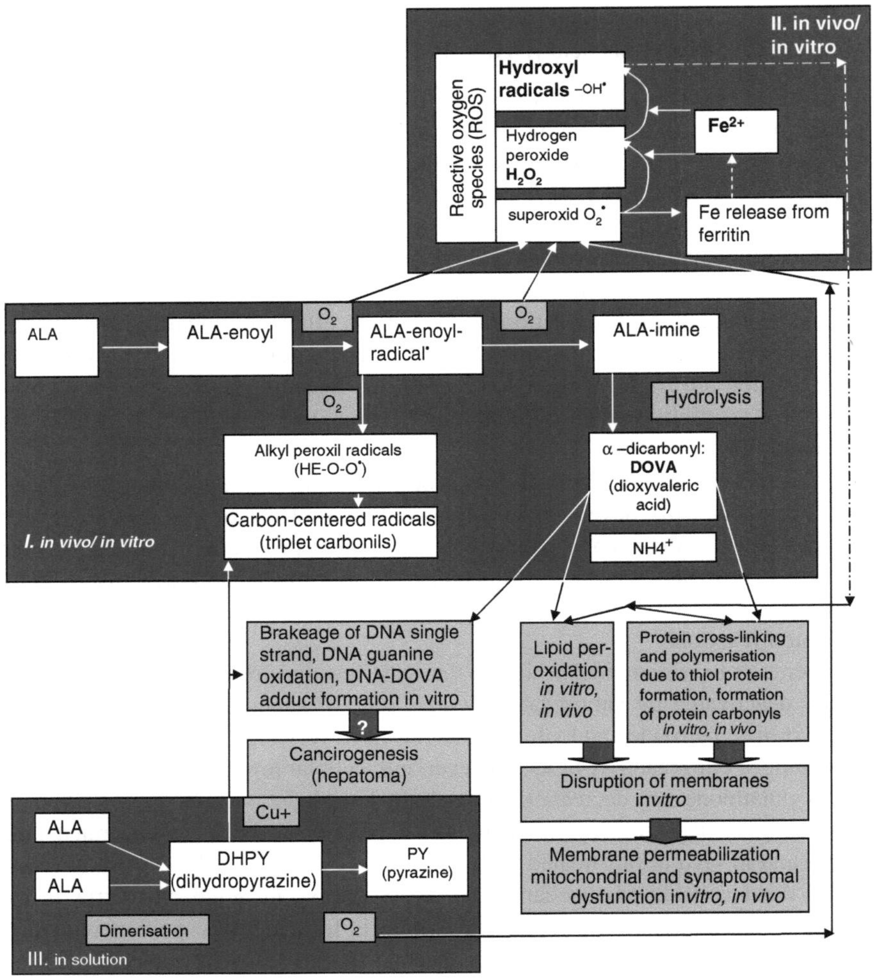

Fig. 4 Three main pathways of oxyradical production from ALA and their potential effects at the biochemical and organelle level. The excess metabolites in ALA-treated rodents are shown in *bold*

4 Oxidative Damage In Vitro and in Animal Models In Vivo Treated with ALA or ALA Synthase Inducers

Markers of OS such as elevated activity of SOD, catalase, glutathione peroxidase, and glutathione reductase in erythrocytes and liver [37], and elevated levels of malondialdehyde with reduced levels of glutathione in the brain have been detected in mice treated with ALA or ALA synthase inducers [38]. It is known that reactive free radicals formed within cells can oxidize biomolecules and lead to DNA damage, lipid peroxidation, and protein oxidation, tissue injury, or cell death [39]. Most frequently, this damage is measured as peroxidized lipid products, carbonyl proteins, and DNA breakage or fragmentation. Recently, ROS have been implicated in attenuated angiogenesis in endothelial cells [40] and in modulating the activation of numerous transcription factors, such as stress-activated protein kinases, heat shock factors, and various pro-inflammatory cytokines [41, 42]. This discovery could have numerous implications in AIP.

OD caused by ALA, ALAS inducers, and pro-oxidants formed from ALA, such as DOVA, has been reported by several investigators [7, 29, 43, 44]. At the biochemical and organelle level, ALA has been shown to cause OD to proteins, as shown by the presence of a fourfold increase of carbonyl proteins in synaptic membrane and whole-brain cortex preparations from ALA-treated rats and a 31% increase of protein sulfhydryl content in synaptic membranes of these animals [43]. It was suggested that protein alterations such as protein thiol cross-linkage are the main cause of the increased membrane permeability with affected function of Ca^{2+}-dependent ion channels promoted by ALA oxidation; this was expressed as an alteration of the transmembrane electrical potential and swelling of the inner membrane of the isolated rat liver mitochondria [43, 45]. Disruption of membrane permeability by OD caused by ALA could be caused by both protein polymerization (pore assembly) and pore opening caused by binding of Ca^{2+} to a number of membrane proteins [28]. A twofold increase of thiobarbituric acid-reactive substance (TBARS) and conjugated diene production in synaptic membranes of rats was detected. Such findings are considered reliable markers for lipid peroxidation [43]. Similar results have been demonstrated in vitro [27].

ALA-induced lipid peroxidation was associated with an increase in liposome permeability and subsequent release of encapsulated carboxyfluorescein [27]. Thus, lipid peroxidation could be an additional mechanism in abnormal protein polymerization causing impaired membrane permeability in ALA-treated rats. In addition, Ca^{2+} uptake was slightly increased, probably a result of the OD to the synaptosome membrane [43].

Acute and chronic administration of ALA synthase inducers or 40 mg/kg ALA to rats has resulted in a significant increase of total intracellular non-heme iron in the liver, spleen, and brain (up to 70% higher than the initial levels). This increase originated from ferritin by both ALA-generated O_2^- and ALA-enoyl radical [43, 44, 46] and was accompanied by OD to tryptophan and cysteine residues of L- and H-subunits of ferritin [44]. Similar results were obtained in vitro at ALA concentrations of 5×10^{-5}–10^{-3} M. In addition, activation of iron-regulating protein (IRP-1) from together with release of iron from ferritin was found [43, 47]. In the brain, the most prominent increase was detected in cortex and striatum [43]. In the plasma of these rats, iron levels were not affected, but ferritin levels were increased, suggesting a general mobilization of iron [43].

The role of iron in chronic OD to the nervous system that was shown to be involved in the pathophysiology of neurodegenerative disorders is believed to occur mainly by metal-promoted membrane lipid peroxidation and oxidative modification of various membrane and membrane-associated proteins [48]. The potential role of iron in reversibility of neurological symptoms in AIP could also be associated with membrane dysfunction, which, however, is more likely to be transient in contrast to the progression and chronicity of this process in neurodegenerative disorders.

Mitochondrial dysfunction in the form of impaired membrane electrical properties and morphology, Ca^{2+} release, and increased respiratory rate was found after OS induced by ALA exposure [41, 45, 49].

Increased levels of DOVA are responsible for formation of adducts with DNA bases, breakage of single-strand DNA, and in vitro promotion of DNA guanine oxidation [29, 50].

OD and a twofold decrease in the density of GABA receptors was detected in brain extracts of ALA- and DOVA-treated rats [24, 43, 51]; this could be an additional mechanism of disturbed GABA efflux in synaptosomes.

Treatment of myelinating cocultures of Schwann cells and sensory neurons with 10^{-3}–10^{-4} M ALA resulted in pronounced increase of protein carbonylation and formation of hydroxynonenal and malondialdehyde. These results are in accordance with reported experiments on liver, spleen, and erythrocytes [30]. It was also shown that Schwann cells are more vulnerable to OS than neurons because of the selective decrease in the expression of mitochondrial respiratory chain proteins in glia but not in neurons [30].

In summary, after treatment with ALA or ALAS inducers in vitro and in vivo, a two- to fourfold increase in factors indicating OS with induced membrane lipid peroxidation and protein modifications has been detected. Increased membrane permeability caused by both protein and lipid oxidation is the main effect of this treatment at the organelle level. Additional minor effects include altered synaptosome and mitochondrial function causing membrane dysfunction. In addition, ALA exposure affects DNA stability.

5 Evidence of Oxidative Stress in Patients with AIP

Twofold elevation of activity of the erythrocyte antioxidant enzymes SOD and glutathione peroxidase has been detected in symptomatic patients with AIP, in contrast to asymptomatic patients in whom only 30% of the normal activity was found [52]; this is currently almost the only direct evidence of increased OS in patients with AIP. Similarly increased SOD, glutathione reductase, and catalase activity in erythrocytes of patients with VP was found [53]. VP is another form of acute porphyria caused by inherited deficiency of protoporphyrinogen oxidase, the seventh enzyme of heme biosynthesis (see Fig. 1). Acute attacks in VP are indistinguishable from those of AIP, and a similar increase of porphyrin precursors in plasma and urine is present [1].

To produce oxyradicals from ALA in experiments, prolonged exposure of relatively high concentrations of ALA is required. These experimental findings suggest that there are only limited potential clinical implications of ALA-induced OD (data listed above). The levels of ALA in reported experiments and in patients with AIP and knockout HMBS (−/−) mice, the animal model of AIP, are compared in Fig. 3. In most of those experiments the ALA plasma level found was higher than in patients with AIP. However, the amount of ALA injected into rats (40 mg/kg) [43] that produced OD was even lower than that given to patients during photodynamic therapy (60 mg/kg) [54, 55] and to a healthy volunteer (50–80 mg/kg) without causing abnormal signs of symptoms [22].

ALA is not the strongest pro-oxidant when compared, for example, with amino acetone accumulation in diabetes mellitus, causing a more pronounced OD to ferritin and isolated mitochondria [7]. The lack of information regarding intracellular levels of ALA in patients with AIP makes it difficult to predict the putative oxidant effect of ALA in vivo [7]. It cannot be excluded that OD occurs at the lower concentrations of ALA or that intracellular concentrations of ALA are higher than extracellular. If OD is responsible for some of the neurological manifestations of AIP, it is most likely to occur in severe attacks with high levels of plasma ALA as there is a correlation between the severity of the attack and the plasma level of ALA [4, 21]. The majority of patients with AIP who have had severe acute attacks with neurological manifestations have constantly elevated plasma ALA levels with an additional two- to threefold increase during an acute attack.

In animals, OD caused by extended administration of ALA was more obvious than when they were treated for relatively shorter longer periods (2 weeks or longer) [29, 43, 44, 46].

Photodynamic therapy of tumors with ALA as an exogenous source of photosensitizing porphyrins did not cause chronic elevation of ALA [56]. In these patients with normal activity of the enzymes involved in heme biosynthesis, ALA does not accumulate in tissues but converts into porphyrins, which are well known to produce free radical OD and destroy the tumor cells only after exposure to light [56].

Because porphyrins can cause OD only after light exposure, photodermatosis after exposure to sunlight is the clinical presentation of cutaneous porphyrias, with up to 100-fold elevation of porphyrins in the circulation [5]. Other organs can be affected in patients with cutaneous porphyrias only when exposed to unfiltered light during open surgery [57] or during photodynamic therapy of tumors [56], because the tumor cells have a higher porphyrin absorption than unaffected tissues. Thus, despite increased levels of porphyrin excretion during acute attacks of AIP [4], it is unlikely that porphyrins per se produce any OD in the nervous system.

The chronically increased ALA levels in the majority of patients with AIP [4], and experimental ALA-induced DNA damage such as fragmentation, breakage, alkylation of the guanine moieties, and formation of adducts with DNA bases [29, 50], could be implicated in the high frequency of liver cancer in these patients [36]. However, no correlation was found between the occurrence of hepatoma and the severity and recurrence of acute porphyric attacks in patients with a number of *HMBS* mutations in whom constantly high levels of ALA were documented [58].

6 Clinical Manifestations of an Acute Porphyric Attack and a Potential Role of Oxidative Damage

Direct neurotoxicity of ALA caused by its transient excess in tissues could explain the reversible, mainly motor and sometimes also sensory, neuropathy, and the autonomic neuropathy and mild mental symptoms, observed during acute attacks of AIP [2]. However, the occurrence of posterior reversible encephalopathy syndrome (PRES) associated with affected blood–brain barrier (BBB) permeability and peripheral nerve demyelination in severe acute attacks suggests that additional mechanisms may be involved, such as transient auto-oxidation of ALA excess or activity of ROS pro-inflammatory cytokines and endothelial factors such as vascular endothelial growth factor (VEGF) precursor. The potential OS in AIP could be related to ALA as it affects membranes and causes transient increase of cell membrane permeability and impaired function of mitochondria.

6.1 Acute Encephalopathy

Acute encephalopathy may appear during an acute attack [59]; it manifests as a combination of mental symptoms, seizures, syndrome of inappropriate secretion of antidiuretic hormone (SIADH), and rarely focal CNS deficits. Only 1% of ALA crosses the intact BBB [3, 60, 61], and thus direct neurotoxicity of ALA does not explain it solely. Acute encephalopathy in AIP can be visualized during PRES with diffusion-weighted (DW) magnetic resonance imaging (MRI) if performed within a few hours from onset [62–73]. The development of PRES in AIP suggests that the BBB was damaged, permitting access of neurotoxins such as ALA into the CNS [74]. The precise mechanism of affected BBB permeability in AIP has not been yet studied. Increased blood pressure is a common finding during an acute attack, but rarely is it elevated to the level of hypertensive crisis (>180/100 mmHg). This finding favors the theory of endothelial toxicity being the main cause of vasogenic edema in AIP whereas hypertension more likely acts as a cofactor for PRES as happens also in eclampsia [74]. OD, which has been shown to affect biological membranes, could be a potential cause of endothelial toxicity and PRES.

Because vasospasm leading to focal ischemia is unusual in AIP with PRES [69, 70, 73], it cannot be "blamed" for the occurrence of PRES, and thus the etiology of this from of encephalopathy in AIP is still obscure, although OS may be involved. The CNS is very sensitive to OD because of a relatively low antioxidant capacity

[38]. Increased levels of ALA in patients with AIF and in the rat model could directly contribute to OD to brain neurons [38].

6.2 *Acute Peripheral Neuropathy*

A possible role of free radicals has previously been suggested in certain neuropathies [75, 76], either directly via lipid peroxidation or indirectly by the activation of proteases and phospholipase A2, which causes damage to Schwann cells and demyelination [75].

Acute porphyric peripheral neuropathy (PNP) is predominantly motor and typically associated with a history of abdominal pain and dysautonomia, encephalopathy, and mild increase in liver transaminase activity [59, 77]. Based on scarce evidence from neurophysiological studies of 39 patients with acute PNP [77–88], its pathology was considered as axonal [89]. Direct neurotoxicity of ALA [3, 18–20] is currently accepted as the neuropathological causative. Although the blood–nerve barrier lowers ALA concentration in the perineural fluid, which is around 30% of that measured in serum [3] (see Fig. 3), its concentration is high enough to be neurotoxic [3, 18, 19] and cause axonal degeneration and polyneuropathy.

However, several lines of evidence support the presence of demyelination in addition to axonopathy as an independent mechanism of PNP in AIP. We have reevaluated previously published neurophysiological studies on patients with AIP during an acute attack and found that the values of the recorded motor conduction velocities mentioned in those publications were significantly slow for age (range, 25–50 m/s; mean, 40 m/s) [78, 79, 81–85]. Prolonged distal motor latencies, and minimal F-wave latencies [79, 82, 83], abnormal temporal dispersion, and presence of conduction block were also noted [79]. Neuropathological studies that have been done on sural nerve biopsy or autopsy have supported both demyelinating [90–92] and axonal [93, 94] changes, usually coexisting in the same nerve.

We have studied six severely affected patients during an acute phase of PNP and found mixed demyelinative and axonal damage in four patients [95]. The most dramatic case, in whom more than 60% decrease of motor conduction velocities was measured a week after the onset of PNP, died within a month from ventilator-associated pneumonia complicated by septicemia. Two additional patients with acute PNP had axonopathy without features of demyelination. This finding indicates that the presence of demyelination may suggest a more severe attack and may support our hypothesis of higher probability of OD with higher levels of ALA. Furthermore, a 1-year-old homozygote patient with AIP and extremely high ALA urinary excretion who died at the age of 3 years had chronic demyelinative PNP documented by neurophysiological and pathological studies. Brain MRI showed deep cerebral white matter abnormalities with relative preservation of the corpus callosum, cerebral gray, and infratentorial structures. These findings could indicate selective cerebral oligodendrocyte injury. Moreover, the infant had clinical features indistinguishable from a progressive neurodegenerative disease of infancy [96].

It was shown that Schwann cells are more vulnerable than neurons to ALA OD [30]. A similar vulnerability of those cells to other pro-oxidants was also observed [75]. Based on those observations, OD could be a reasonable explanation for demyelination in porphyric PNP.

Encephalopathy with PRES and peripheral demyelination occur only in severe acute attacks, which is in accordance with our hypothesis of a potential role of OS as a partial explanation of these conditions.

7 Possible Therapeutic Implications

Because OS could be implicated in severe neurological manifestations of AIP, the use of antioxidants together with heme arginate [16] could be a potential therapeutic approach. Among antioxidants shown in vitro to diminish or prevent OD caused by ALA were melatonin [97, 98], bilirubin [99], and glutathione [97]. However, the possible benefit of these metabolites should first be tested in clinical trials.

The fact that patients with similar high levels of ALA may present with variable severity of attacks and associated neurological symptoms and signs may indicate that there is a variable vulnerability to ROS resulting from polymorphisms in the genes encoding antioxidant enzymes [100] or other factors of ROS defence and cleavage. Thus, patients who are more vulnerable will suffer from greater neuronal damage caused by ALA-derivated free radicals.

8 Conclusions

Direct neurotoxicity caused by transient excess of ALA could explain the mild mental changes and symptoms of motor and autonomic neuropathy with mixed mainly axonal-demyelinative axonal features, confirmed by both neurophysiological and pathological findings, that occur during an attack of AIP. The main gaps in the present understanding of the pathogenesis of neurological manifestations in AIP include the development of porphyric encephalopathy, which manifests as PRES with affected blood–brain permeability, and peripheral nerve demyelination in severe acute attacks, which cannot be directly attributed to ALA accumulation during the attacks. Additional mechanisms should be identified to explain these conditions. The transient OD to cellular organelles resulting from auto-oxidation of excess ALA could contribute to neuropathy and encephalopathy of AIP. A two- to fourfold increase in the factors representing OS and causing dysfunction of biological membranes, mitochondria, and synaptosomes has been demonstrated in vitro in cultures of neurons and Schwann and other cells treated with ALA as well as in rodents treated with ALA or ALAS inducers. In symptomatic patients with AIP, a twofold elevation of erythrocyte SOD and glutathione peroxidase activity has been detected, whereas in asymptomatic patients the level was only 30% higher than normal [52]. Currently, those findings are the only evidence of increased OS in human AIP. Thus, the role of OD in neurological manifestations of AIP is only hypothetical. The higher concentration of ALA in the experimental situation as compared to the relatively lower levels of ALA in the plasma of AIP patients is one of the weak points of this hypothesis. The relatively low oxidative potential of ALA compared to other pro-oxidants is another weak point.

As the intracellular level of ALA and intracellular parameters of OS are yet unknown in patients with AIP, it cannot be excluded that OS could be induced in patients by higher levels of ALA, and thus OD could be a contributor to severe encephalopathy with PRES and peripheral nerve axonal and demyelinative damage in AIP and other forms of acute porphyria with paroxysmal neurological manifestations.

References

1. Kauppinen R. Porphyrias. Lancet. 2005;365:241–52.
2. Pischik E, Kauppinen R. Neurological manifestations of acute intermittent porphyria. Cell Mol Biol (Noisy-le-grand). 2009;55:72–83.
3. Meyer UA, Schuurmans MM, Lindberg RL. Acute porphyrias: pathogenesis of neurological manifestations. Semin Liver Dis. 1998;18:43–52.
4. Kauppinen R, von und zu Fraunberg M. Molecular and biochemical studies of acute intermittent porphyria in 196 patients and their families. Clin Chem. 2002;48:1891–900.
5. Poh-Fitzpatrick MB. Molecular and cellular mechanisms of porphyrin photosensitization. Photodermatology. 1986;3:148–57.
6. von und zu Fraunberg M, Timonen K, Mustajoki P, et al. Clinical and biochemical characteristics and genotype–phenotype correlation in Finnish variegate porphyria patients. Eur J Hum Genet. 2002;10:649–57.
7. Bechara EJ, Dutra F, Cardoso VE, et al. The dual face of endogenous alpha-aminoketones: pro-oxidizing metabolic weapons. Comp Biochem Physiol C Toxicol Pharmacol. 2007;146:88–110.
8. Strand LJ, Felsher BF, Redeker AG, et al. Heme biosynthesis in intermittent acute prophyria: decreased hepatic conversion of porphobilinogen to porphyrins and increased delta aminolevulinic acid synthetase activity. Proc Natl Acad Sci U S A. 1970;67:1315–20.
9. Meyer UA, Strand LJ, Doss M, et al. Intermittent acute porphyria – demonstration of genetic defect in porphobilinogen metabolism. N Engl J Med. 1972;286:1277–82.
10. Sassa S, Solish G, Levere RD, et al. Studies in porphyria. IV. Expression of the gene defect of acute intermittent porphyria in cultured human skin fibroblasts and amniotic cells: prenatal diagnosis of the porphyric trait. J Exp Med. 1975;142:722–31.
11. Sassa S, Zalar GL, Kappas A. Studies in porphyria. VII. Induction of uroporphyrinogen-I synthase and expression of the gene defect of acute intermittent porphyria in mitogen-stimulated human lymphocytes. J Clin Invest. 1978;61:499–508.
12. Fraser DJ, Podvinec M, Kaufmann MR, et al. Drugs mediate the transcriptional activation of the 5-aminolevulinic acid synthase (ALAS1) gene via the chicken xenobiotic-sensing nuclear receptor (CXR). J Biol Chem. 2002;277:34717–26.
13. Handschin C, Lin J, Rhee J, et al. Nutritional regulation of hepatic heme biosynthesis and porphyria through PGC-1alpha. Cell. 2005;122:505–15.
14. Rodgers PA, Stevenson DK. Developmental biology of heme oxygenase. Clin Perinatol. 1990;17:275–91.

15. Hift RJ, Meissner PN. An analysis of 112 acute porphyric attacks in Cape Town, South Africa: evidence that acute intermittent porphyria and variegate porphyria differ in susceptibility and severity. Medicine (Baltim). 2005;84:48–60.
16. Mustajoki P, Nordmann Y. Early administration of heme arginate for acute porphyric attacks. Arch Intern Med. 1993;153:2004–8.
17. Jover R, Hoffmann F, Scheffler-Koch V, et al. Limited heme synthesis in porphobilinogen deaminase-deficient mice impairs transcriptional activation of specific cytochrome P450 genes by phenobarbital. Eur J Biochem. 2000;267:7128–37.
18. Loots JM, Becker DM, Meyer BJ, et al. The effect of porphyrin precursors on monosynaptic reflex activity in the isolated hemisected frog spinal cord. J Neural Transm. 1975;36:71–81.
19. Percy VA, Lamm MC, Taljaard JJ.Delta-aminolaevulinic acid uptake, toxicity, and effect on [^{14}C]gamma-aminobutyric acid uptake into neurons and glia in culture. J Neurochem. 1981;36:69–76.
20. Cutler MG, Mair J, Moore MR. Pharmacological activities of delta-aminolaevulinic acid, protoporphyrin IX and haemin in isolated preparations of rabbit gastric fundus and jejunum J Auton Pharmacol. 1990;10:119–26.
21. Sardh E, Harper P, Andersson DE, et al. Plasma porphobilinogen as a sensitive biomarker to monitor the clinical and therapeutic course of acute intermittent porphyria attacks. Eur J Intern Med. 2009;20:201–7.
22. Mustajoki P, Timonen K, Gorchein A, et al. Sustained high plasma 5-aminolevulinic acid concentration in a volunteer: no porphyric symptoms. Eur J Clin Invest. 1992;22:407–11.
23. Fukuda H, Casas A, Batlle A. Aminolevulinic acid: from its unique biological function to its star role in photodynamic therapy. Int J Biochem Cell Biol. 2005;37:272–6.
24. Adhikari A, Penatti CA, Resende RR, et al. 5-Aminolevulinate and 4,5-dioxovalerate ions decrease GABA(A) receptor density in neuronal cells, synaptosomes and rat brain. Brain Res. 2006;1093:95–104.
25. Brennan MJ, Cantrill RC. Delta-aminolaevulinic acid and amino acid neurotransmitters. Mol Cell Biochem. 1981;38:49–58.
26. Princ FG, Juknat AA, Amitrano AA, et al. Effect of reactive oxygen species promoted by delta-aminolevulinic acid on porphyrin biosynthesis and glucose uptake in rat cerebellum. Gen Pharmacol. 1998;31:143–8.
27. Oteiza PI, Bechara EJ. 5-Aminolevulinic acid induces lipid peroxidation in cardiolipin-rich liposomes. Arch Biochem Biophys. 1993;305:282–7.
28. Vercesi AE, Castilho RF, Meinicke AR, et al. Oxidative damage of mitochondria induced by 5-aminolevulinic acid: role of Ca^{2+} and membrane protein thiols. Biochim Biophys Acta. 1994;1188:86–92.
29. Fraga CG, Onuki J, Lucesoli F, et al. 5-Aminolevulinic acid mediates the in vivo and in vitro formation of 8-hydroxy-2′-deoxyguanosine in DNA. Carcinogenesis (Oxf). 1994;15:2241–4.
30. Felitsyn N, McLeod C, Shroads AL, et al. The heme precursor delta-aminolevulinate blocks peripheral myelin formation. J Neurochem. 2008;106:2068–79.
31. Hunter GA, Rivera E, Ferreira GC. Supraphysiological concentrations of 5-aminolevulinic acid dimerize in solution to produce superoxide radical anions via a protonated dihydropyrazine intermediate. Arch Biochem Biophys. 2005;437:128–37.
32. Onuki J, Rech CM, Medeiros MH, et al. Genotoxicity of 5-aminolevulinic and 4,5-dioxovaleric acids in the Salmonella/microsuspension mutagenicity assay and SOS chromotest. Environ Mol Mutagen. 2002;40:63–70.
33. Velosa AC, Baader WJ, Stevani CV, et al. 1,3-Diene probes for detection of triplet carbonyls in biological systems. Chem Res Toxicol. 2007;20:1162–9.
34. Lee JC, Son YO, Choi KC, et al. Hydrogen peroxide induces apoptosis of BJAB cells due to formation of hydroxyl radicals via intracellular iron-mediated Fenton chemistry in glucose oxidase-mediated oxidative stress. Mol Cells. 2006;22:21–9.
35. Timmins GS, Liu KJ, Bechara EJ, et al. Trapping of free radicals with direct in vivo EPR detection: a comparison of 5,5-dimethyl-1-pyrroline-*N*-oxide and 5-diethoxyphosphoryl-5-

methyl-1-pyrroline-*N*-oxide as spin traps for HO* and SO_4*. Free Radic Biol Med. 1999;27:329–33.
36. Ummus RE, Onuki J, Dornemann D, et al. Measurement of 4,5-dioxovaleric acid by high-performance liquid chromatography and fluorescence detection. J Chromatogr B Biomed Sci Appl. 1999;729:237–43.
37. Lelli SM, San Martin de Viale LC, Mazzetti MB. Response of glucose metabolism enzymes in an acute porphyria model. Role of reactive oxygen species. Toxicology. 2005;216:49–58.
38. Rodriguez JA, Martinez Mdel C, Gerez E, et al. Heme oxygenase, aminolevulinate acid synthetase and the antioxidant system in the brain of mice treated with porphyrinogenic drugs. Cell Mol Biol (Noisy-le-grand). 2005;51:487–94.
39. Cheeseman KH, Slater TF. An introduction to free radical biochemistry. Br Med Bull. 1993;49:481–93.
40. Piao ZH, Yoon SR, Kim MS, et al. VDUP1 potentiates Ras-mediated angiogenesis via ROS production in endothelial cells. Cell Mol Biol (Noisy-le-grand). 2009;55:1096–103.
41. Calabrese V, Lodi R, Tonon C, et al. Oxidative stress, mitochondrial dysfunction and cellular stress response in Friedreich's ataxia. J Neurol Sci. 2005;233:145–62.
42. Mitra S, Abraham E. Participation of superoxide in neutrophil activation and cytokine production. Biochim Biophys Acta. 2006;1762:732–41.
43. Demasi M, Penatti CA, DeLuca R, et al. The prooxidant effect of 5-aminolevulinic acid in the brain tissue of rats: implication in neuropsychiatric manifestation in porphyrias. Free Radic Biol Med. 1996;20:291–9.
44. Rocha ME, Dutra F, Bandy B, et al. Oxidative damage to ferritin by 5-aminolevulinic acid. Arch Biochem Biophys. 2003;409:349–56.
45. Hermes-Lima M, Castilho RF, Valle VG, et al. Calcium-dependent mitochondrial oxidative damage promoted by 5-aminolevulinic acid. Biochim Biophys Acta. 1992;1180:201–6.
46. Rocha ME, Bandy B, Costa CA, et al. Iron mobilization by succinylacetone methyl ester in rats. A model study for hereditary tyrosinemia and porphyrias characterized by 5-aminolevulinic acid overload. Free Radic Res. 2000;32:343–53.
47. Monteiro HP, Abdalla DS, Augusto O, et al. Free radical generation during D-aminolevulinic acid autooxidaation: induction by hemoglobin and connections with porphyrinopathies. Arch Biochem Biophys. 1989;271:206–16.
48. Mattson MP. Metal-catalyzed disruption of membrane protein and lipid signaling in the pathogenesis of neurodegenerative disorders. Ann N Y Acad Sci. 2004;1012:37–50.
49. Hermes-Lima M. How do Ca^{2+} and 5-aminolevulinic acid-derived oxyradicals promote injury to isolated mitochondria? Free Radic Biol Med. 1995;19:381–90.
50. Douki T, Onuki J, Medeiros MH, et al. DNA alkylation by 4,5-dioxovaleric acid, the final oxidation product of 5-aminolevulinic acid. Chem Res Toxicol. 1998;11:150–7.
51. Penatti CA, Bechara EJ, Demasi M. Delta-aminolevulinic acid-induced synaptosomal Ca^{2+} uptake and mitochondrial permeabilization. Arch Biochem Biophys. 1996;335:53–60.
52. Medeiros MH, Marchiori PE, Bechara EJ. Superoxide dismutase, glutathione peroxidase, and catalase activities in the erythrocytes of patients with intermittent acute porphyria. Clin Chem. 1982;28:242–3.
53. Ferrer MD, Tauler P, Sureda A, et al. Enzyme antioxidant defences and oxidative damage in red blood cells of variegate porphyria patients. Redox Rep. 2009;14:69–74.
54. Brown SB. The role of light in the treatment of non-melanoma skin cancer using methyl aminolevulinate. J Dermatol Treat. 2003;14 (suppl 3):11–4.
55. Friesen SA, Hjortland GO, Madsen SJ, et al. 5-Aminolevulinic acid-based photodynamic detection and therapy of brain tumors (review). Int J Oncol. 2002;21:577–82.
56. Frank J, Lambert C, Biesalski HK, et al. Intensified oxidative and nitrosative stress following combined ALA-based photodynamic therapy and local hyperthermia in rat tumors. Int J Cancer. 2003;107:941–8.
57. Wahlin S, Srikanthan N, Hamre B, et al. Protection from phototoxic injury during surgery and endoscopy in erythropoietic protoporphyria. Liver Transplant. 2008;14:1340–6.

58. von und zu Fraunberg M, Pischik E, Udd L, et al. Clinical and biochemical characteristics and genotype-phenotype correlation in 143 Finnish and Russian patients with acute intermittent porphyria. Medicine (Baltim). 2005;84:35–47.
59. Pischik E, Bulyanitsa A, Kazakov V, et al. Clinical features predictive of a poor prognosis in acute porphyria [letter]. J Neurol. 2004;251:1538–41.
60. Percy VA, Shanley BC. Porphyrin precursors in blood, urine and cerebrospinal fluid in acute porphyria. South Afr Med J. 1977;52:219.
61. Gorchein A, Webber R. d-Aminolaevulinic acid in plasma, cerebrospinal fluid, saliva and erythrocytes: studies in normal, uraemic and porphyric subjects. Clin Sci. 1987;72:103–12.
62. King PH, Bragdon AC MRI reveals multiple reversible cerebral lesions in an attack of acute intermittent porphyria. Neurology. 1991;41:1300–2.
63. Aggarwal A, Quint DJ, Lynch JP. MR imaging of porphyric encephalopathy. Am J Roentgenol. 1994;162:1218–20.
64. Kupferschmidt H, Bont A, Schnorf H, et al. Transient cortical blindness and bioccipital brain lesions in two patients with acute intermittent porphyria. Ann Intern Med. 1995;123:598–600.
65. Black KS, Mirsky P, Kalina P, et al. Angiographic demonstration of reversible cerebral vasospasm in porphyric encephalopathy. Am J Neuroradiol. 1995;16:1650–2.
66. Susa S, Daimon M, Morita Y, et al. Acute intermittent porphyria with central pontine myelinolysis and cortical laminar necrosis. Neuroradiology. 1999;41:835–9.
67. Garg MK, Mohaparto AK, Dugal JS, et al. Cortical blindness in acute intermittent porphyria. J Assoc Physicians India. 1999;47:727–9.
68. Utz N, Kinkel B, Hedde JP, et al. MR imaging of acute intermittent porphyria mimicking reversible posterior leukoencephalopathy syndrome. Neuroradiology. 2001;43:1059–62.
69. Yen PS, Chen CJ, Lui CC, et al. Diffusion-weighted magnetic resonance imaging of porphyric encephalopathy: a case report. Eur Neurol. 2002;48:119–21.
70. Celik M, Forta H, Dalkilic T, et al. MRI reveals reversible lesions resembling posterior reversible encephalopathy in porphyria. Neuroradiology. 2002;44:839–41.
71. Engelhardt K, Trinka E, Franz G, et al. Refractory status epilepticus due to acute hepatic porphyria in a pregnant woman: induced abortion as the sole therapeutic option? Eur J Neurol. 2004;11:693–7.
72. Maramattom BV, Zaldivar RA, Glynn SM, et al. Acute intermittent porphyria presenting as a diffuse encephalopathy. Ann Neurol. 2005;57:581–4.
73. Wessels T, Blaes F, Rottger C, et al. Cortical amaurosis and status epilepticus with acute porphyria. Nervenarzt. 2005;76:992–5, 7–8.
74. Neal CR, Hunter AJ, Harper SJ, et al. Plasma from women with severe pre-eclampsia increases microvascular permeability in an animal model in vivo. Clin Sci (Lond). 2004;107:399–405.
75. Smith KJ, Kapoor R, Felts PA. Demyelination: the role of reactive oxygen and nitrogen species. Brain Pathol. 1999;9:69–92.
76. Pop-Busui R, Sima A, Stevens M. Diabetic neuropathy and oxidative stress. Diabetes Metab Res Rev. 2006;22:257–73.
77. Ridley A. The neuropathy of acute intermittent porphyria. Q J Med. 1969;38:307–33.
78. Albers JW, Robertson WC, Daube JR. Electrodiagnostic findings in acute porphyric neuropathy. Muscle Nerve. 1978;1:292–6.
79. Barohn RJ, Sanchez JA, Anderson KE. Acute peripheral neuropathy due to hereditary coproporphyria. Muscle Nerve. 1994;17:793–9.
80. Defanti CA, Sghirlanzoni A, Bottacchi E, et al. Porphyric neuropathy: a clinical, neurophysiological and morphological study. Ital J Neurol Sci. 1985;6:521–6.
81. Flugel KA, Druschky K-F. Electromyogram and nerve conduction in patients with acute intermittent porphyria. J Neurol. 1977;214:265–79.
82. King PH, Petersen NE, Rakhra R, et al. Porphyria presenting with bilateral radial motor neuropathy: evidence of a novel gene mutation. Neurology. 2002;58:1118–21.

83. Maytham DV, Eales L. Electrodiagnostic findings in porphyria. J Lab Clin Med. 1971;17:99–100.
84. Nagler W. Peripheral neuropathy in acute intermittent porphyrias. Arch Phys Med Rehabil. 1971;52:426–31.
85. Reichenmiller HE, Zysno EA. Neuropsychiatrische Strörungen bei vier Fällen von akuter intermittierender Porphyrie. Verhandl Dtsch Gesellsch Med. 1970:748–52.
86. Stein JA, Tschudy DP. Acute intermittent porphyria: a clinical and biochemical study of 46 patients. Medicine (Baltim). 1970;49:1–16.
87. Wochnik-Dyjas D, Niewiadomska M, Kostrzewska E. Porphyric polyneuropathy and its pathogenesis in the light of electrophysiological investigations. J Neurol Sci. 1978;35:243–56.
88. Zimmerman EA, Lovelace R. The etiology of the neuropathy in acute intermittent porphyria. Trans Am Neurol Assoc. 1968;93:294–6.
89. Kasper LD, Braunwald E, Fauci A, et al. Harrison's principles of internal medicine. 16th ed. New York: McGraw-Hill; 2005.
90. Denny-Brown D, Sciarra D. Changes in the nervous system in acute porphyria. Brain. 1945;68:1–16.
91. Gibson JB, Goldberg A. The neuropathology of acute porphyria. J Pathol Bacteriol. 1956;71:495–506.
92. Hierons R. Changes in the nervous system in acute porphyria. Brain. 1957;80:176–92.
93. Anzil AP, Dozhic S. Peripheral nerve changes in porphyric neuropathy: findings in sural nerve biopsy. Acta Neuropathol. 1978;42:121–6.
94. Suarez JI, Cohen ML, Larkin J, et al. Acute intermittent porphyria: clinicopathological correlation. Report of a case and review of the literature. Neurology. 1997;48:1678–83.
95. Pischik E. Neurological manifestations and molecular genetics of acute intermittent porphyria in North Western Russia. Helsinki: University of Helsinki; 2006.
96. Solis C, Martinez-Bermejo A, Naidich TP, et al. Acute intermittent porphyria: studies of the severe homozygous dominant disease provides insights into the neurologic attacks in acute porphyrias. Arch Neurol. 2004;61:1764–70.
97. Juknat AA, Kotler ML, Quaglino A, et al. Necrotic cell death induced by delta-aminolevulinic acid in mouse astrocytes. Protective role of melatonin and other antioxidants. J Pineal Res. 2003;35:1–11.
98. Karbownik M, Reiter RJ. Melatonin protects against oxidative stress caused by delta-aminolevulinic acid: implications for cancer reduction. Cancer Invest. 2002;20:276–86.
99. Noriega GO, Tomaro ML, del Batlle AM. Bilirubin is highly effective in preventing in vivo delta-aminolevulinic acid-induced oxidative cell damage. Biochim Biophys Acta. 2003;1638:173–8.
100. Huang WC, Juang SW, Liu IM, et al. Changes of superoxide dismutase gene expression and activity in the brain of streptozotocin-induced diabetic rats. Neurosci Lett. 1999;275:28–5.

Index

I

J

K

L

M

O

MIX
Papier aus verantwortungsvollen Quellen
Paper from responsible sources
FSC® C105338

If you have any concerns about our products,
you can contact us on
ProductSafety@springernature.com

In case Publisher is established outside the EU,
the EU authorized representative is:
Springer Nature Customer Service Center GmbH
Europaplatz 3, 69115 Heidelberg, Germany

Printed by Libri Plureos GmbH
in Hamburg, Germany